Ergonomics 5.0

In the ever-evolving digital age, technology continues to transform our working methods. Ergonomics 5.0 is emerging as a fundamental approach in this new era, ensuring that interactions between humans and technology are efficient, effective, healthy, safe, and sustainable. This book introduces the transformative potential of Ergonomics 5.0 and its essential role in linking humans and technology.

This book is for practitioners seeking to understand and apply the principles of Ergonomics 5.0 in their work environment to ensure that technology is used ethically, safely, and enjoyably. It brings together leading researchers and professionals in ergonomics and human factors from around the world and covers key principles such as digital intelligence, resilience, and sustainability, showcasing how these elements integrate into Ergonomics 5.0. Through international case studies, research findings, and best practice examples, it offers to the reader how to implement Ergonomics 5.0 in various sectors, enhancing efficiency, productivity, and quality of life for workers.

Ergonomics 5.0: Improving Efficiency, Productivity and Contentment in the Workplace is an ideal read for professionals, academics, and students in ergonomics and human factors, plus designers, architects, engineers, and related professionals.

Ergonomics 5.0
Improving Efficiency, Productivity and Contentment in the Workplace

Edited by Eduardo Ferro dos Santos and Marcelo M. Soares

CRC Press
Taylor & Francis Group
Boca Raton London New York

CRC Press is an imprint of the
Taylor & Francis Group, an **informa** business

Designed cover image: iStock

First edition published 2026
by CRC Press
2385 NW Executive Center Drive, Suite 320, Boca Raton FL 33431

and by CRC Press
4 Park Square, Milton Park, Abingdon, Oxon, OX14 4RN

CRC Press is an imprint of Taylor & Francis Group, LLC

Library of Congress Cataloging-in-Publication Data
Names: Ferro, Eduardo dos Santos editor | Soares, Marcelo Marcio editor
Title: Ergonomics 5.0 : improving efficiency, productivity and contentment in the workplace / edited by Eduardo Ferro dos Santos, and Marcelo M. Soares.
Other titles: Ergonomics five point zero
Description: First edition. | Boca Raton : CRC Press, 2026. | Includes bibliographical references and index. |
Identifiers: LCCN 2025044187 | ISBN 9781032906447 hardback | ISBN 9781032907147 paperback | ISBN 9781003559399 ebook
Subjects: LCSH: Human engineering | Design--Human factors | Product design | LCGFT: Essays
Classification: LCC TA166 .E7166 2026
LC record available at https://lccn.loc.gov/2025044187

ISBN: 9781032906447 (hbk)
ISBN: 9781032907147 (pbk)
ISBN: 9781003559399 (ebk)

DOI: 10.1201/9781003559399

Typeset in Times
by codeMantra

Contents

PART II Implementations

About the Editors

Dr. Eduardo Ferro dos Santos is Associate Professor at the University of São Paulo (Brazil) with a transdisciplinary academic background. He holds a Habilitation in Health, Safety, and Environment (USP), a Postdoctoral degree in Educational Technologies (UNESP), a Ph.D. and M.Sc. in Production Engineering, and undergraduate degrees in Mechatronics Engineering and Physiotherapy. Dr Ferro dos Santos teaches and supervises at undergraduate, graduate, and specialization levels across multiple institutions, focusing on Human Factors Engineering, Educational Technologies, and Sustainability Engineering. Internationally recognized, he received the 2023 IEA/Tsinghua International Prize for his significant contributions to ergonomics. Beyond academia, he provides consulting, training, and project development services, having worked with over 500 organizations in Brazil and abroad. As editor, he demonstrates a strong commitment to advancing ergonomics in shaping sustainable and human-centered industrial futures, aligning scientific rigor with social responsibility and technological progress.

Professor Marcelo Soares is a volunteer professor in the Graduate Program in Design at the Federal University of Pernambuco and a visiting professor at the Federal University of Maranhão, Brazil. He previously served as a full professor at Hunan University and SUSTech in China. He holds a master's degree in Production Engineering from the Federal University of Rio de Janeiro and a Ph.D. from Loughborough University, with postdoctoral work at the University of Central Florida. He served as president of ABERGO for seven years and was a member of the International Ergonomics Association board. Professor Soares has published approximately 50 journal articles, 190+ conference papers, and 110 books/chapters. He authored "Ergodesign for Product Design: A Human-Centered Approach" (in English, Portuguese, and Chinese) and co-edited several handbooks on Human Factors. His research focuses on ergonomics, usability, user experience, and product design. He is a Certified Professional Ergonomist and received the 2023 IEA Fellow award.

Contributors

Maria Victoria Cabrera Aguilera
Department of Mathematics, Faculty of Technology, State University of Rio de Janeiro Resende, Brazil

Vera Lopes de Abreu Lima
Pontifical Catholic University of Rio de Janeiro, Rio de Janeiro, Brazil

Robson José de Moura Barbosa
Postgraduate Program in Production Engineering, School of Industrial Metallurgical Engineering, Fluminense Federal University, Volta Redonda, Brazil

Eric Brangier
Université de Lorraine, Metz, France

Fabiana França
Federal University of Maranhão, São Luis, Brazil

Bernardo Bastos da Fonseca
Department of Production Engineering, Faculty of Technology, State University of Rio de Janeiro, Resende, Brazil

Eduardo Ferro dos Santos
Department of Basic and Environmental Sciences, São Paulo University, Lorena, Brazil

David Hitchcock
David Hitchcock Limited, Hugglescote, UK

Alison Alfred Klein
Postgraduate Program of Mechanical Engineering, Federal University of Parana, Curitiba, Brazil

Sanjeet Kumar
Department of Industrial and Systems Engineering, IIT Kharagpur, Kharagpur, India

Bernard Michez
Ergotec Company, Toulouse, France

Cláudia Mont'Alvão
Pontifical Catholic University of Rio de Janeiro, Rio de Janeiro, Brazil

Julien Nelson
Université Paris Cité, Paris, France

Alexandra Nemery
École de Technologie Supérieure, Montréal, Canada

Christopher Nemeth
Applied Research Associates, Inc, Albuquerque, USA

Maria Lucia Okimoto
Postgraduate Program of Mechanical Engineering, Federal University of Parana, Curitiba, Brazil

Márcia Raissa A.F. Pereira
Postgraduate Program in Ergonomics, Federal University of Pernambuco, Recife, Brazil

Fabiana Raulino
Pontifical Catholic University of São Paulo, São Paulo, Brazil

Pradip Kumar Ray
Department of Industrial and Systems Engineering, IIT Kharagpur, Kharagpur, India

Nilo Antonio de Souza Sampaio
Department of Mathematics, Faculty of Technology, State University of Rio de Janeiro, Resende, Brazil

Martin Schrepp
SAP SE, Walldorf, Germany

Carina Santos Silveira
Federal University of Bahia, Salvador, Brazil

Marcelo M. Soares
Postgraduate Program in Design, Federal University of Maranhão, São Luis, Brazil Postgraduate Program in Design, Federal University of Pernambuco, Recife, Brazil

Raimundo Lopes Diniz
Department of Design and Technology, Federal University of Maranhão, São Luiz, Brazil

Paul Thompson
Harmuk Limited, York, UK

Suzi Maria Carvalho Mariño
Federal University of Bahia, Salvador, Brazil

Matteo Zallio
University of Cambridge, Cambridge, UK

Aleksandar Zunjic
Faculty of Mechanical Engineering, University of Belgrade, Belgrade, Serbia

Preface

Ergonomics 5.0 emerges in response to a critical inflection point in the relationship between technology and labor. Over the past few decades, successive industrial revolutions have redefined the way we work, live, and interact with technology. While much of the discourse still revolves around Industry 4.0, it is undeniable that we have moved forward—we are now experiencing the reality of Industry and Society 5.0.

The most significant evolution lies in the growing awareness that the pursuit of efficiency can no longer be separated from the promotion of human well-being—just as innovation cannot progress without ethical and sustainable foundations. It is in the convergence of digital intelligence, resilience, and sustainability that Ergonomics 5.0 finds its core purpose.

This concept represents more than a new methodological approach; it signals a paradigm shift that once again places the human being at the center of system design and organizational decision-making. It calls on us to rethink not only our tools and processes but also the very values that underpin our productive environments.

This book brings together distinguished researchers and professionals from diverse disciplinary and geographical backgrounds, all united by a shared goal: to enable people to work intelligently, inclusively, ethically, and sustainably—achieving high performance, but in ways that remain deeply human.

This book presents an integrative and recent perspective, encompassing historical, methodological, technical, and ethical aspects of Ergonomics 5.0. We hope this book is informative, but even more, that it is thought-provoking. In this way, we intend for it to inspire readers to contemplate further questions, applications, and potential solutions in creating good ergonomics for all. The chapters presented here offer a comprehensive and forward-looking exploration of ergonomics in this era of technological–human convergence.

The book is comprised of two parts. The first part of this book addresses the principles and concepts of Ergonomics 5.0 and contains eight chapters. These chapters introduce the principles, approaches, and advances of Ergonomics 5.0; explain how ergonomic principles support the emergence of a human-centric orientation in Industry 5.0; show how prospective ergonomics contributes to innovation in emerging and future contexts of use; describe how resilience can be applied in ergonomics in a way to anticipate and model solutions that ensure human efficiency and well-being; explore the fundamental principles of inclusive design in advancing Ergonomics 5.0; discuss the impact of artificial intelligence on the future of workers; examine the user problems in interacting with privacy interfaces and how ergonomics can reduce these issues; and finally, discuss the advantages and limitations to measure the user experience (UX) in Ergonomics 5.0 scenario.

The second part of this book includes some implementations of Ergonomics 5.0 and comprises two chapters with the applications of ergonomics in automotive industry; how ergonomics and artificial intelligence can impact occupational stress and strategies for workers's health and organizational effectiveness; what are the implications, opportunities, and challenges of gamification in evolving Ergonomics

5.0; how to use of an advanced platforms knowing as kinebot and passive skeleton to assess ergonomic practices and activities; the use of a human-centric assessment tool to reduce dissatisfaction, discomfort and injury to contribute to the provision of a human-centric workplace; and finally a report of investigation of the nature and extent of ergonomic challenges encountered in patient handling activities within Indian healthcare systems.

The goal of this book is to contribute to the consolidation of Ergonomics 5.0 as an operational vision—capable of guiding ergonomists, informing policies, and inspiring transformative actions across sectors. We hope this work serves both as a reference and a provocation for scholars, professionals, and decision-makers committed to advancing truly human-centered innovation.

Eduardo Ferro dos Santos
Marcelo M. Soares
Editors

Acknowledgements

We wish to express our deepest gratitude to the contributors whose intellectual generosity, methodological rigor, and visionary perspectives have shaped the substance and character of this work.

Each chapter in this volume reflects not only the expertise of its authors but also a shared commitment to advancing ergonomics in alignment with the complex demands of the contemporary world. From foundational theoretical insights to applied innovations in diverse industrial contexts, your contributions have enriched this book with multidimensional approaches, empirical depth, and practical relevance.

We recognize the effort required to produce high-quality academic work under tight deadlines, often while balancing multiple professional obligations. Their dedication, responsiveness during the editorial process, and openness to constructive dialogue were essential to achieving the coherence and quality we envisioned for this publication.

Moreover, we are particularly grateful for the spirit of collaboration and mutual respect that permeated our exchanges. This collective effort transcends the bounds of disciplinary silos and national borders, exemplifying the transdisciplinary and international cooperation that Ergonomics 5.0 calls for.

May this volume serve not only as a repository of knowledge but as a catalyst for further research, dialogue, and transformation in the fields of human factors, design, and occupational innovation. To all of you—authors, reviewers, and academic partners—our sincere thanks.

Part I

Principles and Concepts

The first part of this volume is dedicated to establishing the conceptual and theoretical foundations of Ergonomics 5.0 in light of the transformative processes introduced by Industry 5.0 and Society 5.0. As technological systems evolve toward higher complexity and autonomy, ergonomics is called upon to reframe its practices, not only to adapt to these new configurations but to shape them in alignment with human values actively.

This section examines how ergonomics, originally focused on optimizing physical and cognitive performance, has expanded to encompass principles of digital intelligence, human centrality, sustainability, and resilience. The chapters herein delve into the philosophical and operational transitions that support this evolution, offering critical perspectives on the changing role of human factors in the design of future-oriented systems.

Together, these contributions underscore the urgent need for a renewed ergonomic paradigm—one that can engage ethically and effectively with the socio-technological systems of the 21st century. This part provides the epistemological and methodological scaffolding for the applied discussions that follow in Part II, where principles meet practice.

DOI: 10.1201/9781003559399-1

1 Ergonomics 5.0
Principles and Perspectives

Eduardo Ferro dos Santos and Marcelo M. Soares

1.1 BACKGROUND

The discourse surrounding Industry 4.0, which revolutionized the industrial sector by integrating technologies such as automation, the Internet of Things, artificial intelligence, and big data, remains substantial. However, technological advances have progressed further. The current trajectory marks the advent of Industry 5.0, a paradigm prioritizing human-centricity and sustainability as core drivers of technological transformation.

This chapter examines the historical evolution of industrial revolutions, culminating in Era 5.0, and highlights the pivotal roles of human centrality and sustainability in reshaping production processes, as well as the interplay between technology and society. In this context, the concept of Ergonomics 5.0 emerges as a progressive framework that aligns with the advancements of Industry 5.0 and amplifies them through solutions that integrate technological innovation, human well-being, and environmental stewardship. By fostering a synergy between digital intelligence (DI), resilience, and sustainability, Ergonomics 5.0 introduces an integrative approach that redefines ergonomic practices to address the unique challenges of this era.

Accordingly, this chapter emphasizes the importance of reimagining human-technology interactions, advocating for a future where technological progress is aligned with inclusion, balance, and shared prosperity.

1.1.1 The Concept of Eras

Eras are significant historical periods characterized by transformative changes in various domains, including culture, technology, politics, economics, and society. These divisions help us understand the dynamics of each period and trace an evolutionary trajectory of events.

In the industrial context, the era concept is also used to describe technological and organizational advancements that have reshaped how we produce, work, and consume. Figure 1.1 illustrates the evolution of industrial eras, from the first revolution (Industry 1.0) to the fifth (Industry 5.0), highlighting the defining characteristics of each phase.

Industry 1.0, the initial phase of industrial evolution, was characterized by the mechanization of production through the use of steam-powered machinery. The energy harnessed from steam revolutionized industry and transportation. This era

DOI: 10.1201/9781003559399-2

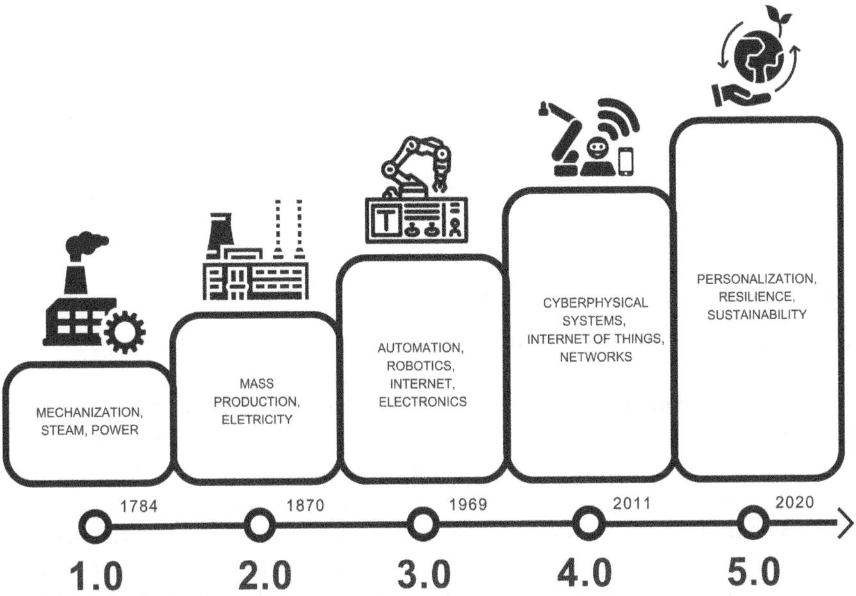

FIGURE 1.1 Evolution of industrial eras.

Source: The authors.

was characterized by a shift from manual and artisanal production processes to mechanized systems, which significantly boosted productivity.

During the second phase, a significant transformation occurred with the advent of electricity, marking the emergence of **Industry 2.0**. This era enabled mass production by developing electrical systems and assembly lines, as exemplified by Henry Ford's innovative production line. This marked the beginning of large-scale manufacturing, allowing the production of goods in vast quantities at reduced costs.

The era of **Industry 3.0** brought computing to the forefront of industrial processes. Production became controlled by computational systems, enhancing factory efficiency and precision. This period, known as the automation era, saw automated systems replacing human labor in many industrial sectors.

The much-discussed **Industry 4.0** was defined by integrating systems, particularly cyber-physical systems and connective networks. During this era, factories became highly interconnected and intelligent, facilitating machine-to-machine communication and automating complex processes. The complete digitalization of industrial environments has enabled greater flexibility and efficiency in production.

However, we are now in the era of **Industry 5.0**, which began in Europe in 2020 (Breque et al., 2021). This phase emphasizes human-centricity, resilience, and sustainability. Industries now focus on creating personalized products, swiftly adapting to consumer needs and market changes. Sustainability has become a priority, with production processes designed to minimize environmental impacts and promote social well-being. Unlike previous eras, Industry 5.0 transcends a sole focus on efficiency and technology. This new phase integrates human capabilities with technological advancements, promoting more sustainable, personalized, and resilient production.

This advancement is not merely technological but represents a shift in mindset. It highlights the need to strike a balance between technological progress, human well-being, and environmental stewardship. This era addresses contemporary societal demands for innovation, as well as social and environmental responsibility. It is pivotal to combine cutting-edge technology with a more human and sustainable perspective. More than ever, it is time to advance into an era characterized by balance, innovation, and inclusion, aligned with the principles of ergonomics.

1.1.2 Key Pillars of Industry 5.0

The key pillars of Industry 5.0 are:

- **Human-Centricity**: Industry 5.0 values human skills in creating value and promoting harmonious collaboration between humans and machines. Technologies like collaborative robots (cobots) are designed to work alongside workers, complementing rather than replacing their abilities.
- **Sustainability:** Our production is oriented toward reducing environmental impact, reusing resources, and enhancing energy efficiency. Concepts, such as the circular economy, are gaining prominence and encouraging recycling and material reuse.
- **Resilience:** Industrial systems are designed to adapt to crises, market changes, and natural disasters. This includes more robust supply chains and technologies for predicting and mitigating risks.

The vision for Industry 5.0 did not originate in Europe but in Japan, with the introduction of Society 5.0, which was rapidly adopted in Europe for industrial applications. Developed in Japan, Society 5.0 proposes a social revolution based on the strategic use of technology to address societal challenges and enhance quality of life (Deguchi et al., 2020). Unlike previous phases of social organization, Society 5.0 transcends mere digitalization by focusing on harmony between technology and human well-being. The goal is to create an inclusive, sustainable, and human-centered society where technology serves as a means, not an end in itself.

Inspired by the humanistic principles of Society 5.0, Industry 5.0 builds upon the concept of Industry 4.0. While Industry 4.0 is predominantly technology-driven, emphasizing automation and efficiency (Xu et al., 2021), Industry 5.0 prioritizes value creation by fostering collaboration between humans and intelligent machines to address demands for personalization, creativity, and sustainability (Leng et al., 2022). This model aims to strike a balance between technological advancements and environmental respect, adopting practices that promote economic circularity and resilience.

Human centrality is a shared and essential element in the concepts of Society 5.0 and Industry 5.0, demonstrating that technological development cannot be detached from its social context. As an integral part of society, industry plays a vital role in driving social transformation. Conversely, social transformations often drive new industrial revolutions, creating a cycle of interdependence (Huang et al., 2022). Integrating these concepts under Era 5.0 gives us a comprehensive vision of the future, where technology and humanism intertwine to create a balanced and sustainable environment.

Speculations about the next era, Era 6.0, anticipate a transformative global scenario by 2050 (Chourasia et al., 2022). This phase addresses current fears about job displacement due to technology, emphasizing the need for antifragility in businesses and lifelong education to enhance societal resilience amid ongoing changes.

It is paramount to recognize the advancements of Industry 5.0, as Europe and Asia have already done. Continued discussions on Industry 4.0 led us to focus on the digitalization and automation of industrial processes, with optimizing efficiency and productivity as top priorities. However, this model no longer fits the current context. At its core, Industry 4.0 aligns with profit maximization and economic efficiency within traditional business models, often at the expense of underlying environmental threats.

Industry 5.0 highlights the need for a more balanced and human approach to technology application. It recognizes that pursuing efficiency and profit solely may perpetuate or exacerbate social and environmental issues. Beyond efficiency and automation, technologies must be developed and applied to promote social well-being, environmental sustainability, and equity.

Era 5.0 proposes a balance between technological development and human quality of life, addressing global challenges in a more holistic manner. As we enter these new eras, it is crucial to maintain dialogue on how technology can be shaped to serve humanity better, rather than the other way around.

This alignment with human centrality, sustainability, and inclusion makes the transition to this era essential. Industry 5.0 is intrinsically linked to ergonomics, which plays a crucial role in adapting technologies to human needs. From this new era, the concept of Ergonomics 5.0 emerges.

1.1.3 Ergonomics 5.0

We live in the Cyber-Physical Era, a period marked by the growing integration of the physical and digital worlds. This era is defined not merely by the presence of advanced digital technologies but also by their capacity to continuously interact with the real world in a connected and intelligent manner.

The Cyber-Physical Era serves as the technological foundation underpinning concepts such as Industry 5.0 and Society 5.0. In Industry 5.0, cyber-physical systems enable seamless collaboration between humans and machines, facilitating product personalization and promoting more sustainable processes. Similarly, in Society 5.0, these systems address societal challenges and improve the quality of life through intelligent and connected services. Within this dynamic landscape, the concepts of Era 5.0 emphasize the need for humans to remain at the center of all processes. The challenge now lies in ensuring this integration is ethical, inclusive, and sustainable, fostering a future where technology genuinely serves humanity.

In alignment with these concepts, the term Ergonomics 5.0 emerges to provide ergonomists with a perspective on the evolution of technology. It offers a fundamental approach to ensure that interactions between humans and technology are efficient, effective, safe, healthy, and sustainable.

Ergonomics 5.0 represents the era of integrating technological advances with human values, aligning efficiency, well-being, and sustainability in a highly connected and automated world (Santos, 2024). It is based on three key principles: DI, Resilience, and Sustainability. It responds to accelerated social and technological transformations, particularly following the COVID-19 pandemic and the widespread adoption of artificial intelligence.

Ergonomics 5.0 can be defined as a contemporary approach to the interaction between humans and other system elements that integrates technological advances with human values, aligning efficiency, well-being, and sustainability in a highly connected and automated world. Based on the principles of DI, Resilience, and Sustainability, Ergonomics 5.0 recognizes the centrality of the human being in interaction with emerging and disruptive technologies, promoting ethical, safe, and pleasurable experiences. It also incorporates practices that minimize environmental, social, and economic impacts, creating balanced solutions that harmoniously prioritize individual and collective well-being alongside technological progress.

Ergonomics 5.0 is an advanced concept emerging in the context of Industry 5.0 and Society 5.0. It proposes a systematic integration of emerging technologies and human values. Based on the DI, Resilience, and Sustainability pillars, Ergonomics 5.0 seeks to optimize the interaction between humans and technological systems, guaranteeing operational efficiency, safety, and well-being.

Historically, ergonomics has often been constrained by a regulatory approach focused on compliance with outdated standards. Even when revised, such standards are frequently perceived as obsolete due to the slow pace of regulatory processes. However, ergonomics cannot be limited to regulatory norms.

With the rise of the digital era, the nature of work has undergone drastic changes. Tasks once performed manually are now executed through computers, mobile devices, and automated systems. Ergonomics 5.0 responds to this transformation, recognizing that human-machine interaction has evolved significantly and must no longer be approached with inertia. It serves not only as a foundation for advancing ergonomics but also as a guide for ergonomists, who often remain confined to observation and the application of outdated analytical techniques based on tools published decades ago.

Ergonomics 5.0 integrates technological advancements with human values, aligning efficiency, well-being, and sustainability in a highly connected and automated world. Its foundation is built on three core principles: DI, Resilience, and Sustainability.

1.1.4 Digital Intelligence

Humans remain at the center of interaction with emerging and disruptive technologies, ensuring that these tools are employed in ways that are:

- **Ethical:** Respecting human limits and needs while promoting accessibility and inclusion.

- **Efficient**: Optimizing processes and minimizing unnecessary effort.
- **Safe:** Mitigating risks associated with the use of new technologies.
- **Enjoyable:** Providing positive and satisfying experiences that enhance well-being and foster emotional connections.

  This principle underscores the importance of designing user-centered interfaces and systems that facilitate tasks and enrich the human experience.

1.1.5 Resilience

Resilient ergonomics transcends reactive problem-solving by anticipating challenges and optimizing processes to sustain human well-being and efficiency in the face of constant change. Actions must be highly adaptable to keep pace with rapid technological advancements and evolving human needs.

Resilience in ergonomics is not limited to responsiveness; it is expressed in the capacity to anticipate and model solutions that ensure human efficiency and well-being in an environment of relentless innovation and challenges.

1.1.6 Sustainability

Sustainability is a central principle of human interaction within the systems they inhabit, acknowledging the interdependence between:

- **Economic Impacts**: Promoting ergonomic solutions that optimize costs and enhance productivity.
- **Environmental Impacts:** Minimize the use of natural resources and reduce waste in product and process design.
- **Social Impacts**: Ensuring ergonomic practices benefit communities and promote equity, health, and quality of life.

  This principle emphasizes the importance of balancing technological advancements with responsible practices, thereby creating solutions that respect both the planet and its inhabitants.

By combining these principles, Ergonomics 5.0 provides a forward-thinking framework for aligning technological progress with humanity's values. This will pave the way for a more balanced, sustainable, and human-centered future.

Ergonomics 5.0 represents more than a mere technical evolution. Emphasizing DI, resilience, and sustainability provides a pathway to harmoniously integrate technology and humanity, ensuring that technological progress serves the well-being and collective prosperity of society.

The evolution of ergonomics over time reflects humanity's ability to adapt environments, tools, and systems to meet the needs of different eras. The progression from Ergonomics 1.0 to 5.0 illustrates a trajectory that begins with a focus on physical performance and culminates in the integration of technology and sustainability in the digital age.

1.1.7 Ergonomics 1.0—Optimizing Physical Performance

Ergonomics 1.0 is rooted in fundamental human needs. It applies subsistence principles to develop tools and artifacts that enhance performance in daily tasks. The primary goal is to increase efficiency and reduce physical effort in survival activities such as farming and hunting. This phase laid the foundation for ergonomics as an applied science, albeit in an intuitive and empirical manner. Primitive tools, such as the lever and the wheel, exemplify this era, optimizing the application of human force.

1.1.8 Ergonomics 2.0—Human Labor Optimization

During Ergonomics 2.0, applications expanded into military and industrial contexts. In wartime and mass production, the focus was on optimizing human labor, creating "super soldiers," or adapting factory workstations. This era introduced scientific tools to study human variables, marking the beginning of ergonomics as a distinct scientific discipline through the measurement and analysis of human performance. An example is the aircraft cockpit, which is optimized for ease of control and error reduction under high-pressure conditions.

1.1.9 Ergonomics 3.0—Standards and Protocols

The third phase marked a significant leap in formalizing standards, guidelines, and protocols for ergonomic analysis. Regulations began encompassing physical, cognitive, and organizational domains. This era established ergonomics as a standardized practice integrated into organizational management, including the development of regulatory norms and ergonomic management programs.

1.1.10 Ergonomics 4.0—Psychodynamics of Work

With Ergonomics 4.0, activity analysis became a central focus. Moving beyond purely protocol-driven evaluations, this phase sought to understand and transform work at its core, incorporating subjective and psychosocial aspects through the study of work psychodynamics and activity analysis. This era expanded the scope of ergonomics, linking mental health and work dynamics to productivity and well-being—for example, organizational ergonomics studies aimed to improve worker satisfaction and reduce stress-related impacts.

1.1.11 Ergonomics 5.0—Digital Intelligence, Resilience, and Sustainability

The most recent evolution, Ergonomics 5.0, emerges within the context of Industry 5.0 and Society 5.0. This phase responds to accelerated social and technological transformations, particularly following the COVID-19 pandemic and the widespread adoption of artificial intelligence. Ergonomics 5.0 addresses the challenges of the digital age, transforming them into opportunities to create healthier, safer, and more balanced work environments. It bridges the past and future, positioning humans at the center of all innovations.

Ergonomics 5.0 incorporates advanced concepts such as captology, hedonomics, and eudaimonia, aligning technology, well-being, and human purpose in a harmonious framework:

- **Captology:** The study of how technology influences and shapes human behavior is vital in designing ergonomic systems that optimize tasks and guide users toward healthy and efficient interactions (Fogg, 2003). This is crucial for developing tools and interfaces that encourage conscious choices, minimize distractions, and reduce cognitive overload, making work and technological interactions smoother and more productive.
- **Hedonomics**: Focused on designing pleasurable and satisfying experiences during interactions with products and systems. Hedonomics encompasses human satisfaction beyond health, safety, and performance (Hancock et al., 2005). This is particularly significant in environments where human-technology interaction is constant, such as cyber-physical systems and digital interfaces, ensuring individuals feel connected and valued in their tasks.
- **Eudaimonia:** It transcends immediate satisfaction, emphasizing personal growth and fulfillment. It manifests in creating environments and systems that facilitate work and foster human development, enabling individuals to achieve their potential and find meaning in their daily activities (Seaborn et al., 2019). This approach links individual well-being to collective impact, promoting positive outcomes in workplaces and society.

Combining these concepts, Ergonomics 5.0 proposes an integrated model that optimizes human performance and fosters enriching and sustainable experiences. This approach is crucial for addressing the challenges of the digital era and building a future where technology and humanism are deeply intertwined.

1.2 DIGITAL INTELLIGENCE

DI, defined as the ability to understand, reason, and learn using digital information (Rahman et al., 2021), plays a crucial role in Ergonomics 5.0. This approach integrates emerging and disruptive technologies to optimize human interactions, technology, and the workplace. It aims to improve efficiency and ensure worker well-being in an increasingly digital and interconnected world.

Soares (2021) defines Human-Centered Design as the process of creating products and systems based on the general characteristics and peculiarities of human psychology and perception. However, the concept of Human-Centered Design is strongly associated with Industry 5.0. DI is incorporated as a key principle within Ergonomics 5.0. Ergonomics, by definition, is already human-centered. DI enhances this foundation by introducing technological tools that facilitate more intuitive, accessible, and effective interactions, seamlessly integrating technology into the workplace without compromising human well-being.

DI is a transformative force in Ergonomics 5.0, bridging science, technology, and humanism. Its application extends beyond work optimization, fostering healthier, more interactive, and sustainable work environments. By equipping ergonomists with advanced digital tools, DI ensures that the future of work becomes more efficient, human, and inclusive.

This competence is indispensable for ergonomists and workers today, where technology is central to nearly all human activities.

1.2.1 For the Ergonomist

Ergonomists require DI to effectively use emerging technologies and advanced software capable of collecting and processing large volumes of data. With these tools, they can analyze patterns in behavior, posture, movement, and environmental factors affecting worker health and performance.

DI also enables ergonomists to develop digital models and simulations to test and validate solutions before implementation. This skill is critical for designing human-centered workspaces, tools, and interfaces. Ergonomists proficient in DI can create more intuitive and efficient digital systems, fostering seamless interactions between humans and machines. This ensures that technology is an ally in the workplace, enhancing performance without compromising worker well-being.

Additionally, DI empowers ergonomists to share knowledge more effectively by utilizing virtual reality, augmented reality, and interactive simulations. These tools make training sessions more accessible and impactful, enabling workers to navigate complex digital environments effectively.

1.2.2 For the Worker

DI emerges as a tool for workers' adaptation and growth. In the modern workplace, employees increasingly interact with digital systems, intelligent machines, and connected devices. DI equips them to operate these technologies safely and efficiently, minimizing errors and maximizing performance. For instance, machine operators must understand how to adjust the parameters of digital equipment to complete tasks without overexerting themselves or making errors.

DI also helps workers adapt to constant workplace changes, such as the introduction of new digital tools or the transition to hybrid work models. This adaptability is critical in a rapidly evolving world.

1.2.3 A Shared Necessity

Ergonomists and workers require DI to navigate the contemporary workplace, which is characterized by rapid technological transformations. For ergonomists, DI amplifies their ability to develop human-centered solutions. It also gives workers the skills to operate, adapt, and thrive in a digital world.

In Ergonomics 5.0, DI serves as a foundation for innovation and a bridge between human capabilities and technological advancements. It ensures an efficient, inclusive, and future-ready workplace.

1.3 RESILIENCE

Resilience is the ability to face adversity, overcome obstacles, cope with stress, pressure, or trauma, and adapt positively to changes and challenges. It is the skill to recover, learn from difficult experiences, and continue growing even in adverse circumstances (Pinheiro, 2004).

Resilient individuals do more than "cope" with challenges; they demonstrate creativity in problem-solving, remain calm under pressure, and emerge stronger and more adaptable. This ability is not fixed or purely innate but can be developed and refined throughout life.

Resilience is an indispensable quality in both personal and professional contexts. It enables individuals to overcome difficulties, adapt to changes, and thrive even in adverse conditions. In a dynamic and constantly evolving world, particularly in Era 5.0, where advanced technologies, artificial intelligence, and rapid transformations reshape work environments and ways of life, resilience has become a fundamental pillar.

Ergonomics 5.0 acknowledges that modern work environments are characterized by complex challenges, many of which arise from the rapid pace of technological advancements. This evolution has a profound impact on our daily tasks, requiring professionals and organizations to demonstrate flexibility and agility to adapt.

Resilient professionals can effectively and safely integrate new technologies into the workplace. They adapt to changes and view them as opportunities for innovation and process improvement. This mindset is essential in high-complexity environments where stress and pressure are inherent.

Furthermore, resilience directly contributes to mental and emotional well-being, enabling professionals to maintain high performance levels without compromising health. This balance benefits individuals and fosters healthier and more productive workplaces, where ergonomic challenges are viewed as opportunities for learning and growth.

The COVID-19 pandemic underscored the importance of resilience at individual, organizational, and societal levels. The ability to quickly adapt to unexpected changes, such as social isolation, supply chain disruptions, and the need for digital transformation, was crucial to maintaining operations, protecting public health, and sustaining the economy.

In this context, resilience was not merely about surviving the crisis but also learning, adapting, and emerging stronger. Companies and professionals who demonstrated resilience could restructure processes, adopt new technologies, and transform work models with agility, ensuring their relevance even in times of uncertainty.

Resilience is even more critical in Era 5.0. Today's challenges are increasingly complex and interconnected. Accelerated technological advancements, climate change, and global social and economic transformations demand continuous adaptation, lifelong learning, and innovation during crises.

Resilience is about resisting change and thriving in dynamic and uncertain environments. It involves transforming obstacles into opportunities for growth and evolution. This skill is fundamental to advancing ergonomic principles in Era 5.0, where human-technology interactions must be continually refined to meet the needs of well-being and sustainability.

Resilient professionals are willing to embrace change and innovate. They recognize challenges as opportunities for learning and understand that every experience, even adverse ones, contributes to individual growth. This continuous pursuit of development benefits the professional and strengthens teams and organizations.

For instance, when implementing new technologies, resilient professionals demonstrate creativity in problem-solving and openness to testing new approaches that may enhance the work experience. This attitude is vital for the evolution of ergonomics, which must adapt to the constant changes of the digital world.

In Ergonomics 5.0, resilience is an invaluable investment. It ensures that workplaces are safe, healthy, dynamic, and adaptable. In an uncertain future, resilience will be the key to thriving, promoting individual, organizational, and societal growth in harmony with technology and human well-being.

Resilience ensures that ergonomic practices align with the demands of a fast-paced world, transforming workplaces into environments that foster innovation, adaptability, and holistic growth. This critical competence bridges the gap between challenges and opportunities, paving the way for a future where technology and humanity coexist in balance and mutual enrichment.

1.4 SUSTAINABILITY

Sustainability has become a fundamental pillar in contemporary society, guiding decisions across environmental, social, and economic domains. This concept is intrinsically linked to the triple challenge of sustainability, which integrates these three dimensions and seeks solutions to meet the needs of the present without compromising the ability of future generations to meet their own.

In Ergonomics 5.0, sustainability assumes a central role, recognizing that interactions between humans, technology, and the workplace must be guided by principles ensuring balance and harmony. More than an isolated practice, sustainability in ergonomics represents a holistic approach that aims to enhance worker well-being, organizational productivity, and promote a more inclusive, ethical, and sustainable future for all stakeholders.

Ergonomics 5.0 adopts a comprehensive approach to sustainability, integrating economic, environmental, and social aspects into its practices and objectives. This integration fosters a more balanced working model that considers both the short—term and long-term impacts of human-technology interactions.

1.4.1 Economic: Sustainable Economic Impact

Economic viability is a critical component of sustainability in Ergonomics 5.0. Ergonomic changes must be evaluated in terms of their initial implementation costs and the long-term benefits they provide. This includes:

- Investment in ergonomic technologies enhances productivity and efficiency while ensuring worker safety and health.

- Reduction of operational costs, such as occupational health expenses, absenteeism, and employee turnover.
- Improved competitiveness in the global market by valuing human capital and optimizing processes.

Adopting a sustainable economic perspective is crucial for organizations to balance the costs of ergonomic changes with the long-term financial and social benefits achieved.

1.4.2 Environment: Responsible Environmental Preservation

Adopting advanced technologies can often increase the consumption of natural resources and generate waste. In Ergonomics 5.0, environmental sustainability is promoted through responsible practices, including:

- Developing low-impact projects that prioritize recyclable materials and energy-efficient technologies.
- Reducing waste in production processes by utilizing technological solutions that optimize resource usage.
- Sustainable design that considers the entire lifecycle of products—from conception to disposal—minimizing environmental impacts.

This approach ensures that technological progress and innovation become allies of environmental preservation rather than exacerbating ecological issues.

1.4.3 Society: Worker Well-Being and Inclusion

Worker well-being is a fundamental aspect of sustainability in Ergonomics 5.0. This entails creating working conditions that prevent health risks and promote quality of life and job satisfaction. Key initiatives include:

- Safe and healthy work environments minimize stress, fatigue, and ergonomic risks.
- Ethical and social responsibility, incorporating moral values into developing and implementing new technologies.
- Respect for diversity and inclusion, ensuring that ergonomic solutions are accessible and relevant for all workers, regardless of their characteristics or contexts.

This social dimension of sustainability underscores a commitment to creating a work environment that respects human dignity and promotes equity.

Sustainability in Ergonomics 5.0 transcends conventional practices by aligning technological innovation with the principles of environmental preservation, social responsibility, and economic balance. This integrated approach ensures that workplaces not only meet the demands of the present but also contribute to a future where human and technological progress coexist harmoniously, fostering well-being and prosperity for individuals, organizations, and society.

1.5 CONCLUSION

Introducing a concept as transformative as Ergonomics 5.0 is no simple task. It is, in fact, a monumental challenge. For Ergonomics 5.0 to move beyond being a mere concept and become a tangible practice, more than theoretical actions are required. It demands a cultural shift—a transition to a new era in a society that often clings to tradition.

Culture, by definition, is a set of values, beliefs, norms, practices, and behaviors shared by a group. It shapes how people perceive the world, make decisions, and interact. Culture is akin to a living, complex, and interconnected system that defines the identity of a group. It establishes the foundations upon which individuals and groups act and interpret the world around them (Costa, 2023).

Changing culture requires persistence, strategy, and, above all, a collective effort. In the realm of ergonomics, resistance often goes beyond ignorance. It is deeply rooted in the fears and uncertainties that accompany any significant change. People frequently fear the unfamiliar, question the necessity of innovation, and cling to practices that, although outdated, still feel safe and familiar.

Despite its apparent solidity, culture is not an unyielding rock. It is a dynamic system that constantly evolves in response to social, technological, economic, and political changes. This dynamism presents an opportunity. Advancements in Industry 5.0 and Society 5.0 are profoundly transforming financial and social environments. Ergonomics can—and should—"ride this wave" of transformation.

Ergonomics 5.0 is not a fleeting trend or a temporary concept destined to fade over time. It represents a paradigm shift—a new way of thinking and acting in the workplace, grounded in the integration of humans and technology. However, to realize its potential, more is needed than technical adjustments or short-term policies. What is required is an embedded ergonomic culture.

An ergonomic culture goes beyond implementing adjustable workstations or offering occasional training. It is a comprehensive organizational approach that prioritizes worker health, safety, and comfort in all decisions. This entails recognizing that employee well-being is not just an ethical obligation but a critical component of productivity, innovation, and organizational success.

In organizations with a well-established ergonomic culture:

- Work environments are designed to minimize risks and maximize comfort.
- Healthy practices are encouraged and seamlessly integrated into employees' daily routines.
- Leaders and managers collaborate with ergonomists to create ideal working conditions.
- Ergonomics is viewed as an investment, not a cost, with proven returns in productivity and engagement.

This vision must be clear and, more importantly, memorable. It is essential to inspire everyone within the organization to incorporate ergonomics into the corporate DNA.

Promoting an ergonomic culture is rewarding, but it is not without obstacles. One of the biggest challenges is resistance to change. People may question the relevance of ergonomics, view initial costs as prohibitive, or underestimate its long-term benefits.

A lack of knowledge about ergonomics can hinder acceptance. Many still see it as an "extra" or secondary aspect rather than an essential component of organizational success. Resource limitations, especially in small and medium-sized enterprises, pose significant challenges, as tight budgets often restrict long-term initiatives.

However, these challenges are not insurmountable. With committed leadership, continuous education, and clear strategies, ergonomics can be transformed into an organizational priority. Leaders play a pivotal role: They inspire, set examples, and mobilize the necessary resources to overcome barriers.

The benefits of an ergonomic culture are profound and far-reaching. It improves workers' physical and mental well-being, reduces costs associated with absenteeism and occupational illnesses, and enhances productivity and engagement. Beyond these advantages, an ergonomic culture fosters resilience, preparing organizations to meet the challenges of the digital and technological age.

The COVID-19 pandemic provided a clear example of how organizations with robust and adaptive cultures were able to reinvent themselves during an unprecedented crisis. Ergonomics played a vital role in this process, helping companies reorganize their work environments, implement hybrid models, and prioritize employees' physical and mental health. This highlights that fostering an ergonomic culture is not just a strategic choice but a necessity for long-term organizational sustainability.

An ergonomic culture cannot be overlooked in Ergonomics 5.0, where the interaction between humans and technology is increasingly prevalent. It must evolve in tandem with the demands of modern work. This evolution entails creating space for innovation, embracing transformation, and placing humans at the center of all decisions.

Promoting an effective discourse on Ergonomics 5.0 is a collective effort, a shift in mindset, organizational alignment, and a genuine commitment to well-being and progress. By embedding this culture within organizations, Ergonomics 5.0 can redefine workplaces, ensuring they are safe, sustainable, and centered on human values.

REFERENCES

Breque, M., De Nul, L., & Petridis, A. (2021). Industry 5.0: Towards a sustainable, human-centric and resilient European industry, Publications Office of the European Union.

Chourasia, S., Tyagi, A., Pandey, S. M., Walia, R. S., & Murtaza, Q. (2022). Sustainability of Industry 6.0 in a global perspective: Benefits and challenges. Mapan, 37(2), 443–452.

Costa, L. M. (2023). A cultura organizacional e seus impactos na formação das estratégias empresariais. Revista De Gestão E Secretariado, 14(11), 19662–19680.

Deguchi, A., Hirai, C., Matsuoka, H., Nakano, T., Oshima, K., Tai, M., & Tani, S. (2020). What is society 5.0. Society, 5, 1–24.

Fogg, B. J. (2003). Persuasive technology: Using computers to change what we think and do. Morgan Kaufmann.

Hancock, P. A., Pepe, A. A., & Murphy, L. L. (2005). Hedonomics: The power of positive and pleasurable ergonomics. Ergonomics in Design, 13(1), 8–14.

Huang, S., Wang, B., Li, X., Zheng, P., Mourtzis, D., & Wang, L. (2022). Industry 5.0 and Society 5.0—Comparison, complementation and co-evolution. Journal of Manufacturing Systems, 64, 424–428.

Leng, J., Sha, W., Wang, B., Zheng, P., Zhuang, C., Liu, Q., … & Wang, L. (2022). Industry 5.0: Prospect and retrospect. Journal of Manufacturing Systems, 65, 279–295.

Pinheiro, D. P. N. (2004). A resiliência em discussão. Psicologia em estudo, 9, 67–75.

Rahman, T., Amalia, A., & Aziz, Z. (2021). From digital literacy to digital intelligence. In 4th International Conference on Sustainable Innovation 2020–Social, Humanity, and Education (ICoSIHESS 2020) (pp. 154–159). Atlantis Press.

Santos, E. F. (2024). Ergonomia e Novas Tecnologias. SP: Editora Virtuos.

Seaborn, K., Pennefather, P., & Fels, D. I. (2019). Eudaimonia and Hedonia in the design and evaluation of a cooperative game for psychosocial well-being. Human–Computer Interaction, 35(4), 289–337.

Soares, M. M. (2021). Ergodesign methodology for product design: A human-centered approach (1st ed.). CRC Press. https://doi.org/10.1201/9781003214793

Xu, X., Lu, Y., Vogel-Heuser, B., & Wang, L. (2021). Industry 4.0 and Industry 5.0—Inception, conception and perception. Journal of Manufacturing Systems, 61, 530–535.

2 Ergonomics 5.0 for Human-Centric Approach of Industry 5.0

Aleksandar Zunjic and Bernard Michez

2.1 INTRODUCTION

The European Commission introduced Industry 5.0 (Ghobakhloo et al., 2022; Madsen & Berg, 2021). The idea and vision of Industry 5.0 was introduced in a policy brief released in January 2021 (European Commission, 2021; Alves et al., 2023). The new industrial paradigm aims to go beyond the Industry 4.0 philosophy because it includes human-centered, sustainability, and resilience dimensions, in order to further improve European industry (Briken et al., 2023; Van Erp et al., 2024). This chapter is concerned with how to achieve, by application of ergonomics, one of the three major objectives of Industry 5.0, the human-centric approach (Adel, 2022; Alves et al., 2023).

The aim of this chapter is to illustrate how ergonomic principles support the emergence of a human-centric orientation in Industry 5.0. Special emphasis will be laid on the role of ergonomics in the successful incorporation of humans and technology in the modern industrial systems. Also, important aspects of Ergonomics 5.0 that need more work to be done for the successful implementation of this approach in the future will be discussed.

2.2 HUMAN-CENTRIC ERGONOMIC PRINCIPLES FOR INDUSTRY 5.0

Industry 5.0 is the new paradigm that puts humans upfront in the industrial systems focusing beyond technology-centricity of Industry 4.0. The Federation of European Ergonomics Societies (FEES) has formulated four principles to achieve the human-centric approach of Industry 5.0 through ergonomics. These principles accentuate designing industrial systems in a way that takes into account human capabilities and limitations, recognizes the importance of workers' expertise, which also fosters collaborative design processes, as well as keeps in mind that technology cannot solve all the problems.

Below, each principle will be presented in detail and supported by recent scientific literature. For each principle, we will present examples from practice, which show what consequences in terms of human, material, and economic loss can appear when

DOI: 10.1201/9781003559399-3

the principles are not met. After these examples, we will present our case studies from the industry that will also point out what kind of losses can be expected if companies striving to implement Industry 5.0 do not apply them.

2.2.1 Principle One: Design Equipment Considering the Operator's Capabilities and Limitations

The emphasis of Industry 5.0 is design's foundational ergonomic approach that mandates human inclusion in the design of machines, workstations, and installations, with due attention to human physical and cognitive emerged and latent capabilities and limits. Designing "fitting the task to the human" is important to have technology enhancing the ability of humans rather than overpowering them. For example, Kadir and Broberg (2021) show how a human-centered design (HCD) in industrial work systems can generate (and "realize") high worker well-being and high system performance by a systematic infusion of human factor knowledge into design activities. They provide a guideline on how ergonomic factors (such as anthropometric data, cognitive load) are addressed in the phases of system development to ensure that the tools fit human capabilities and limitations. This leads to improved production as the HCDs prevent fatigue, mistakes, and accidents in the operation.

The human-centric nature of Industry 5.0 has made this principle more relevant than before. Previous instances from Industry 4.0 showed that an implementation that is based too much on technology did not account for human factors, and therefore, led to unsatisfactory results (Alves et al., 2023). In contrast, Industry 5.0 frameworks insist that new technologies be developed for and with the human operator in mind. Longo et al. (2020) also highlight the need for human-machine interaction to be developed in a manner that human strengths, such as creativity, flexibility, and ethical decision-making, are augmented upon, and human limitations, like cognitive overload and physical stress, are recognized. This implies that machines and interfaces should be ergonomic: controls or displays should be accessible and understandable, forces or loads on people should be within the range of human strength, as well as work tasks involving machines should be safe and in accordance with the physical and cognitive capabilities of workers.

Current research also states that for human-centric design, beyond ensuring basic safety and comfort, it is necessary to cope with cognitive workload and psychological well-being to perform optimally in high levels of automation (Trstenjak et al., 2025). This could include concepts related to adaptive interfaces that can change the level of load according to the operator's abilities, or assistive technologies (supplementary elements such as an exoskeleton or augmented reality prompts) that would enhance the operator's particular tasks and abilities without replacing the operator's position (Žižić et al., 2022). Such approaches seek to increase the operator's ability, for example, through wearable sensors and AI to extend sensory perception or decision support systems to aid cognitive tasks while always within the scope of human acceptance and control. As Žižić et al. (2022) explain, Industry 5.0 explicitly complements the Industry 4.0 paradigm with orientation to the worker, valuing the operator's skills and ensuring technology is adapted to people, not the other way around. In addition, in order to incorporate the human operator into the design of a system, one must recognize the variability of human characteristics, including the reality of physical dimensions, cognitive skills, and experience, and design flexible work systems that can be accommodated to such diversity.

If the design of systems within the frame of Industry 5.0 adopts ergonomic principles, for example, ensuring controls within the 5th percentile female to 95th percentile male reach envelope, offering adjustable equipment (Pheasant & Haslegrave, 2005), interfaces with natural mappings for easy use (Norman, 2013), and reducing cognitive loads within human capacity (Hart & Staveland, 1988), the operators will be able to comfortably and efficiently work together with advanced technologies in the Industry 5.0 system. Following this principle leads to production systems that are more user-friendly, safer, and more productive as supported design helps operators perform better with fewer errors and interruptions (Kadir & Broberg, 2021).

Designing equipment and installations in Industry 5.0 must start with the human user in mind. The human-centric design strategy has considerable proofs in ergonomics, where we can find that systems that were designed with human capacities in mind have had superior outcomes (Kadir & Broberg, 2021; Žižić et al., 2022). Recognizing the human abilities and constraints, as well as the constraints of technology, Industry 5.0 can enhance the optimal collaboration between humans and machines. As one review correctly notes, human-centric approach places human needs at the heart of the production process, asking what technology can do for workers and how it can be useful (Žižić et al., 2022). This ethos ensures that technological innovation in Industry 5.0 truly serves its end-users: the human operators.

2.2.1.1 Three Mile Island Nuclear Accident

The control room of the Three Mile Island reactor had poorly designed instruments that failed to take into consideration the needs of the operators. The ultimate fault of the accident lay with a critical valve that was stuck in the open position. However, the control panel design only told the operators that the power was off to the valve – not that the valve was open. This convinced operators that the coolant supply was adequate when, in fact, the fluid was draining away. The faulty and poorly thought-out human-factors-related design did not allow the operators to investigate the phenomenon, and contributed to a partial core meltdown. The investigators ultimately concluded that deficiencies in poorly designed instrumentation and interface – not merely "human error" – contributed to the accident (World Nuclear Association, 2022). The nuclear industry, in fact, has made major changes in standards for control room design to account for operator needs. This case directly illustrates the consequences of failing to design for human cognitive capacities and needs: operators were overwhelmed by alarm floods and poor presentation and organization of information, which prevented timely intervention.

2.2.1.2 Therac-25 Radiation Therapy Overdoses

The Therac-25 was a computerized medical accelerator that had meliorated human-computer interaction poorly in its design. The machine was complicated with little feedback for the operator, and had very limited operator error safety controls. Human operator gross errors from very rapid and repeated keystrokes created a race condition where safety was no longer valid, and full-body patients were irradiated with incorrect doses assuming great doses of radiation. (Leveson & Turner, 1993). There were six accidents because the machine failed to foresee the possibility of human errors without the existence of safety nets, demonstrating the danger of ignoring the operator in design. The investigation concluded that there were better HCD strategies to communicate error messages, error feedback, safety locks, and it would have likely avoided these errors and fatalities (Leveson & Turner, 1993). Throughout this investigation, the machine and regulatory standards were changed to include human factors. This case study presented here shows that designing for complex systems without understanding human error and adequate feedback can end in tragedy, and it demonstrates the necessity of human factor considerations for the design of systems.

2.2.1.3 Air Inter Flight 148 Autopilot Crash

A design fault with the autopilot interface of the Airbus A320 made a contribution to this accident in the mountains near Strasbourg. The mode control panel had a single display window to select one of two different descent modes (either vertical speed or flight path angle), and it had ambiguous inputs. In poor weather, the crew selected "33" not realizing it was actually a 3,300 ft/min descent rate, which was an even steeper and more dangerous descent than their intended 3.3 degree descent angle. User interface design flaws failed to separate modes or units properly from each other and did not provide pilot-specific feedback, thereby creating conditions for mode errors that led aircraft into uncontrollable terrain collapse. Furthermore, the investigators determined that cockpit displays with improved pilot-friendly presentation could have prevented the error (Admiral Cloudberg, 2019). A total of 87 people lost their lives in the Airbus A320 airplane accident that occurred on Mount Ste-Odile (Ludes et al., 1994). Following the accident, Airbus implemented changes to the autopilot interface through the addition of "V/S" and "FPA" display labels and four-digit screens to prevent such confusion. The fatal airplane accident showed both the fatal consequences of lacking human-focused design principles in control systems as well as the significance of cockpit interface design that considers human cognitive processes.

Well-analyzed real-world incidents confirm that systems designed without human performance factors in mind result in critical design weaknesses. Human factors leading to tragic system failures have been proven through multiple documented incidents such as Three Mile Island alarms being overwhelmed without access to vital information and Therac-25 having unergonomic user interfaces, and Air Inter Flight 148 with ambiguous autopilot modes. Ergonomic design principles need to be integrated into systems development to guarantee that designs operate properly to support users physically, cognitively and behaviorally thus ensuring operational performance and safety.

2.2.1.4 Change of Displays in a Power Plant

With the evolution of technology, the old analog displays in the control room of an aging power plant needed to be switched to digital technology. This change was made mainly because the old equipment could no longer be supported with spare parts. There were about 50 screens in the control room showing information concerning a number of systems.

A new digital display was needed, and the challenge was to make the new digital display the same as the old analog one. The supplier successfully reproduced all the existing displays but added one new element – a blinking white square in the top right corner of the screen. This square showed that the screen was working and refreshing the data.

Ergonomics experts who immediately noticed a problem noticed the blinking white square could interfere with the detection of other, vitally important, blinking information like that from visual alarms. This distraction means that operators could miss critical alarms in a room with about 50 screens. A simulation was done that includes and excludes blinking squares to prove this issue. Next, to confirm if the blinking indicator indeed did make alarm detection more difficult, results were measured.

There were a number of solutions that the experts on ergonomics together with the supplier considered. With the understanding of human vision physiology, we moved away from a blinking object to a rotating object. This solution was effective because:

- Blinking is particularly sensitive in peripheral vision.
- Rotation is less "intrusive" for peripheral vision.
- Because the object is rotating, it was easy for operators to be able to check the screen's functionality while also noticing any new blinking alarms.

This case shows how important it is to design for humans and the human factors when designing an interface. A small change in the least expected way can affect work efficiency and safety.

2.2.2 Principle Two: Utilize Operator Expertise Before Implementing Changes

Under the second principle, it is necessary to know and to use the enormous knowledge accumulated by those who have worked on the workshop and have labored in industry before a change is tried in an industrial process. That is, the expertise and experience of the workers who actually work the machines and systems must be proactively drawn upon by organizations. These operators hold a great deal of tacit knowledge about the work, knowledge of where things are going wrong, what workarounds they're using, how what left the shop could have been improved, which may be invisible to engineers or managers who are far from the workshop. Using this experiential information can greatly improve the success of new technology roll-outs (Johnson et al., 2019).

In the field of manufacturing, research has indicated that experienced operators acquire skills and "know-how" that are essential for guaranteeing quality and regarding of managing the variability of processes (Johnson et al., 2019). For example, these authors found that human inspectors in aerospace manufacturing had superior abilities in detecting product defects due to years of accumulated tacit knowledge. However, such expertise was frequently not documented in a formal manner and was exposed to the risk of being neglected in the implementation of new automated inspection systems. Johnson et al. (2019) showed that by capturing and scrutinizing operators' tacit knowledge by task analysis and interviews, companies would preserve that expertise and integrate it into the design of assistive technologies. Thus, we create advanced automation that reinforces the nuanced observations derived from workers' expertise.

The more general lesson is that before any change is implemented – be that a process re-design, a new machine, or a software upgrade – the people operating the existing process, who know it best, should be consulted. Those workers can call attention to potential problems and make suggestions based on what has, or hasn't, worked in practice. In the Industry 5.0 dimension, this means inclusive innovation, i.e., the recognition that operators are experts of their own domain and they should also be involved in the change initiatives (Orso et al., 2022).

One method of making use of operator knowledge is participatory design or participatory ergonomics, wherein workers are actively involved in the identification of issues and the formation of solutions in their work environment. There is empirical evidence supporting the fact that participatory approaches result in more effective and acceptable interventions. For example, a recent investigation by Orso et al. (2022) presents the way of work of operators to create, with researchers, by interviews and video-analysis of work tasks, the conception of innovation via an "employee-centric innovation" inspired approach, as a lever of Industry 5.0 transitions. Besides producing valuable data regarding workflow bottlenecks and safety issues, the participatory design process also provided real operational context to the proposed changes (Orso et al., 2022). When employees contribute to technological enhancements, their ownership grows, making them less resistant to change and enabling the utilization of their relevant expertise for solution improvement. These findings corroborate that from organizational research on new technology adoption that employee involvement serves as a positive mediator for success.

Research conducted by Tortorella et al. (2018) among Brazilian manufacturers showed that manufacturers achieving better industrial performance results from Industry 4.0 implementations followed approaches based on both employee involvement and continuous improvement practices. In other words, engaging workers and using their ideas enhanced the benefits of advanced manufacturing tech, whereas neglecting the human element could undermine those benefits (Tortorella et al., 2018). Experienced operators develop the ability to predict system-related problems that designers and developers may have overlooked through their direct use of machinery or workflow modification. Soliciting this input early allows for changes to be made to the design or implementation plan before significant mistakes or rework have to be done. In addition, this principle has a motivational component: taking into account workers' expertise shows respect, as people are more likely to be more engaged when their efforts are valued.

When their suggestions are put into practice, it establishes a positive feedback loop, and operators are encouraged to increasingly come up with positive inputs (Cunha et al., 2022). In fact, a meta-analysis by Cunha et al. (2022) observed that much of the Industry 4.0 development is still too techno-centric and that more "work-centered", participatory approaches are recommended in order to make visible workers' knowledge and concerns. Adopting operator expertise means these technological shifts are based in reality – the messy, complex reality of jobs, and not some best possible world abstraction. It eliminates the difference between "work-as-imagined" by planners and "work-as-done" by doers, which should lead to more seamless introductions of Industry 5.0 advances to the market.

Optimizing the skills of operators is a means for promoting continuous improvement. Workers on the ground are the first to run into issues and come up with improvisations or improvements. Companies that set up processes to gather those insights (like regular team debriefs, suggestion boxes, or Kaizen circles) tend to be more agile and innovative. Indeed, companies that transform operators into problem-solvers may improve their competitiveness. Conversely, not taking advantage of operator expertise may result in "reinventing the wheel" or making adjustments that prevent a working informal practice from being effective.

At the other end of the scale, lack of worker input can lead to failure of new systems because of unexpected problems in their design, problems that those closest to the work could have anticipated. An illustrative example is presented by Johnson et al. (2019): an automated inspection system was first missing certain subtle defects that veteran human inspectors regularly picked up by feel or sound; only when the engineers went to these inspectors and found out their methods were they able to retrofit the automation to consistently feel those defects. Indeed, this is a reminder that Industry 5.0 technologies (such as AI, sensors, robotics, etc.) need to be accompanied by human field savvy for optimal performance.

Before making any workplace change, Industry 5.0 advocates should "learn from the shop floor". Engaging operators early allows organizations to design changes that are workable and beneficial in practice. The literature consistently supports that worker participation leads to better outcomes: improved system performance, higher acceptance of new technology, and sustained continuous improvement (Tortorella et al., 2018). By valuing and utilizing the huge expertise of their workforce, companies can avoid costly missteps and co-create innovations that truly enhance both productivity and worker well-being – fulfilling the human-centric promise of Industry 5.0 (Cunha et al., 2022).

2.2.2.1 Challenger Space Shuttle Disaster

NASA, with its contractor, pushed ahead with the Challenger launch despite receiving numerous engineer warnings about the adverse impact of cold weather on the solid booster O-ring seals. Multiple veteran Morton Thiokol engineers presented management with evidence showing that cold launch temperatures would cause O-ring failure, threatening the safety of the rocket (Rogers Commission, 1986). The management team insisted on prioritizing their scheduling plans above the engineer professional expertise and chose to override engineering expertise. The decision-makers launched the shuttle in conditions that engineers had warned were unsafe, resulting in the O-ring failure, according to the engineer's predictions, with an explosion occurring 73 seconds after liftoff which killed all seven crew members. Expert engineering predictions received scant attention from authorities because the official investigation found this human failure to be an essential cause of the accident (Rogers Commission, 1986). NASA created extensive system changes to maintain operator/engineer expert input during critical go/no-go safety decisions so similar safety expertise failures would not occur again. This tragic case verifies the priority of respecting technical knowledge at points of operation over organization-driven pressures and shows organizations should incorporate operator expertise directly into their decision frameworks.

2.2.2.2 BP Texas City Refinery Explosion

Before this refinery disaster operators together with technicians maintained informal awareness of the distillation tower and its blowdown system though they had no official documentation about this knowledge. The level indicator showed evidence of becoming stuck while operators were aware that the blowdown drum operated using an outdated model prone to overfilling. Management denied or dismissed concerns about high liquid levels and alarming contents, which several experienced operators reported before starting the unit (U.S. Chemical Safety Board, 2007). The unit startup operation led to tower overfilling, which caused hydrocarbons to vent to an atmospheric stack resulting in fatalities of 15 workers and injuries to 180 people. The BP (British Petroleum) management team failed during investigations to incorporate operator skill sets and experience-based learning from existing incidents when making procedural revisions and cutting costs. Safety protocols and extra regulations were created by BP following the tragedy while management began to depend more heavily on worker knowledge, which led an outside panel to confirm operator experience should be essential for process change management. The case shows that managers who fail to grasp experiential shop-floor knowledge before updating procedures leave behind uncontrolled risks that cause fatal accidents as they bypass their responsibility to include operator expertise in change procedures.

2.2.2.3 Deepwater Horizon Oil Rig Blowout

The drilling supervisors and crew members on the Deepwater Horizon drilling rig possessed good understanding of operational behavior, which went unheeded when crucial decisions about minor procedures were made. BP (British Petroleum) modified the well completion schedule to boost efficiency through changes in the verification of cement integrity processes and wellbore temporary sealing without consulting the experienced drill crew about potential outcomes. The drill crew members voiced their concerns about a puzzling negative pressure test outcome at their last rig meeting before a shore-based BP manager disregarded their warnings and authorized drilling operations (National Commission, 2011). The decision to ignore drill crew experts and their recommendation for additional verification led to a plan that focused on equipment repair and heavyweight drilling mud replacement with less dense seawater slurry. The disaster, driven by misjudged information, ended in a fatal gas surge after the final decision was made. This explosion in the well led not only to the death of 11 workers, but also to an environmental oil spill catastrophe. The National Commission (2011) determined that human error in the blowout stemmed from BP operators who failed to incorporate their predictions and observations into their management of the failure. This historical event showed various industries that subject matter experts need to be involved in operations before any process or procedural changes are made. The new safety regulations for offshore operations demanded operators to conduct risk assessments before making any operational changes. The blowout event shows that operators should have used their environmental understanding and combined different protective measures, but they failed, which proves how important technical decisions are in high-risk operations.

The examined disaster cases show clear evidence that organizations that dismiss frontline staff knowledge create a gap between strategic choices and real operational requirements. Preventable disasters at Challenger, Texas City, and Deepwater Horizon were caused by disregarding both operator warnings and standing expertise. Successful change management practices require built-in systems to preserve and merge operator feedback because essential live operational know-how is needed to guide crucial safety choices.

2.2.2.4 Human Factor in Modernizing the Production System

This example comes from the food industry. The production concerns curdled cheese. The work of the operators consisted of taking the product with a ladle in a 40 kg basin and filling 1 kg pots for marketing. The filled pots are brought by a conveyor belt to a machine for the application of the antifungal blotter and the lid. The marking is the next step, and then the crating. The production line was old, subject to breakdowns, and it must be renovated with new equipment.

The HFE (human factors/ergonomics) specialist was mobilized when the project had already started, and several decisions had been made. In particular, the new depositing machine for the blotter and the lid had a maximum capacity of 5000 pots per hour. This matched the calculated average hourly production and aligned with the projected requirements of the facility. However, observations made on the production activity showed that the production varied between 3000 and 7000 pots/hour.

Indeed, the challenge for the teams was to process all the tonnage planned for the day, at the risk of having to throw away what had not been packaged. The speed variation was linked to several factors:

- Machine hazards and breakdowns. This factor was important with the equipment before the change, but it was necessary to anticipate that the new machines would also have breakdowns.
- Interruptions for adjustments. Maintenance sometimes had to intervene to adjust parameters and bring the different equipment in coherence.
- The decline in worker performance observed during the half-day between 10 a.m. and its peak at noon could have resulted from blood sugar depletion, which lowered their overall energy levels.
- The team had a set of internal rules, and relationships between supervisors and operators, and the team required a varying amount of time (for the behavior the team engaged in) to clean the production line at the shift change, based on events that occurred throughout the shift.

At the stage of highlighting the defect linked to the design, the risk of high loss in production was established. The project owner requested a redesign of the blotting paper and cover installation machine to incorporate the range of adjustments identified during testing.

The experience of the operators and their adaptation to the requirements of the work is decisive here. The design error was due to:

- The failure to take into account the operating methods in the existing system. This is a common argument heard by HFE specialists: "Why analyze current practices with outdated equipment when everything will be replaced?"
- The capacities and limitations of the operators, and the variation of efficiency caused by the circadian rhythms.
- The collective functioning of the team, the variability of related times, and the durations related to "ancillary" activities.

It is obvious from this case study that the importance of bringing the operator in along with making the change. Average performance was used as the basis for the design of the new equipment, and little consideration was given to the variance of production, which was most well-known to the operators. Ignore the benefit of their experience, their working styles, their biorhythms, and the way that they work and interact as a team, and you have a design flaw that made us more vulnerable to lost production. Corrections weren't made until real field conditions were seen.

2.2.3 Principle Three: Include Teams in the Design Process

Whereas the second principle concerns the knowledge of individual operators, the third principle broadens the view and stresses the role of design and problem-solving as a team activity. This means it acknowledges that the successful design of complex industrial systems needs to have input from groups, not one user group or one special department. This involves teams enrolling multidisciplinary groups (e.g., operators, engineers, supervisors, maintenance operators, safety scientists, and even end-users) in activities such as brainstorming, prototyping, and decision-making in designing the system. For instance, an engineering design team that works in isolation could develop a highly automated solution that, in principle, maximizes throughput; but when employees on an assembly team are engaged in the design process, they might highlight ergonomic hazards or maintenance challenges with such a design, resulting in a revised design that is safer and more maintainable. The co-creative iteration leads to a design that is robust both in terms of technology and human aspects, which indeed is one of the key targets of human-centric Industry 5.0 systems.

Moreover, involving teams in design gets all of them to have the cost of ownership and extends acceptance of the new system. When entire teams are involved in developing a new operational process or process tool, they are more likely to appreciate it and put it into use. The design process is more like a dialogue rather than a decree from above. From a psychological standpoint, neighborhood engagement meets workers' demands for autonomy and competence, thereby raising motivation. It seems that cross-functional and team-based design organization practices lead in terms of innovation performance. In manufacturing, a study revealed mediation effects of high employee involvement – typically defined by continuous improvement team meetings – on the link between adopting new technologies and performance gains (Tortorella et al., 2018). This implies that the more the adoption of Industry 5.0 technologies was carried out by a more pervasive form of participatory teams (instead of being enforced by management decisions), the higher the gain in productivity and quality.

On the other hand, the presence of inclusiveness and diversity in the design teams facilitates the assembly of teams with diverse perspectives and skills helping innovation and effective problem-solving (Trstenjak et al., 2025). On the other hand, the presence of inclusiveness and diversity in the design teams facilitates the assembly of teams with diverse perspectives and skills helping innovation and effective problem-solving (Trstenjak et al., 2025). In practice, during designing a new production line, this might mean that operators' practical advice is assembled with engineers where technical feasibility is guaranteed, managers(grid) where business goals are met, and ergonomists where human well-being is devoted. Collectively such teams design solutions using various methods of workshops, simulations, or virtual reality prototyping to test design options (Orso et al., 2022). This joint process results in a more socio-technical design, that is, a system that optimizes the social (human) and the technical aspects at the same time.

Inclusiveness of the teams also means that many industrial tasks are done by teams and therefore the design should consider team dynamics, not just human-machine interaction. Suppose multiple workers and collaborative robots cooperate in a manufacturing cell, in which case it is important to involve the whole team in the layout and workflow design. They can assist in deciding how tasks are divided between humans and robots, how information will be shared among team members, and so on. This can be done by simulating team interaction during the design phase to identify possible coordination problems early. Furthermore, teams can be instrumental in developing contingency procedures: experienced teams, like many individuals, typically have ideas for dealing with common problems (say, breakdown of a machine or need for an urgent order) that a designer may not have anticipated. Including those teams in scenario planning ensures that the new system will serve rather than undermine such critical team responses.

Instead of a technology-disrupting approach, an Industry 5.0 approach would enable people, recalibrating to be human-centric in responding to demands already faced by businesses. For example, research into human-robot collaboration highlights that smoothly and resiliently working in human-machine teamwork is a design challenge that should take into account how humans communicate and trust their machine partners (Kaasinen et al., 2022). This is due to inserting human team members in the design process stating the robot's roles and behavior that take effect in co-designing the human automation teamwork protocols.

Additionally, with a team-based design, there is inherently a system of checks and balances. When it comes to critical requirements, every member of a design team can be his own hero: one can worship at the altar of safety; another can deify efficiency; and so on. The team and its members are able to find solutions that answer to these needs, with a reasonable degree of satisfaction, after a healthy conflict. This allows for a decreased chance of system failure due to design flaws. A security engineer on the team might argue for more guarding or emergency stops that a production engineer might not have included to a higher degree. The resulting final design provides better protection to the workers, yet is still able to meet the production requirements. Inclusive design teams are also more likely to lead to incorporating universal design principles for users to limit the exclusion of a work group (e.g., aging workers, workers with different abilities or skill levels) upon the employees serving as members on the design teams to consider this during the design process. Once more, this speaks to the Industry 5.0 vision of more inclusive and sustainable workplaces.

Since Industry 5.0 initiatives draw on the collective intelligence of multidisciplinary teams, one can ensure new systems are not only technologically advanced but also practical, safe, and embraced by the workforce. It actually makes the design phase a microcosm of Industry 5.0 – human-centric – technology development with people, not only to people and not only for people. This empowers the team, and it's also more likely that after the system has been deployed, the team will work together well to preserve it. This yields smoother adoption of innovations (Orso et al., 2022) and continuous team-driven improvements over time for the organization.

2.2.3.1 London Ambulance Service Dispatch System Failure

The London Ambulance Service introduced a new computer-aided dispatch (CAD) system without adequately involving the dispatchers, call handlers, and ambulance crews who formed the response team. The development process took place under strict deadlines while end-users were not adequately consulted about their actual business procedures and workload requirements (Beynon-Davies, 1999). Upon launch, the poorly designed system could not handle the call volume or correctly prioritize incidents, and its user interface confused operators. This resulted in dispatch delays of hours, units going to the wrong address, or not at all. As many as 20–30 lives were lost due to the disarray in the first days of implementation (Beynon-Davies, 1999). It has been ascertained that the failure stemmed from the complete exclusion of the user (team) in the design of the system; the ambulance teams' knowledge of the city, priority setting, and communication effectiveness during high-pressure situations were not incorporated by the developers. The end result was the complete and utter failure of the project. It became a textbook case in IT project management of why all of the team (end-users, dispatchers, medics) must participate during the design and testing process for safety-critical systems. This is a failure that if operational teams are not involved in design, the systems that will be developed will be operationally wrong, and therefore asserting why there is a need to embrace participatory, multisectoral design.

2.2.3.2 FBI Virtual Case File Project

The "Virtual Case File", commonly known as VCF, was designed to revolutionize case management for the field agents and other personnel of the FBI across the country, and it took a considerable time and money to implement. Unfortunately, the VCF initiative was not so effective due to ineffective requirements development and almost negligible contribution from the field agents. The developers and managers failed to involve the teams of FBI agents who were going to use the software in their work on a daily basis and gain an understanding of the investigation process (FBI, 2005). This end product did not fit the needs of the agents, which was complex, exhibited poor performance, and delivered an incomplete product, from the perspective of the agents who had to use it. Unfortunately, there was no feedback from the end-users during the development of the design. There were last alterations in requirements at the end of the project development cycle, resulting in a runaway project. In 2005, after over $100 million and five years had been put into the VCF, the initiative was shut down in an official report, and one of the reasons highlighted was that the user teams had not been involved in the planning and design processes, and hence a product that the agents would not use (FBI, 2005). The FBI then had to start over (with the "Sentinel" project), this time by a process of taking a huge and convoluted requirement and translating it into a plethora of end-user feedback to avoid repeating the aforementioned errors. It also shows that complex environments must be created with the help of end-users, and not through top management's idea of an appropriate need.

2.2.3.3 Boeing 737 MAX MCAS Development

MCAS (Maneuvering Characteristics Augmentation System) was a software system included to automatically push the plane's nose down sides to very specific situations, and Boeing included it with hardly any feedback from test pilots or airline crews about any issues. The flight crews who would be operating the aircraft had little involvement in the design work carried out through isolated groups in the company, which led to especially dangerous problems with MCAS that nobody at Boeing had foreseen (U.S. House Committee, 2020). MCAS relies on a single sensor and can automatically push the nose of the plane down multiple times if it malfunctions. During testing and certification, there was no thought given to what kind of responses flight crews would have if MCAS failed in flight! Their failed responses in two crashes in 2018 and 2019 involving the 737 MAX, which resulted in the deaths of 346 individuals, came from the unexpected failure of MCAS that no one, including the pilots, fully understood from the pilot training. The investigations showed that the company's approach of siloed engineering has led directly to the exact flaws in the design of MCAS that put the planes into a crash situation and caused the loss of life. The worldwide grounding of the 737 MAX fleet and redesign came from these tragic failures to have all stakeholders present, especially end-users. This case gives a dramatic view of why advanced technologies require a multidisciplinary view in the design, the view of who operates a product matters, and we must keep hearing from them again in the review of designs, because there is always danger and unknowns that one group may not be aware of.

These examples of failure highlight that the exclusion of various operation teams in both design and implementation caused systems to be misaligned with the operational realities of system use. We saw this with the London Ambulance CAD collapse, FBI VCF (Virtual Case File) failure, and Boeing 737 MAX design defects; the failure to include operational teams resulted in unsuitable and operationally ineffective responses and rejections of technology, and therefore, safety threats. The time needed for multidisciplinary team players needs to be recognized as early engagement is critical for the successful design of resilient, effective, and accepted systems.

2.2.3.4 Modernization of Software Systems in the Telecommunications Sector

This case study comes from the field of telecommunications, more precisely from sales agencies for a provider of these services. The project concerns the renovation of management, invoicing, and recovery applications. The existing applications are extremely complex, written in a programming language that hardly anyone understands anymore, and all previous attempts to replace them with more modern languages have failed. As a result, the chosen solution was to keep the legacy applications and build a modern IT overlay using up-to-date tools. The HFE specialist was called in after several commercial agencies refused to implement the new tool. The analysis quickly revealed the cause of the refusal. The new software had only been tested in one agency, where agents had fixed workstations. When they arrived in the morning, they connected their computers to various databases, which took around 30 minutes. Once this was done, everything functioned normally throughout the day as customers arrived. However, the agencies that rejected the new software had a very different setup: the computers were placed in the customer service area for "self-service" use, and agents would sit down at them as needed, depending on customer flow. But the 30-minute connection time was unacceptable – both for the agents and the customers. Skipping this step wasn't an option either, because every transaction had to be logged. At this stage of development, it was not possible to find a fully satisfactory solution, as the choices of technologies and software packages were irreversible. Still, the "self-service" model could be partially maintained by installing more powerful computers in the sales area, and by establishing multiple connections in the morning before customer arrival (several agents logged into each computer in advance).

2.2.4 Principle Four: Technology Cannot Solve All Problems – It May Lead to Delusion

The fourth principle is a warning about technological solutionism, the assumption that whatever problems there are, they can all be resolved by new technology. We need, according to FEES, to recognize that technology, despite the development, has inherent boundaries and that the automaton or algorithm is not infallible, and that the use of automation or algorithms can lead to dangerous outcomes or "delusions" of invulnerability. In the context of Industry 5.0, this principle serves as a reminder that even in Industry 5.0, we still have the need for human insight, oversight, and ethical judgment and cannot solve human or social organizational problems with technology alone.

In history, there have been many times when the overreliance on automation resulted in failure or brought new problems. Bainbridge (1983) in a classic paper entitled "Ironies of Automation" noted that as our machines become more and more sophisticated and autonomous, the more human monitoring and handling of rare anomalies becomes possible but also less likely for reasons of eroding human skills and vigilance in those very situations. These warnings, however, remain relevant decades later as stated by Strauch (2018) that fundamental "ironies of automation" identified by Bainbridge have not been resolved even in the modern automated systems. However, an irony is that automation often transforms the nature of human problems rather than attenuating them – for example, rather than manual routine, operators are confronted by the problems of supervising complex automation, maintaining situation awareness, and intervening correctly when the automation hands over control (Strauch, 2018).

Organizations that simply trust technology without human supervision risk major system failures that result from conditions outside the programming scope. The domain of transportation research revealed that trusting automation too much leads drivers and pilots to lose awareness of their situation along with slower responses that elevate accident risks (Papadimitriou et al., 2020). Drivers who were complacent with a highly automated driving system did not react to dangers in sufficient time, and as a result had a high crash rate, because they thought that the system could take care of everything, so they didn't remain engaged (Papadimitriou et al., 2020). These authors quantified this: if automation is relied upon too heavily, there is an increase in reaction time, a decrease in situational awareness and an increase in accident rate. Such evidence justifies the need for continuous vigilance about technology and human supervision regardless of how sophisticated technology is.

In Industry 5.0 factories, we might also picture such a case: if an operator decides that a smart machine can never be wrong, perhaps they stop looking at the outputs from it, but if a sensor fails or an AI makes some unexpected mistake, the human oversight is missing and an error, a defect, or even a safety hazard is not noticed. Automated systems are subject to automation bias, a well-known phenomenon in which users will accept automated system recommendations, even when they are erroneous, in such high-reliability environments as aviation and medicine (Mosier et al., 1998; Parasuraman & Manzey, 2010). For this reason, a human-centric approach requires that technology is a tool for helping humans, not to solve all problems with an autonomous capability.

In spite of this, the automation of catching certain issues using human judgment and common sense, as well as the adaptability that goes with it, is still very important. However, as the FEES white paper points out, some Industry 5.0 discussions assume an underlying belief that all needed transformations for sustainability and social well-being will simply come with advanced technological solutions, which "history has shown countless times that this is unrealistic", and purely technical solutions can generally not be realized without interdisciplinary, human-centered approaches (Michez et al., 2025). In other words, without integrating human and organizational factors, the implementation of technology that is bordering on cutting-edge is likely to yield suboptimal results or unintended consequences.

Embracing the limitations of technology means a bunch of specific practices. Firstly, it implies humans are in the loop of critical decision-making and control. As AI and autonomous systems get more sophisticated, there is a need for human oversight, which prevent a disaster. A good example of this is quality inspection in a factory. In this case, AI should be used, but there should be a human quality engineer periodically reviewing the AI's judgments, especially in borderline cases because the AI might systematically miss some type of defect.

The second point is that it requires investing in human training and readiness. On the other hand, if operators know all about the failure modes and limitations of the technology, they can be vigilant and intervene when necessary (Parasuraman & Manzey, 2010). A study in aviation demonstrated that pilots with high trust in an autopilot system monitored less than those trained to remain skeptical and their attention was also less (Papadimitriou et al., 2020). Therefore, we must succeed in creating an organizational culture that prefers the alerting of humans over blind faith in machines.

Third, accepting technical limitations also deals with the human, non-technical side of issues. Many of our workplace grievances arise from human communication, or workflow needs that are sociocultural or socio-economic, for example, low productivity might be due to poor team coordination, or fatigue, and would not be fixable by just upgrading a machine. So, instead of believing that a new technology would have resolved these issues, HCD would tackle these issues with ergonomics job design, better work-rest schedule, and potentially other means like training. According to Parasuraman and Riley (1997), understanding the factors related to human use of automation suggests that the key to reducing the risks of misuse, disuse, and overuse of automation is to strike a balance between trust in technology and operator engagement.

However, this principle isn't antitechnological at all; it is what can be referred to as optimally skeptical or rationalistic in using gadgets. Nevertheless, AI, IoT, and robotics technologies are key enablers for Industry 5.0 specifications for efficiency and innovation while smart integration is an important factor. For example, for an organization that wants to schedule nurses for its hospitals, using an AI tool may neglect the exhaustion of such workers since an AI does not care about their fatigue, and thus, a human manager is necessary to set boundaries that protect nurses from being overworked.

For the European Commission, Industry 5.0 is a shift in thinking beyond technology that places human values at the front and center of technology deployment (Žižić et al., 2022). An organization must continuously ask itself about the limits of technological capabilities. What could possibly go wrong? Do we consider the influence of people on the situation? Intensive questioning of their systems allows organizations to logically assess what may be a delusion about technological capability. However, as Michez et al. (2025) point out, relying solely on technological innovations to achieve Industry 5.0 goals is unrealistic, as these need to be integrated into social, organizational and human factors approaches in order for technology to effectively support human needs. The authors caution against overlooking the need for social change in favor of solely technical fixes as this offers itself up to be counterproductive.

Another component of understanding technology limits is the ability to use technology with consideration to ethical and safe use. Many Industrial 5.0 systems of automation (e.g., decision-making tools powered by AI) may bring ethical dilemmas or inherent biases. There is a danger in unexceptionally trusting technology, and in this example, unethical behaviors may occur (e.g., a bias in an algorithm unfairly distributing work or risk). Humans are responsible for managing and correcting the issues caused by technology. Strauch (2018) emphasizes that human operators must remain connected as active participants, prepared to intervene when automation fails, falters, or behaves in unpredictable and/or unconventional ways. By implementing automation in operational design with operational transparency (e.g., explainable AI; manual override that is intuitive; alerts that are simple to comprehend), operators can utilize this trust in the way it is intended – not too much (over-possessing the trust) but also not too little (not using automated processes when it is beneficial).

The calibrated trust aims to use the strengths of technology (speed, precision, massive data crunching) and relies on humans for the strengths of humans (judgment in novel situations, moral issues, creative problem solving). Papadimitriou et al. (2020) remind us that distrust of automation may actually decrease reaction time since the operator is still mentally engaged – some skeptics say our senses, or being prepared. The purpose is to maintain a calibrated trust so that humans and technology support one another.

Thus, Principle Four concludes that technology is not an end in itself, but is a means to an end. Industry 5.0 should make use of advancing technology but with an eye toward employing it as part of a more extensive socio-technical solution. Lacking human vigilance (in blind faith), faith in an automated or digital system is an illusion of safety and can create its own failures (Strauch, 2018). By understanding what technology can do (and what it cannot do), organizations can implement the human checks, contingencies, and complements that represent a realistic, human-aware method of deploying technology in the workplace that actually serves humanity's interests.

To paraphrase, "guns don't kill people, people do", an equivalent expression would be that technology won't solve a problem, but rather people working with technology. Thus, amidst the human-centeredness of the Industry 5.0 enterprise where human insight and control will always be designed into the mix, we will not be misled into believing that complex industrial issues can simply be resolved with technology (Michez et al., 2025). So the most successful Industry 5.0 implementations are those that pair technology with human wisdom, intuition, and flexibility in the knowledge that both are necessary and interdependent.

2.2.4.1 Boeing 737 MAX and MCAS Automation

Boeing incorporated the MCAS automated system into the new 737 MAX to make it handle like the older models of the 737, thinking that the problem of aircraft stability would be solved by technology. Boeing believed that this software fix would ensure that pilots would never encounter this handling issue in the first place, and so they did not prioritize pilot training and disclosure about MCAS (U.S. House Committee, 2020). This overconfidence in technology was rather illusional. In service, a faulty sensor activated MCAS, and the automation forced two aircraft into fatal dives (Lion Air 610 and Ethiopian 302). Since the pilots were not aware of MCAS's behavior, they could not act against it. The crashes showed that technology used as a replacement for proper design of the aircraft and pilot training can be disastrous Herkert et al. (2020). The automated system did not take into consideration all the failure possibilities and with the human operators not fully informed, the safety net was missing. This led to increased focus on Boeing and the FAA (Federal Aviation Administration): the solution is not just in the software update, but in the sensors, in the alerts, and in the training – a reminder that technical solutions must be accompanied by human-centered ones (U.S. House Committee, 2020). This case is a classic example of what is referred to as the "technology delusion" – the notion that technology can solve all problems of human factors integration and training.

2.2.4.2 Tesla Autopilot Overreliance Fatal Crash

A Tesla Model S car on a "Autopilot" (Level 2 driver-assist technology) crashed into a tractor-trailer at highway intersection speed, and the driver of the car died.

The driver had activated the automation on a straight road and was over-reliant on the system, and did not pay attention to the road. The car's camera system failed to identify the white truck against a bright background and did not apply the brakes at all. The driver was over-reliant on the system and had let the system take over most of the control and responsibilities of driving and expected the system to handle all the situations (NTSB, 2017). Tesla's design enabled this complacency – it did not limit Autopilot use to proper conditions or guarantee the driver's awareness. The National Transportation Safety Board concluded that the probable cause of the accident was the failure of the truck driver to yield right of way and the Tesla driver's inattention due to overreliance on the system that allowed him to have his hands off the wheel for an extended period. This case is a good example of what can be called the "technology delusion". This is because the belief that the car could drive itself led to tragedy. Therefore, NTSB encouraged manufacturers to incorporate measures (such as monitoring the driver and limiting the use of the system) to ensure that partial automation is not considered as full autonomy (NTSB, 2017). It was evident that human drivers cannot be left out and that technology alone, without such human factor, cannot eliminate all the accidents. This case shows that partial automation systems must be used with the supervision of human beings and that relying on technology is dangerous.

2.2.4.3 Uber Self-Driving Test Car Pedestrian Fatality

In this incident, an Uber test vehicle operating in autonomous mode struck and killed a pedestrian crossing a street at night. Uber's experimental self-driving technology was touted as the solution to human driver errors, but in practice, it had serious gaps. Onboard sensors did detect the woman pushing a bicycle, but the software logic repeatedly reclassified her (as a vehicle, then a bicycle, etc.) and never decided on an emergency stop. Uber had also reduced the supervising human role from a two-person team to a single safety driver, expecting the automation to manage most situations. Unfortunately, that lone operator became distracted (streaming a video on her phone) and did not intervene in time (NTSB, 2019). The investigation found Uber's development program had a poor safety culture: it placed excessive trust in the self-driving system and lacked sufficient safeguards or operator training (NTSB, 2019). The technology was simply not mature enough to handle all scenarios (e.g., it was not programmed to recognize jaywalkers crossing mid-block). Critically, Uber assumed the advanced technology would compensate for having an inattentive human, which proved fatally wrong. The NTSB assigned blame to Uber ATG's inadequate risk management and the safety driver's failure to oversee the automation. This tragedy demonstrated that cutting-edge automation cannot solve every problem on its own – rigorous safety processes, operator vigilance, and system limitations must be addressed. In response, Uber suspended and overhauled its autonomous testing, and the industry refocused on requiring attentive human backup even as the technology evolves (NTSB, 2019). This fatality emphasizes that autonomous systems must still be embedded within robust human supervision frameworks, reinforcing that technology alone is insufficient to guarantee safety.

The analyzed incidents show that while using technology, there is a risk of relying too much on the technology and not enough on people. From Boeing's MAX to Tesla's Autopilot, to Uber's ATG testing, systems faltered when human elements were overlooked and when the idea of the system being perfect was not questioned.

These examples show that the technology has to be wrapped into non-technology solutions, which will keep the factors that include the needed attention, responsibility, and moral decision-making in supporting the sustainable elements of safety and performance.

2.2.4.4 Trust in Technology in Metallurgical Chemistry

This case study comes from the field of production in metallurgical chemistry. The production plant was old; the project consisted of completely renovating it. Among the topics covered by the design team was the one of designing the new supervision system. A gap in design principles appeared between the HFE specialist and the project's IT specialist.

Considering the sensitivity of the production concerned, the IT specialist suggested that there were three possibilities for the production parameters:

- Instructions and the measured values are within the quality and safety rules. The operator and the algorithms have nothing to do except monitor the smooth running of the operations.
- The measured values are outside the quality requirements. The system issues alarms to the operator so that he can identify the drift and initiate corrective actions.
- The measured values are about to or are getting outside the safety requirements. In this case, the algorithms take over and launch commands to maintain the process in safe conditions.

The HFE specialist argued that he agreed on all three levels except the last one. In the event of leaving the security perimeter, the system had to warn the operator of its takeover before launching the commands. This condition was central to ensuring the operator's trust in the machine, and beyond that, there is no machine, sensor, or electronic card that is safe from breakdowns or defects. For the operator to play his role as ultimate regulator, it was necessary that the information be given before the action and that he be able to stop any action according to the cases.

The solution to this situation required the solicitation of the project owner and the search for cases where the machine could be faulty and issue commands wrongly. Several cases were identified where the algorithms could be "trapped" by the information from the sensors. Among the important findings, there was one where the design team admitted that it was impossible to foresee all the possible cases of deviation of the machines. As a result, the project owner decided to keep the operator in the position of the ultimate regulator, taking into account even his participation in the control of emergency situations that are too fast (such as the emergency stop of the machine).

2.3 UNDERSTANDING THE ERGONOMIC BASIS OF THE PRINCIPLES

The four principles presented in this chapter can be seen as ergonomic since they are a direct reflection of ergonomics' fundamental philosophy: adapting work to the human and respecting human capacities and limitations, as well as providing safety, comfort, and efficiency of workers within work systems. The first principle that requires the bearing of the human operator in the design of the equipment is a direct out of the ergonomic notion of "fitting the task to the human". According to this principle, it is crucial to take into account workers' physical and cognitive limitations and potential when creating the work environment and assignments. The ergonomic essence of its nature is reflected in the way its consistent use avoids physical exhaustion and mental overcrowding, error and accident risk, and increases working comfort – an aim that is at the heart of ergonomic design.

The second principle, which mandates the use of operator expertise before introducing changes, is also perfectly in accordance with ergonomics thinking. The role of workers' practical knowledge about their tasks is well understood by ergonomics, because workers are constantly confronted by real-world scenarios that pure engineers and managers could easily ignore. This principle is ergonomic because it involves a participatory orientation with a scope for workers not to be passive users of a system but active contributors of ideas that would improve the applicability and amenability of ergonomic interventions.

The third principle is the one that endorses the inclusion of teams in the design process, and it is also closely related to ergonomics due to both the participatory ergonomics and the multidisciplinary approach. Ergonomics by its nature promotes the cooperation of professionals in various fields (e.g., engineers, workers, supervisors, safety specialists) since only the collaboration of professionals enables creating systems adapted to the needs of users. The principle directly implements the ergonomic approach, which promotes teamwork in detecting the risks, constructing, and testing solutions, thus leading to safer, more effective, and user-friendly systems.

The fourth principle, which cautions on the limitations of technology, also possesses a clear ergonomic dimension. Ergonomics understands that technology can never be the final answer to issues at workplaces, particularly ones that entail human interaction, decision-making, and ethical issues. The applied principle is ergonomic as it emphasizes the necessity of constant human control and involvement of workers in training related to the technologies adopted. By focusing on the role of human beings in monitoring, understanding, and the use of technology, the principle connotes the core of ergonomic belief that technology should serve human beings, but not the reverse way. Figure 2.1 presents that these four principles stem from ergonomics, as well as that they contribute to the achievement of the human-centric approach of Industry 5.0.

2.3.1 Ergonomics 5.0 for Industry 5.0

It is beyond doubt that ergonomics can take people closer to the goals of Industry 5.0. This chapter examines ways in which the human-centric approach in Industry 5.0 can be achieved. These four ergonomic principles mentioned above are important determinants when we want to design toward human-centered outcomes. However, it is also essential to consider those aspects of ergonomics that should be developed and changed for the better in the future. That is how ergonomic knowledge could be applied in practical situations more successfully in order to have a human-centric approach in Industry 5.0. In this part, we will mention some of the main spheres

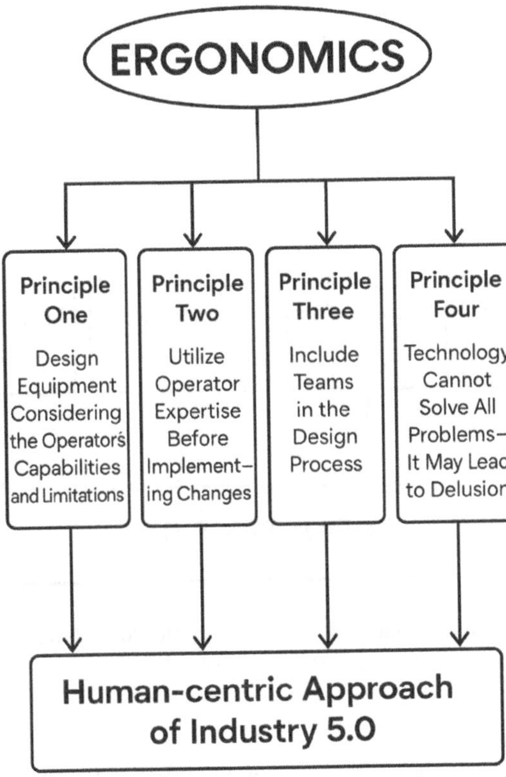

FIGURE 2.1 The four principles originate from ergonomics, and their proper application contributes to achieving the human-centric approach of Industry 5.0.

of Ergonomics 5.0, which are extremely important for making Industry 5.0 occur (from the human-centric perspective). The areas are indicated in Figure 2.2, and their importance to Industry 5.0 will be discussed below.

2.3.2 Participative Ergonomics for Ergonomics 5.0

Participatory ergonomics is defined by employee involvement in the design, planning, and improvement of their work and work environment. The rationale behind such an approach is the idea that workers, based on their daily interaction with tasks, can provide important insights that would result in better safety, higher efficiency, and increased acceptability. Haines et al. (2002) define participatory ergonomics as a process whereby the workers, supervisors, and managers are part of the process of identifying and implementing ergonomic improvements.

Participatory ergonomics is the foundation for the second and third principles in the human-centric ergonomic approach to Industry 5.0. The second principle highlights the need to engage operator expertise prior to changes being made because it requires a participatory approach to utilize the unwritten and unrecorded experiential knowledge that is critical for system functionality (Johnson et al., 2019). The

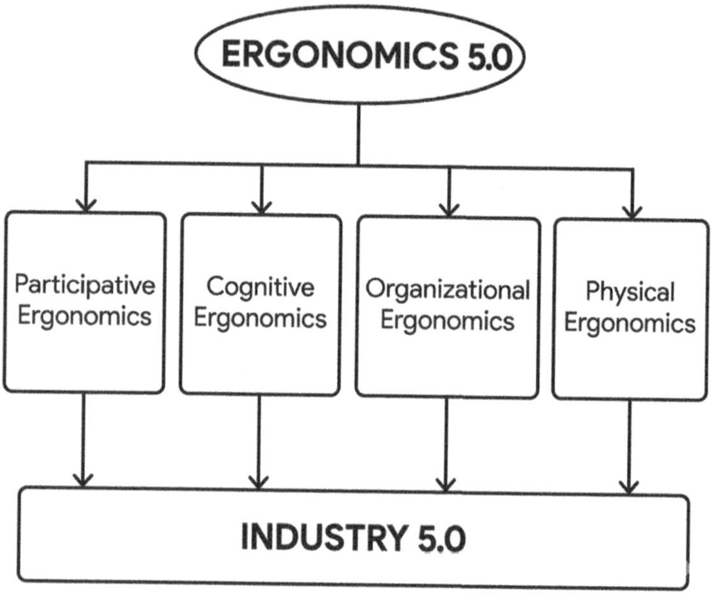

FIGURE 2.2 Areas of Ergonomics 5.0 that especially should be developed in the future in order to achieve the human-centric approach of Industry 5.0.

third principle focuses on the involvement of teams in the design process during the challenge, which once again requires participatory ergonomics because of its ability to effectively integrate various expert knowledge (operators, engineers, supervisors, ergonomists, etc.) at the stages of solution development and testing.

Participatory ergonomics in a human-centered approach is essentially based on: it perceives human beings as essential ingredients in a system rather than mere users of it. Providing employees the chance to be involved in the decision-making about their work results in higher levels of autonomy, belonging, and control, and accordingly will increase engagement and satisfaction. As a result, a design is implemented that combines technical and social aspects by taking into account the context and needs of the systems' actual users. That is why the primary idea of the human-centered approach is implemented – systems are built together with people, not only for them (Alves et al., 2023).

On a larger scale, participatory ergonomics is part of the principles of Ergonomics 5.0 and Industry 5.0. Each paradigm is concerned with developing just, accommodating, and enduring technology, and the application of participatory techniques in the design process can enhance commitment, hence increasing acceptance and resilience in work systems. Michez et al. (2025) reinforce this viewpoint by saying that our focus lies not only in creating technology that is functional thanks to operator participation and stakeholder engagement, but also in fostering sustainable transformations in work practices and human values.

The development of participatory ergonomics to a point where it is a fully human-centric approach is an ongoing process of methodological and institutional development. First of all, it was mentioned above that there is a necessity to develop and implement strong, sustainable methodologies for participatory design, such as digital workshops, video analyses, simulations, and virtual co-creation. Then, it is necessary to train engineers, managers, and people who make decisions in participatory methods, and how to use them efficiently. Furthermore, technological solutions should allow workers operating in various regions or on alternative shifts to engage in co-creation activities. In addition, attempts should be made to involve workers who might be marginalized, including older people, staff with disabilities, and workers with poor digital skills, in the collaborative design activities. Moreover, developing multiple assessment procedures will aid in quantifying the value of the participatory processes in increasing productivity, safety, and the quality of work experience.

2.3.3 Cognitive Ergonomics for Ergonomics 5.0

Cognitive ergonomics is a branch of ergonomics that takes cognition to mean mental processes of the worker in conjunction with the work system. The International Ergonomics Association (IEA) defines cognitive ergonomics as the mental processes (perception, memory, reasoning, and motor response) that should affect interactions between people and other elements of the system (IEA, 2000). Cognitive ergonomics concerns how information arrives, is interpreted, and is used to make decisions and act, on the basis of competence, in complex work systems.

Taking into account the four ergonomics principles, cognitive ergonomics is especially important for the first, which is concerned with designing equipment that considers the cognitive and physical abilities and limitations of the operators. In addition to physical work characteristics, these abilities include cognitive aspects. Human attention misalignment with interfaces, limits of decision-making, or tolerance to overload with interfaces, automated systems, or alarms may lead to significant problems such as operational errors. Systems that are based on cognitive ergonomic principles can be very sophisticated in terms of technology, but they can be operationally inefficient and extremely dangerous. Systems that were always able to consider these factors will probably make fewer errors and achieve overall system reliability.

At the core of a human-centric approach, cognitive ergonomics promotes the design of technologies and work procedures that are specific to individual thought and behavior rather than basing on hypothetical assumptions by emphasizing the difference between perceived and real work approaches. The analysis of the cognitive workload that operators experience from automated systems (such as the number of alarms and interaction with screens) contributes to better decision outcomes and a better understanding of the operational environment. Also, user interfaces based on identifiable mental models encourage more user approval, reduce training needs, and increase engagement and focus. These considerations follow the basic principles of the human-centric approach: Adaptation of technology to human capacity is better than the reverse.

Cognitive ergonomics is vital in the creation of Ergonomics 5.0 and its role within Industry 5.0 because, given that automated systems (Industry 4.0) must be replaced by cooperative systems (Industry 5.0), the understanding and appreciation of In cases when humans and technologies coexist, and make decisions together, it is vitally important that a design embraces the actual human thinking processes. When the operator's capability to stay mentally tuned in and understand the borders of automation is ignored, phenomena such as "automation bias" or disconnect from oversight might happen which would undermine the system's resilience as well as increase risks.

To better add to a human-centric approach, getting a head start on supporting cognitive ergonomics in the future, key areas need to be improved. First, adaptive interfaces should be implemented; they will dynamically respond to the operator's cognitive load, for example, reducing the information on display in response to high stress. Second, the implementation of decision-support systems should be broad enough that it would not take over the controlling role while supplying other information and warnings without harming the autonomy of a human. Third, there should be encouragement of the creation of educational programs, which combine engineering design with original ideas of cognitive sciences: the bridge between technology and its users needs to be established at the conceptual phase of system development. Finally, measures of the cognitive workload have to be developed to measure in real-world operations, so as to modify solutions to actual requirements instead of idealized user models.

2.3.4 Organizational Ergonomics for Ergonomics 5.0

As one of the central spheres of ergonomics, organizational ergonomics is considered the scientific field aimed at optimizing socio-technical systems (organizational structures, policies, processes) (Bau et al., 2012). To explain, it strives at humanizing the work organization – discussing such issues as communication, teamwork, participatory design, organizational culture (Bau et al., 2012). Work organization must be designed so as to truly support people. This is exactly where organizational ergonomics becomes vital – thanks to good processes and policies, it guarantees that better technologies will result in safer and more comfortable, and efficient work without the strain and mistakes (Trstenjak et al., 2025).

Organizational ergonomics becomes very important when in relation to the four ergonomics principles that support the human-centric approach of Industry 5.0, principle three requires the inclusion of teams in the design process. Competent application of such a principle presupposes the existence of a system of organization enabling cooperation, two-way contact, and decision-making according to joint knowledge.

Introduction of new technologies alone is not enough for a system to be completely human-centered – organizational changes are necessary to ensure that technology responds to users and people are involved in defining work processes. This demonstrates the necessity of organizational ergonomics. With participatory employee involvement in work design and enhancement in teamwork and communication, organizational ergonomics guarantees that the work system is really designed according to human needs.

In practice, if modern (e.g., collaborative) technologies are to be implemented, it is important to take into account such factors as employee training, adaptation of organizational policies, and change readiness. Any engineering efforts that undertake to develop elaborate technological initiatives without this organizational approach are likely to be met with resistance or fail to give the desired performance. Organizational ergonomics contributes directly to Industry 5.0 vision realization as overcoming these barriers – through investments in employee skill development, a partnership approach toward the implementation of new solutions, and the implementation of human-centric design, guarantees successful people and technology integration.

Improvements in the area of organizational ergonomics will contribute to an even greater human-oriented approach in the future. One direction of development is the introduction of objective methods for the control of labor workload and well-being (sensors, data analysis) for workflow optimization based on human needs. The other direction is the creation of flexible training and work organization strategies for different groups of employees, for instance, the inclusion of the generations through individual training of the older employees in order to keep them productive in a team-based technology environment. Lastly, future developments in organizational ergonomics should particularly concern themselves with psychosocial factors. There is a need in the study of the effects of modern high-tech environments (automation, working with robots, digitalization) on mental health, and the introduction of support measures to lessen stress and increase the well-being of employees. These development threats will help keep the human-centric approach at the center of industrial work evolution in the new stages.

2.3.5 Physical Ergonomics for Ergonomics 5.0

Physical ergonomics is the branch of ergonomics that studies human anatomical, anthropometrical, physiological, and biomechanical characteristics relative to physical activity (Donisi et al., 2022). In other words, it carries out exploration of the ways in which the traits and abilities of the human body affect the interaction of people with the work environment, tools, and tasks. Physical ergonomics aims at arriving at a point where tasks will be adapted to human beings so that, through work design, excessive physical pressure, discomfort, and potential dangers can be reduced in favor of comfort and efficiency. Based on this definition, physical ergonomics is interested in material handling, proper working postures, repetitive movements, prevention of musculoskeletal disorders, and ergonomic design of workplaces, as well as work-related physical safety and health (Donisi et al., 2022).

Physical ergonomics directly meets at least one of the four established principles – the principle that calls for equipment and work environment to be designed based on the operator's abilities and limitations. By means of application of anthropometric, biomechanical, and physical ability data, physical ergonomics ensures that machines, tools, and work processes fit human build and strength, a basic element of the first principle. This eliminates fatigue, injuries, and errors as well as concurrently enhances system productivity. Physical ergonomics also indirectly supports the fourth principle. In focusing on human limitations and needs, it confirms the fact that the absence of ergonomic requirements typically results in technological initiatives failing – for example, new automated equipment will not be ergonomically adapted, and as a result, it will introduce further strain or danger for workers.

Physical ergonomics is necessary for true human-centric solution design because it guarantees the integration of human physical needs in the design of work. Through ergonomic shaping of tasks and tools, physical strain and fatigue decrease, directly resulting in a positive impact on the health and satisfaction of employees. Healthier and more satisfied workers then work better, commit fewer errors, and are more innovative for the sake of both employers and the macro socio-technical system. What this means is that physical ergonomics helps to guarantee that the "human center" in the "human-centric" approach is not just an ideology but an entity present in reality, as the environment where people complete their work should be designed around people. Without valid physical ergonomics, human-centered projects would not be fully implemented, because if one fails to take into account physical limitations and the comfort of the labor, one compromises their health and comfort, and thereby their motivation, too.

Physical ergonomics is a major factor in human-machine interaction harmonization. For instance, if workplaces and interfaces of machines are developed with reference to human dimensions and reach, if collaborative robots and exoskeletons with an ergonomic assessment are introduced, technology can become an instrument of extension of human capacities of the body. According to recent research, the success of new technologies relies upon parallel improvement of ergonomic working conditions. If technology changes are implemented with insufficient regard to ergonomics, its use of technology is sub-optimal and may also create new risks to workers, but if ergonomics is improved while technology is neglected, potential benefits cannot be achieved fully. Thus, the synergy between micro-ergonomics (physical and mental work elements) and macro-ergonomics (system organization) is given primary focus in relating Industry 5.0 solutions (Trstenjak et al., 2025). Physical ergonomics is an important part of this framework – thus, technological progress is balanced with human work environment modifications so future factories are truly human-centric. In other words, Industry 5.0 tenets are transformed into tangible design prescriptions that protect workers' health, yet improve their outcomes as well, which is the essence of ergonomic contributions to the 5.0 paradigm.

In the future, the human-centric approach through several complementary directions should be additionally motivated by the development of physical ergonomics. One pathway is the increased use of modern technologies for the discovery and resolution of ergonomic problems. Wearable sensors, smart devices, and work monitoring systems would be able to measure in real time the individuals' body posture, muscle strain, and movements of workers; this way, situations of risk could be identified and eliminated before the occurrence of injury or overload. In addition, the use of digital human models and simulations (e.g., virtual and augmented reality) integration into the workplace design will allow for testing of various design options and their optimization from the ergonomic point of view prior to their use. This enables the embedding of physical ergonomics into future system design from the outset rather than after the design has been made. Another important innovation is the personalization of ergonomic solutions – individual adaptation of tools, equipment, and work procedures to workers' characteristics (e.g., the fine-tuning of workstation height to individual anthropometry and the alteration of work speed to physical condition). This is of great consequence, especially with the aging workforce and diversity among the employees. Personalization of work environments can improve not only the length of careers but also overall well-being.

Finally, the future of physical ergonomics is its merger with other aspects of ergonomics – cognitive and organizational. A holistic human-centered approach requires to consider physical strain, mental stress, and organizational work factors at the same time. The creation of hybrid methodologies that combine the evaluation of physical and cognitive job requirements, as well as a culture of participatory worker involvement in developing safer and more pleasant workplaces, will support pursuing full Ergonomics 5.0. This integrative approach will guarantee that the aims of Industry 5.0 are achieved in reality, that is, through workplaces, which at the same time are productive and adapted to human needs.

2.4 CONCLUSION

Ergonomics is of high importance in realizing the human-centric approach of Industry 5.0. The studied ergonomic principles clearly show that successful human-technology integration presumes a deep understanding and respect for human abilities and limitations, and active participation of operators and multidisciplinary teams in the implementation of new solutions. The given practical examples and case studies serve to confirm that the disregard of ergonomic principles all too often causes serious consequences such as safety incidents, production loss, and lower efficiency of the system. In the future, it will be necessary to continue developing such aspects as participatory, cognitive, organizational, and physical ergonomics as integral parts of Ergonomics 5.0. This will help make sure that ergonomics as a scientific discipline does not lag behind fast technological developments and support adequate realization of the human-centric vision of Industry 5.0. Only in this manner can the industry reach its potential – with technological superiority and human peculiarities, but in an environment which is sustainable, productive, and safe.

REFERENCES

Adel, A. (2022). Future of Industry 5.0 in society: Human-centric solutions, challenges and prospective research areas. Journal of Cloud Computing: Advances, Systems and Applications, 11, Article 40. https://doi.org/10.1186/s13677-022-00314-5

Admiral Cloudberg. (2019, August 31). The crash of Air Inter Flight 148: Analysis. Medium (Analysis of BEA investigation conclusions on Flight 148's UI design and pilot factors).

Alves, J., Lima, T. M., & Gaspar, P. D. (2023). Is Industry 5.0 a human-centred approach? A systematic review. Processes, 11(1), 193. https://doi.org/10.3390/pr11010193

Bainbridge, L. (1983). Ironies of automation. Automatica, 19(6), 775–779.

Bau, L. M., Farias, J. P., Buso, S. A., & Passero, C. R. (2012). Organizational ergonomics of occupational health methods and processes in a Brazilian oil refinery. Work (Reading, Mass.), 41(Suppl 1), 2817–2821. https://doi.org/10.3233/WOR-2012-0529-2817

Beynon-Davies, P. (1999). Human error and information systems failure: The case of the London Ambulance Service computer-aided dispatch system. Interacting with Computers, 11(6), 699–720. https://doi.org/10.1016/S0953-5438(98)00050-2

Briken, K., Moore, J., Scholarios, D., Rose, E., & Sherlock, A. (2023). Industry 5 and the human in human-centric manufacturing. Sensors, 23(14), 6416. https://doi.org/10.3390/s23146416

Cunha, L., Silva, P., & Maggioli, S. (2022). Exploring the status of the human operator in Industry 4.0: A systematic review. Frontiers in Psychology, 13, 9530396. https://doi.org/10.3389/fpsyg.2022.928139

Donisi, L., Cesarelli, G., Pisani, N., Ponsiglione, A. M., Ricciardi, C., & Capodaglio, E. (2022). Wearable sensors and artificial intelligence for physical ergonomics: A systematic review of literature. Diagnostics, 12(12), 3048. https://doi.org/10.3390/diagnostics12123048

European Commission. (2021). Industry 5.0: Towards a sustainable, human-centric and resilient European industry (Policy brief). Directorate-General for Research and Innovation, European Commission.

FBI. (2005). The FBI's management of the trilogy information technology modernization project, Audit Report No. 05–07.

Ghobakhloo, M., Iranmanesh, M., Mubarak, M. F., Mubarik, M., Rejeb, A., & Nilashi, M. (2022). Identifying Industry 5.0 contributions to sustainable development: A strategy roadmap for delivering sustainability values. Sustainable Production and Consumption, 33, 716–737. https://doi.org/10.1016/j.spc.2022.08.003

Haines, H., Wilson, J. R., Vink, P., & Koningsveld, E. (2002). Validating a framework for participatory ergonomics (PE). Ergonomics, 45(6), 309–327. https://doi.org/10.1080/00140130210156352

Hart, S. G., & Staveland, L. E. (1988). Development of NASA-TLX (Task Load Index): Results of empirical and theoretical research. In P. A. Hancock & N. Meshkati (Eds.), Human mental workload (pp. 139–183). North-Holland.

Herkert, J., Borenstein, J., & Miller, K. (2020). The Boeing 737 MAX: Lessons for engineering ethics. Science and Engineering Ethics, 26, 2957–2974. https://doi.org/10.1007/s11948-020-00252-y

International Ergonomics Association (IEA). (2000). What is Ergonomics (HFE)? Retrieved from https://iea.cc/what-is-ergonomics/

Johnson, T. L., Fletcher, S. R., Baker, W., & Charles, R. L. (2019). How and why we need to capture tacit knowledge in manufacturing: Case studies of visual inspection. Applied Ergonomics, 74, 1–9. https://doi.org/10.1016/j.apergo.2018.07.016

Kaasinen, E., Anttila, A.-H., Heikkilä, P., Laarni, J., Koskinen, H., & Väätänen, A. (2022). Smooth and resilient human–Machine teamwork as an Industry 5.0 design challenge. Sustainability, 14(5), 2773. https://doi.org/10.3390/su14052773

Kadir, B. A., & Broberg, O. (2021). Human-centred design of work systems in Industry 4.0: Challenges and opportunities. Applied Ergonomics, 92, 103334. https://doi.org/10.1016/j.apergo.2020.103334

Leveson, N. G., & Turner, C. (1993). An investigation of the Therac-25 accidents. IEEE Computer, 26(7), 18–41. https://doi.org/10.1109/MC.1993.274941

Longo, F., Padovano, A., & Umbrello, S. (2020). Value-oriented and ethical technology engineering in Industry 5.0: A human-centric perspective for the design of the factory of the future. Applied Sciences, 10(12), 4182. https://doi.org/10.3390/app10124182

Ludes, B., Tracqui, A., Pfitzinger, H., Kintz, P., Levy, E., Disteldorf, M., Hutt, J. M., Kaess, B., Haag, R., Memheld, B., Kaempf, C., Friederich, E., Evenot, E., & Mangin, P. (1994). Medico-legal investigations of the Airbus A320 crash upon Mount Ste-Odile, France. Journal of Forensic Sciences, 39(5), 1147–1152.

Madsen, D. Ø., & Berg, T. (2021). An exploratory bibliometric analysis of the birth and emergence of Industry 5.0. Applied System Innovation, 4(4), 87. https://doi.org/10.3390/asi4040087

Michez, B., Etienne, P., Žunjić, A., Ferreira, P., & Szabó, G. (2025). What ergonomics and human factors return of experience tells to Industry 5.0. In Proceedings of the 10th International Ergonomics Conference (ERGONOMICS 2024) (pp. 33–43). Springer Series in Design and Innovation.

Mosier, K. L., Skitka, L. J., Heers, S., & Burdick, M. (1998). Automation bias: Decision making and performance in high-tech cockpits. The International Journal of Aviation Psychology, 8(1), 1–20. https://doi.org/10.1207/s15327108ijap0801_1

National Commission on the BP Deepwater Horizon Oil Spill and Offshore Drilling. (2011). Deep water: The gulf oil disaster and the future of offshore drilling. U.S. Government and National Commission on the BP Deepwater Horizon Oil Spill and Offshore Drilling, Washington, D.C.

National Transportation Safety Board (NTSB). (2017). Collision between a car operating with automated vehicle control systems and a tractor-semitrailer truck near Williston, Florida, May 7, 2016 (NTSB/HAR-17/02).

National Transportation Safety Board (NTSB). (2019). Collision between vehicle controlled by developmental automated driving system and pedestrian, Tempe, Arizona, March 18, 2018 (NTSB/HAR-19/03).

Norman, D. A. (2013). The design of everyday things: Revised and expanded edition. Basic Books.

Orso, V., Ziviani, R., Bacchiega, G., Bondani, G., Spagnolli, A., & Gamberini, L. (2022). Employee-centric innovation: Integrating participatory design and video-analysis to foster the transition to Industry 5.0. Computers & Industrial Engineering, 173, 108661. https://doi.org/10.1016/j.cie.2022.108661

Papadimitriou, E., Schneider, C., Aguinaga Tello, J., Damen, W., Lomba Vrouenraets, M., & ten Broeke, A. (2020). Transport safety and human factors in the era of automation: What can transport modes learn from each other? Accident Analysis and Prevention, 144, Article 105656. https://doi.org/10.1016/j.aap.2020.105656

Parasuraman, R., & Manzey, D. H. (2010). Complacency and bias in human use of automation: An attentional integration. Human Factors, 52(3), 381–410. https://doi.org/10.1177/0018720810376055

Parasuraman, R., & Riley, V. (1997). Humans and automation: Use, misuse, disuse, abuse. Human Factors, 39(2), 230–253. https://doi.org/10.1518/001872097778543886

Pheasant, S., & Haslegrave, C. M. (2005). Bodyspace: Anthropometry, ergonomics and the design of work (3rd ed.). CRC Press.

Rogers Commission. (1986). Report of the presidential commission on the space shuttle challenger accident. https://ntrs.nasa.gov/citations/19860015255

Strauch, B. (2018). Ironies of automation: Still unresolved after all these years. IEEE Transactions on Human-Machine Systems, 48(5), 419–433. https://doi.org/10.1109/THMS.2018.2844243

Tortorella, G.L., Miorando, R.F., Caiado, R.G., Nascimento, D.M., & Portioli Staudacher, A. (2018). The mediating effect of employees' involvement on the relationship between Industry 4.0 and operational performance improvement. Total Quality Management & Business Excellence, 32, 119–133.

Trstenjak, M., Benešová, A., Opetuk, T., & Cajner, H. (2025). Human factors and ergonomics in Industry 5.0—A systematic literature review. Applied Sciences, 15(4), 2123. https://doi.org/10.3390/app15042123

U.S. Chemical Safety Board (CSB). (2007). Investigation report: Refinery explosion and fire, BP Texas City, March 23, 2005 (Report No. 2005-04-I-TX).

U.S. House Committee on Transportation and Infrastructure. (2020). Final committee report: The design, development, and certification of the Boeing 737 MAX.

Van Erp, T., Pereira Carvalho, N. G., & Gerolamo, M. C. (2024). Industry 5.0: A new strategy framework for sustainability management and beyond. Journal of Cleaner Production, 461, 142271. https://doi.org/10.1016/j.jclepro.2023.142271

World Nuclear Association. (2022). Three Mile Island accident. Retrieved from https://world-nuclear.org

Žižić, M. C., Mladineo, M., Gjeldum, N., & Celent, L. (2022). From Industry 4.0 towards Industry 5.0: A review and analysis of paradigm shift for the people, organization and technology. Energies, 15(14), 5221. https://doi.org/10.3390/en15145221

3 Challenging Innovation with Prospective Ergonomics

Julien Nelson and Eric Brangier

3.1 INTRODUCTION

Ever since the appearance of the concept of innovation in the literature, it has been seen as a primary engine of economic growth (Schumpeter, 1934). The importance of technological progress is often emphasized in these works (Flichy, 2007). Yet, invention does not equate innovation. When assessing whether or not an innovation is a success, one must also consider whether or not it has been adopted by the surrounding social system (Rogers, 1962), i.e., the market. This decision depends in part, from the point of view of end users, on the novelty and usefulness of the technology (Davis, 1989; Nielsen, 1993). In this context, ergonomics is often seen as a means to improve the acceptability of innovations to their intended "end users".

This view, however, underplays the power of innovative technologies to transform human activities. The current speed of innovation and technological disruptions is profoundly transforming our workplaces. Every day, new technologies emerge, accelerating progress in various fields such as mobility, transportation, artificial intelligence, healthcare, and energy. Innovation cycles are shortening, forcing companies to quickly adapt to remain competitive. Technological disruptions, on the other hand, are reshaping traditional models, redefining industries, and creating new markets. These changes not only challenge established companies but also alter workers' behaviors. In this fast-paced environment, agility becomes essential, and the ability to anticipate the effects of future technologies on human activity is crucial for survival in a competitive economy.

Prospective ergonomics is an emerging form of ergonomic intervention. It provides a crucial contribution to innovation by anticipating future user needs and facilitating the design of products and services that align with technological advancements (Robert & Brangier, 2009, 2012, 2024). Unlike traditional ergonomics, which focuses on improving existing tools, prospective ergonomics projects itself into emerging and future contexts of use. It allows innovative organizations to anticipate the future impact of emerging technologies - both in positive terms (e.g., identifying promising new avenues for the design of goods and services that offer radical improvements in usefulness and user experience) and negative terms (e.g., identifying emerging human factors concerns surrounding a specific technology) By placing humans at

 DOI: 10.1201/9781003559399-4

the center of the design process, prospective ergonomics supports a user-centered approach, enabling innovations that better meet expectations, reduce the risks of innovation failure, and anticipate future uses.

This chapter will present an overview of a new challenge for the future of ergonomics: to support innovation with prospective ergonomics in the development of Industry 5.0. Its goal is first, to explain the needs of innovation in Industry 5.0, second to encourage researchers generally interested in the future to orient their research toward the creation of new methods or the adaptation of existing ones, and toward validating these methods in the field so as to know how to work efficiently with the future; third, to inform the practitioners about the existence and characteristics of methods for dealing with the future, and help them to select the best methods for their innovation projects.

3.2 ERGONOMICS AND INDUSTRY 5.0

Industrial development is always part of an evolution of economic and political projects to modernize production tools. Five fairly conventional phases can be distinguished, beginning with the industrial revolution (Figure 3.1).

As with previous stages (e.g., Schwab, 2016), the rise of Industry 5.0 is inseparable from that of specific Key Enabling Technologies (KETs), particularly in the field of Artificial Intelligence (AI) and robotics. However, a specific feature of Industry 5.0 is that it is also connected with a different view of society. Whereas "Society 4.0", i.e., the Information society, relied on the multiplication of sources of information and on new means to process this information for value creation, Society 5.0 is openly more focused on human aspects (Figure 3.2). It is therefore a transition taking multiple forms: digital, environmental, energy, social, economic, and obviously, productive. This new stage of industrial development carries enormous expectations for

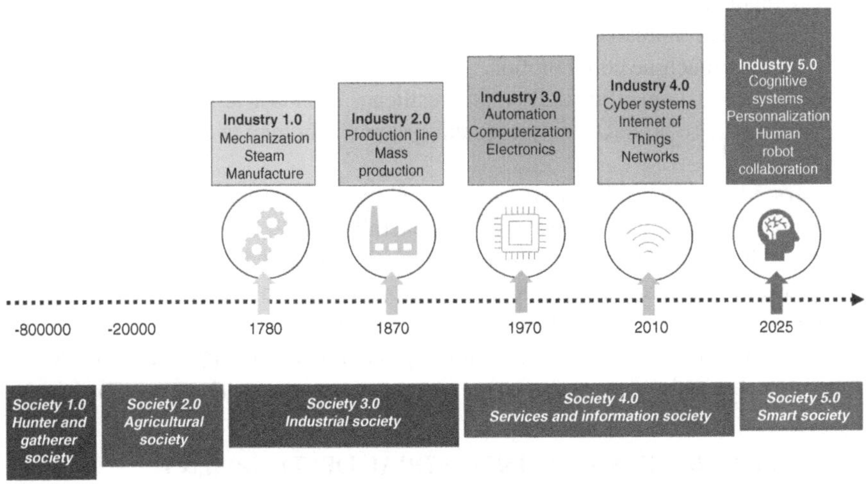

FIGURE 3.1 The five major stages of industrial development and the kind of societies.

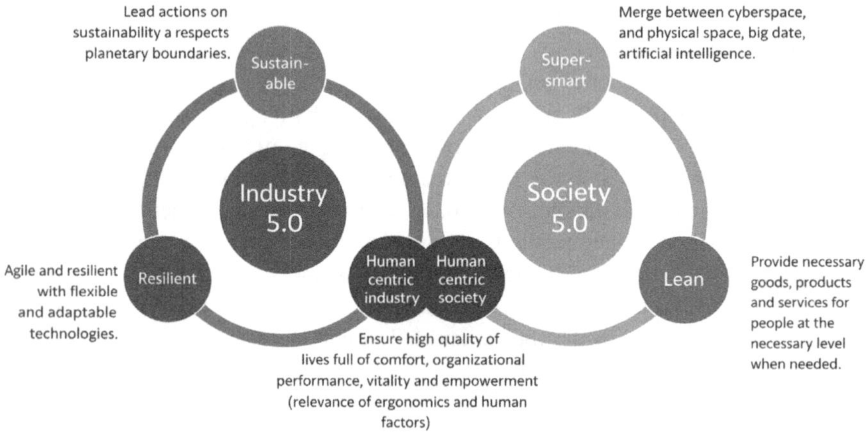

FIGURE 3.2 Relation between Industry 5.0 and Society 5.0 (inspired by Huang et al., 2022).

innovation and leads ergonomics to evolve toward anticipating future human needs and behaviors.

Industry 5.0 and Society 5.0 include characteristics closely linked to ergonomics, in the sense that they emphasize the human dimensions of work and life. The economy must work for humans, and innovation will be at the service of humans and society. Industry must help develop workers' skills while increasing business competitiveness. The human-centered approach is therefore at the heart of the 5.0 approach, including for digital technologies and artificial intelligence. This new industry therefore aims to:

- Create sustainable and human-centered industrial processes,
- Develop human creativity to solve unprecedented problems that machines cannot effectively address,
- Combine human skills with technological advances centered on human-machine collaboration,
- Ensure sustainability, agility, and resilience of industrial activities,
- Personalize products, services, and systems to human needs.

Industry 5.0 therefore places human and technical performance as the keystone of tomorrow's productivity. Industry 5.0 is generally considered a more complete version of Industry 4.0, in the sense that it is not only concerned with promoting productivity, working conditions, and economic growth – which is already a considerable challenge in itself – but also places confidence in the ability of these KETs to become a resilient provider of prosperity. Technology adoption is an opportunity for a more human-centric industry.

3.3 ERGONOMICS AND INDUSTRIAL DEVELOPMENT

Ergonomics and human factors engineering (HFE) emerged during the Second World War, with concerns for the aeronautical and military sectors. It then spread to

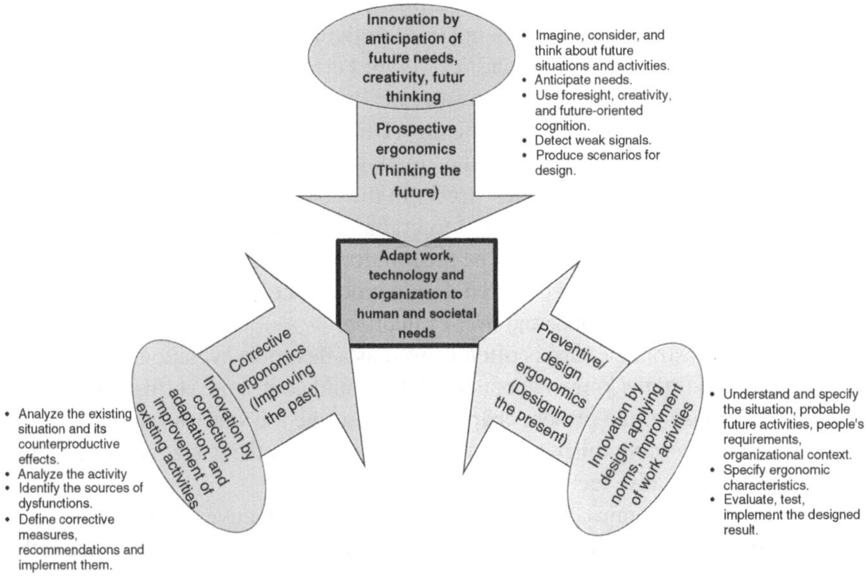

FIGURE 3.3 Ergonomics to collect, analyze, and interpret data on people and activities to design the future of work.

all areas of society. The development of the discipline has followed that of the industry. Overall, the connections between HFE and systems design have followed three stages (Robert & Brangier, 2012 ; Figure 3.3):

HFE was first considered as a form of applied psychology that could serve to understand system failures in terms of design flaws from a user-centered perspective (Chapanis et al., 1949). This approach is known as *corrective ergonomics* (de Montmollin, 1967). Although corrective ergonomics has greatly contributed to the development of the profession, it has some major drawbacks. First, it implies that the system is already designed and implemented, and most likely, that it presents some design flaws that have had adverse effects on user performance and/or well-being. Second, in such cases, the expenses related to systems design to this point can be considered "sunk costs", as the design project has yielded a suboptimal solution. To avoid this, and following the emergence of project-based design management (Midler, 1995; Wheelwright & Clark, 1992), one solution has been to include HFE professionals as members of project teams, so that design flaws can be detected and corrected before system implementation. This is known as *design ergonomics*[1] (de Montmollin, 1967).

Prospective ergonomics is a fairly new mode of ergonomic intervention, particularly suited to some challenges of innovation design. It is defined as "the part of ergonomics that attempts to anticipate future needs and activities so as to create new artifacts that will be useful and provide positive user experience" (Robert & Brangier, 2009, p. 165). Its ambition is to contribute to defining the orientations of innovative design before the beginning of design projects (Liem, 2017; Nelson et al., 2013; Robert & Brangier, 2024).

Indeed, HFE practitioners have often expressed concern, perhaps even frustration, with the idea that their discipline is not used to the full extent of its transformative potential (e.g., Dul et al., 2012; Hendrick, 1996; Perrow, 1983). One way to remedy this is to find new ways for ergonomics to intervene not just in the implementation, but in the *definition* of company strategies (Dul & Neumann, 2009). In line with Porter (1996), we define strategy as "the creation of a unique and valuable position, involving a different set of activities" (p. 3). In the context of design strategy, we can identify two main expectations related to ergonomics involvement. First, choosing which concepts should or should not be greenlit for design and development projects; and second, planning the user-centered design process, i.e., defining which resources will be needed for prototyping and user evaluation.

Although this strategic turn is often viewed as a desirable evolution for ergonomics, surprisingly little is known about how HFE could contribute at this key strategic level. The purpose of the present chapter is to describe the state of current research and practice concerning this issue, specifically as regards the design of systems featuring highly innovative technologies in Industry 5.0.

Prospective ergonomics, far more than design and corrective ergonomics, is concerned with anticipating the future of human activities. One notable difference is that it purposely aims to "jump into the unknown" and address situations that do not yet exist: Can we imagine products that don't yet exist and that will satisfy tomorrow's needs? What will tomorrow's services be? Can we anticipate people's expectations?

As noted above, Industry 5.0 places a strong emphasis on innovative technologies to explore new sources for value generation. Whereas conventional forms of ergonomic intervention are well suited to producing *incremental* innovation, i.e., discrete improvements of existing systems, prospective ergonomics focuses more on radically new, *disruptive* technologies as a means to renew human needs (Norman, 2010). Beyond the disruption of a technological state of affairs, it poses the question of disrupting the structure and meaning of human activities to generate value to users and society (Norman & Verganti, 2014). The question of harnessing the potential of disruptive technologies for company and societal growth has long been posed by technologists and management scholars (e.g., Christensen, 1997). Prospective ergonomics aims to address this challenge by placing human activity, specifically *future* human activity, firmly at "center stage". Furthermore, it does not view future activity merely as something to be anticipated, but more as something to be *designed*.

In recent years, management research has done much work to characterize the various organizational forms that may lead to successful innovations. We argue that ergonomics research should focus much more on how the discipline can contribute to these various paths to innovation success. As creativity is often viewed as a precursor of innovation (Cropley & Cropley, 2015), how can HFE contribute to improved performance in creative work? Some work has focused on the ergonomics of workstation design for creative workers (e.g., Dul, 2019). However, in the present case, prospective ergonomics is concerned more with using creativity techniques to anticipate future use. The questions that this situation poses are not dissimilar to those that HFE was confronted with at the onset of design ergonomics: what form should this anticipation take? What stakeholders should HFE practitioners be conversing with, and when?

3.4 ERGONOMICS AT THE FUZZY FRONT-END OF HUMAN CENTRIC INNOVATION

The argument in favor of using HFE to support innovative design usually runs as follows: ergonomists are charged with collecting data "in situation" to help characterize user needs; these needs serve as a basis for the production of requirements for the design of a new system; these requirements are then used for generating design solutions, which then lead to the production of a prototype; finally, these prototypes are subjected to iterative user evaluation. The only difference between corrective and design ergonomics seems to be in the "distance" between the situations of user activity being analyzed and the future situations of use of the system being designed. In the first case, corrective ergonomics, the situation that is analyzed – i.e., the *situation of action of reference* (Daniellou, 2004) – involves the use of the very system that is being redesigned. In the second case, design ergonomics, ergonomists still rely on identifying situations of reference for analysis, but there is often a wish that the future system will be somewhat different from the existing situation.

The problem arises, therefore, when the design brief does not make it possible to clearly identify a population of end users and/or a situation of reference for analysis (Brangier & Bastien, 2006; Nelson et al., 2013). This is particularly the case when a company's innovation strategy is strongly reliant on producing technological novelty (Buisine et al., 2021; Jaruzelski et al., 2014). Although in some cases, such novelty may be valued *per se* (Hooge, 2010), the ideal situation is that the product development process yields a system that is useful to its users. In other words, this is an inversion of the typical position of ergonomics: the goal is not to identify the optimal technological solution to address user needs, but to identify the best possible application of a given technology, i.e., the one that provides the most value to its users (Verganti, 2011). Only once this has been done, can any resources be allocated to the user-centered design process.

The fuzzy front-end of innovation, or FFE, refers to "all actions that are taken between the first consideration of an opportunity and the decision whether or not to start product development" (Gassmann & Schweitzer, 2014, p. 5). Acknowledging the possibility that ergonomics might have a role to play in defining innovation design strategy entails admitting the need for ergonomists to identify key stakeholders at the FFE and learn the bases for a productive dialog (Brangier et al., 2017).

One key consideration in doing so is the change in the status of user needs. In conventional – i.e., corrective and design – ergonomics, user needs refer to a psychological experience in users arising from the use of an existing system, and which can be inferred from analyzing their interactions with – and perceptions of – said system. Whether the system succeeds or fails in addressing user needs will determine the tone of the user experience,[2] and therefore the identification of *unmet* user needs can serve as a basis for defining the design requirements for a new system. In *prospective* ergonomics, the project has not yet begun and the FFE stakeholders are examining opportunities for innovation, defined as "situations in which new goods, services, raw materials, and organizing methods can be introduced and sold at greater than their cost of production" (Shane & Venkataraman, 2000, p. 220). Ultimately, although it is important to consider the "bottom line" – i.e., company profit – the goal of UCD is first and foremost to achieve acceptance of a new product or technology in its end users, and this depends in part on their judgments of the *usefulness* of the technology for achieving their goals. In UCD, it is for users to make judgments of this kind. In the FFE, it is for suitably mandated decision-makers to examine the potential future utility of a technology in order to greenlight a project (Loup-Escande et al., 2020). Only later on will these hypotheses be verified in the user-centered design process.

3.5 CONCEPTUAL BACKGROUND FOR INNOVATION IN HUMAN-CENTRIC INDUSTRY

Brangier and Robert (2014) and Robert and Brangier (2024) have proposed four key foundations for prospective ergonomics:

- The ergonomic foundation: prospective ergonomics naturally remains focused on the main objectives of HFE, i.e., to ensure optimal performance, well-being and user experience by analyzing activities (Karwowski, 2005).
- The prospective foundation: prospective ergonomics should be grounded in collaborations with other disciplines, with the methodological tools necessary for identifying the trends that are likely to shape human activity in the future. These trends may be technological, sociological, demographic, etc.
- The creative foundation: creativity is defined here as the ability to produce work that is novel and suited to the goals of the task for which it is produced (Runco & Jaeger, 2012). In this instance, prospective ergonomics should rely on the methods and tools of creative design for generating novel concepts for system design that provide user-relevant applications of innovative technologies.
- The future thinking foundation: more than other forms of ergonomic intervention, prospective ergonomics relies on ergonomists, FFE stakeholders, and others to imagine future situations of use. Proposing methods to do so requires developing knowledge of the cognitive processes underlying such projections (Colin et al., 2021).

3.6 FUTURE ACTIVITY

Arguably, the ability to formulate and critically examine descriptions of future activity is at the heart of any ergonomic intervention. We have already outlined above (see Section 3) that the analysis of existing situations of reference can serve as a basis for defining the requirements relative to a new system. This is only one half of the equation however, as designing the new system will require producing descriptions of expected future situations of human-system interactions (Robertson, 2004; Rosson & Carroll, 2002). Pinsky and Theureau (1984) emphasized this as a key challenge for ergonomics: how can one be expected to generate valid descriptions of future situations of use based only on the analysis of existing situations? The conventional approach to this problem appears to be through a gradual reduction in the degrees of freedom when producing such descriptions. Faced with an uncertain future, it is possible to implement simulations of future situations of activity to collect reasonably accurate descriptions of the future (Barcellini et al., 2015; Daniellou, 2007; Maline, 1994; van Belleghem, 2018, 2021).[3]

This line of thinking is representative of what Brangier, Vivian et al. (2019) call "probable activity". Descriptions of future activity are essential for determining the quality of design solutions from a user-centered perspective. In this sense, a user test carried out on a prototype can be seen as a means to produce data concerning one specific class of future situations: what would happen if this prototype were to be selected for development and commercialization? It is important to acknowledge that these descriptions are inherently limited. However, much effort is expended in imagining future situations of activity, it will never be possible to completely predict all possible contingencies. Yet, in some cases, ergonomists may wish to anticipate as many different outcomes as possible as a way to make the system "future-proof" ("requisite imagination"; Adamski & Westrum, 2003). In such cases, there is an incentive to expand the range of futures that the ergonomist can anticipate. Brangier, Vivian et al. (2019) call this approach the "possible future".

What approach does prospective ergonomics have for future activities? Because prospective ergonomics aims to intervene before corrective and design ergonomics, it would be tempting to say that it operates at a different time scale – that it aims to anticipate human activities "from further back", or further ahead, than its two counterparts. Certainly, some authors have attempted to apply a prospective ergonomics framework to designing systems with a very long "shelf life", e.g., in aeronautics (Ruault et al., 2014). However, it would be short-sighted to differentiate prospective ergonomics based on this criterion alone. Indeed, it often deals with situations where the immaturity of a technology is the main obstacle to realizing its promise to users (Loup-Escande & Loup, 2021), but where it is complicated to set a clear date when sufficient maturity will be reached. For example, when will interactive windows be a reality in mass transportation (Buisine et al., 2016)? Or do interactive tabletop interfaces provide true support for real-time collaboration (Nelson et al., 2012)?

Rather than reasoning in terms of time frames, it seems more fruitful to reason in terms of the *level of maturity of a design concept*. We use this term in reference to C-K design theory (Hatchuel & Weil, 2009): a concept is a starting point for the design process. Specifically, a design concept may include sufficient information to identify "reference situations" for user research – e.g., a project focusing on designing "a partially autonomous vehicle for maintaining the autonomy of elderly drivers" would probably begin with a careful analysis of problematic driving situations for this audience – or it may not – e.g., "a system making use of artificial intelligence (AI) for improving healthcare outcomes". Whether a concept is mature or not is not a binary matter. For example, "anticipating the effects of the latest developments in AI on the activity of soldiers in the military sector" (Lauton et al., 2024) is a perfectly acceptable question for prospective ergonomics, but it would probably need to be broken down into several more specific problems (e.g., what developments are we talking about? What types of military operations? Etc.). In reference to C-K theory, the purpose of prospective ergonomics here is to *plan future knowledge expansions* in human factors, to be addressed either as fundamental research questions or within a user-centered design project.

3.6.1 Prospective and Foresight

This latter example illustrates the second main characteristic of the future as viewed in prospective ergonomics. Ergonomics is first and foremost a profession of design. In Simon's (1996) terms, its purpose is to "change existing situations into preferred ones" – or alternately, to prevent undesirable situations from taking place. In this sense, it shares much with two related fields, *prospective* (Berger, 1957; Godet, 1986) and *strategic foresight* as they pertain to the broader field of Futures studies. As Poli (2022) notes, the latter comprises several related terms: "*forecasting* deals with data extrapolation, *foresight* with the visualization of possible futures, and *anticipation* with their translation into action" (p. 578).

Forecasting echoes our previous points on the various approaches to future activity in ergonomics (see Section 4.1). Since ergonomics views human activity firmly as a situated phenomenon, forecasting may focus not just on divining future forms of human activity, but on identifying trends in the evolution of the context in which these activities take place. As Liebl and Schwarz (2010) note, the term is used in two ways. One refers to statistical analyses of time series, particularly to the long-term component of these models. The other meaning "is mainly used to address the intuitive feeling that something is important" (p. 314, our emphasis). They further point out that a "quantitative approach of the future [corresponding to the first meaning] is not appropriate for strategic issue management using weak signals".

Concerning forecasting, ergonomics already uses statistical models to project future values from quantitative datasets, e.g., in anthropometry (Barbé et al., 2014). It is also possible to infer trends from qualitative data, although this usually relies on the involvement of domain experts. In addition, experts are also able to identify "weak signals". Initially theorized by Ansoff (1975) simply as "symptoms of possible change in the future", weak signals have been the subject of many successive expansions. Coffman (1997) proposes a seven-point definition:

> a weak signal is (i) an idea or trend that will affect the business or the business environment; (ii) new and surprising from the signal receiver's vantage point although others may also perceive it; (iii) sometimes difficult to track down amid other noise and signals; (iv) a threat or opportunity to an organization; (v) often scoffed at by people who 'know'; (vi) usually has a substantial lag time before it will mature and become mainstream; and (vii) therefore represents an opportunity to learn, grow and evolve.

This definition highlights the fact that weak signals are often initially confused with "noise", with their information value only being recognized *a posteriori*. However, some domain experts are able to recognize them, at the time of occurrence, as markers of strategically important phenomena. Several types of "signals" might fit this definition for prospective ergonomics: a trending hashtag on social networks, the unexpected success of a Kickstarter campaign, a public event marking the beginnings of a new social movement, or even the appearance of a new word designating an emerging social practice. Some authors (e.g., Brangier, Brangier et al., 2019) have proposed interview methods specifically aimed toward this.

Concerning foresight and anticipation, prospective ergonomics, like design ergonomics, relies on scenarios to describe situations of future use. Broadly defined, scenarios are "hypothetical sequences of events constructed for the purpose of focusing attention on causal processes and decision points" (Kahn & Wiener, 1967, p. 6). The two types of scenarios differ based on the types of decision they support (Lauton et al., submitted-b):

> Scenarios in design ergonomics aim to help project team members to generate and assess potential *solutions* to a design problem. In the early stages, they are usually based on the analysis of real-world, existing situations of human activity and aim to illustrate problems with existing systems. In later stages, they aim to describe and evaluate possible solutions, specifically their benefits in terms of usability and/or user experience.[4]

Scenarios in prospective ergonomics aim to help stakeholders at the FFE of innovation to select which projects to greenlight. This usually relies on formulating hypotheses concerning the projected future utility of a possible design concept. Their goal is not to summarize the results of user research, but to trigger avenues for future user research. It is important to note that these scenarios do not necessarily claim to exhibit predictive validity, but need to be sufficiently convincing to trigger the decision to launch a project (Lauton et al., submitted-a).

3.6.2 Future Thinking – Future Oriented Cognition

To imagine innovative products for Industry 5.0, prospective ergonomics questions the future. It seeks to identify future needs and activities that products and systems will need to address. It also aims to anticipate new needs and activities that will emerge from the use of new technical or organizational devices. To support this task of future thinking, prospective ergonomics invokes future-oriented cognition (Martin et al., 2021).

Future-oriented cognition is our ability to imagine and simulate an experience that has occurred, could have occurred, or might occur. According to D'Argembeau (2016), future-oriented cognition appears to be based on mnemonic systems or processes, particularly episodic memory, i.e., memory related to lived events. Indeed, the neuronal activity observed during tasks aimed at envisioning the future and tasks aimed at remembering past events reveals the activation of similar brain regions. Future-oriented cognition is thus the human capacity to mentally project oneself into the future, to imagine and simulate a realistic experience (Colin et al., 2021). This ability allows individuals to anticipate future events.

This capacity we have to imagine future events and experiences constitutes an element for the ergonomist to integrate into their approach and support through their methodology. They can use and promote it to anticipate future needs and activities, particularly by relying on principles to collect data about the future by soliciting projection into the future. The ergonomist will seek, through various methods (see Colin et al., 2021; Robert & Brangier, 2024), to guide the mental construction of the context of the future event, and to use projection supports to facilitate the evocation of future events. These can include ideation techniques, diagrams representing activity, sentences to complete, experience journals, guided scenarios, personas, etc.

3.6.3 Creativity

One of the first challenges when considering the use of idea generation techniques for prospective ergonomics is to define exactly what kind of "idea" we are interested in. These ideas may concern, for example, hypothesized user needs, ideas for system functions that are thought to be novel and adapted, or concepts of design solutions (Buisine et al., 2016).

Given this diversity, it is perhaps not surprising that the work concerning this creative foundation has been rich and varied, albeit somewhat scattered. Users of existing systems have long been considered a source of creative ideas for innovation (see Hewing, 2014, for a review). One problem is that users may have poor knowledge of the technology's usefulness to their own activity, and it becomes difficult to express, or even imagine, their own user needs (Robertson & Robertson, 2012).

As we have noted above (see Section 4.1), one situation that represents a particular challenge for ergonomics is when the initial design brief does not include a clear statement concerning the identity of end users. In those instances, idea generation may be carried out by designers themselves. In such cases, idea generation also relies on the use of intermediate objects of design. Personas (Cooper, 1999) have proven a valuable source of inspiration for designers to generate ideas regarding user needs and useful functions for innovative technologies (e.g., Bornet & Brangier, 2016). More recent work has focused on the encouraging contributions of virtual environments, which allow designers to embody user avatars (Bonnardel & Pichot, 2020; Buisine et al., 2016).

This line of work typically focuses on assessing idea generation as a creative task, using classical measures of creative performance such as idea fluency or originality (Kaufman et al., 2008). Eventually, however, these ideas must be put together into coherent stories or scenarios of use, so that designers and stakeholders at the FFE of innovation will be able to process them. Some authors have proposed the use of creativity techniques to generate scenarios, seen as combinations of elementary features (Moon & Han, 2016; Nelson et al., 2012, 2013, 2014). However, research concerning how scenarios of use may illustrate initial concepts and be processed at the FFE is far sparser. *Processing*, here, encompasses two separate activities: *idea evaluation* refers to judging ideas along one or more dimensions, comparing them to a benchmark or to each other; while idea selection involves some form of decision-making, typically retaining one idea for further development or for implementation (Rietzschel et al., 2019). Although creative *ideas* are well known to be judged according to the central dimensions of originality and suitability (e.g., Runco & Charles, 1993), other authors have proposed more complete taxonomies. In addition, and perhaps more importantly, little is known about the second stage of idea selection at the FFE. Cognitive ergonomics has done great work characterizing the cognitive activity of design (Barcellini et al., 2014; Visser, 2006) with the potential effect of improving the integration of HFE in design processes; comparatively little is known in HFE about the processes and activities that take place at the FFE of innovation. To our mind, this will be an essential challenge for prospective ergonomics in the years to come.

3.7 METHODOLOGICAL BACKGROUND FOR INNOVATION IN HUMAN CENTRIC INDUSTRY

A classification of methods for prospective ergonomics is based on the type of construction that results from the data collection. Robert and Brangier (2019) made a distinction between *descriptive* and *projective* methods. Descriptive methods aim to accurately report the events: what happened or what is going on. Projective methods aim to help people think about the future and project themselves into the future.

3.7.1 Descriptive Methods

Initial work in the field placed prospective ergonomics strongly in the field of technology-push innovation strategies, in such a way that the design concept did not include any information concerning the (methodologically crucial) matter of who the end users are and what activities the future system should aim to support. In such cases, conventional methods for the description of human activities are of little use (Brangier & Bastien, 2006). As noted above, however, the maturity of a design brief is not a binary matter (see Section 4.1). Prospective ergonomics is also used in situations where the brief is well-defined enough to identify broad *domains* of activity, if not clear situations of reference. In this case, it is still important for ergonomists to generate knowledge about current activities and user needs with its usual methods – interviews, focus groups, ethnographic observation, etc. – but with a broader and more structured approach. For example, it is suggested that the major physical and societal transformations occurring in the era of the Anthropocene are likely to lead to major evolutions in human activities (Boisadan & Buisine, 2025). What form will human cities take on in the later Anthropocene, and how will they operate? It seems that such a reflection might begin with studying exactly what activities are structurally important in the running of a city today, before going on to imagine what things might look like tomorrow[5]: will these activities still exist, and will new essential activities have appeared? How might their interrelations have evolved?

3.7.2 Projective Methods

The emerging interest of prospective ergonomics for future-oriented cognition (see Section 4.3) is borne from a practical consideration: to propose methods allowing participants in a prospective ergonomics intervention to imagine future situations of human activity, which we term *projective methods*. One method that stands out is the Guided Imaginary Projection (GIP; Allinc et al., 2019; Maniga et al., 2024), which "aims to have future users 'live' an experience of the service in an imaginary way, as if they were using it". To achieve this, it is necessary for the users to begin with accessing as much information as possible about the novel technology involved *and* to have a clear starting point in terms of what future situation of activity they are to project themselves into. The prospective ergonomist may then rely on specific interview methods to help participants construct and access descriptions of their own future envisioned experience (e.g., Vermersch, 2015). It is worth noting that these projections are hypothetical, like most of the claims prospective ergonomics deals with at the FFE of innovation, and that these hypotheses should be clearly stated and verified once the user-centered design process has begun in earnest.

Two things should be noted about the use of projective methods in prospective ergonomics. First, the materials used in these methods should include a mix of details that are stabilized and of elements that are not yet defined. For example, Maniga et al. (2024), while working on an urban redevelopment project in Italy,

TABLE 3.1
Some methods for prospective ergonomics and innovation (from Robert & Brangier, 2019)

Content	Descriptive	Projective
Verbal	Task analysis. Activity analysis. Interview. Shadowing. User journey may. Card sorting. Probing. Experience sampling. Counterfactual thinking…	Prospective interview. Guided imaginary projection. Expert community staff. Future thinking task. Prospective scenarios. Personas. Prospective story-telling. Sentences to complete. Needs and uses ideation interview…
Figurative (static/ dynamic)	Drawings. Sketches. Photos. Films…	Avatars. Drawings. Sketches. Personas. Story-board. Animation…

used hand-drawn sketches of areas of interest in this project. Interestingly, the sketch showed the objects present in these environments clearly, but the users were represented as blacked-out figures. In other words, the "use frame" (in the sense of Nelson et al., 2012) should remain at least partly open-ended. Second, one should not forget that stakeholders at the FFE, not just users, should be able to engage in these projections. This is quite a risky proposition. Indeed, in a design project, it is generally not considered desirable for stakeholders in a design project to base their decisions on their own projections, rather than on empirical data based on user research ("I-methodology"; Akrich, 1995). At the FFE, on the other hand, one could argue that this is unavoidable: not only is such empirical data often inaccessible, it is these very stakeholders who will defend a given design concept when deciding which projects will be the focus of future development. Rather than avoiding these projections altogether, it would be more fruitful for prospective ergonomics to study how they are constructed in the mind of decision-makers, and how they impinge on their subsequent decisions.

The integration of these two methodological orientations enables the construction of a comparative framework that juxtaposes, on the one hand, the descriptive and the projective, and on the other, the verbal and the figurative (Table 3.1).

3.8 PRACTICAL BACKGROUND FOR INNOVATION IN HUMAN CENTRIC INDUSTRY: AN EXTENDED VIEW OF THE ROLE OF ERGONOMICS IN INNOVATION DESIGN

Prospective ergonomics aims to guide the involvement of ergonomics in the early stages of innovation design. Where innovation is often seen, even today, as an accidental phenomenon, some conceptualizations tend to view innovation as a planned process. The aim of the game is not to achieve success in one project, but to generate as much value as possible from whatever knowledge has been generated by Research functions – and conversely, to guide research toward interesting "pockets" of knowledge (Le Masson et al., 2010).

In Section 4.1, we introduced the term *concept* from C-K design theory. This theory suggests that the value generated in design projects comes not just from the knowledge they generate, but from the concepts, i.e., unusual starting points for design, that are produced within. This duality is especially visible in the case of ergonomics' contribution to innovation design. Its methods make it possible to generate knowledge (e.g., concerning user needs, the evaluation of design solutions, etc.). Increasingly, however, the literature also acknowledges its contribution to the generation of new concepts, i.e., to creative design. Both these types of processes – knowledge generation and concept generation – are necessary for innovation. Yet, with rare exceptions, the literature treats them as being separate.[6]

One notable exception, the KCP workshop (Hatchuel & Weil, 2009), consists in dividing the work into three successive stages: (1) the K (Knowledge) stage aims to "build upon internal and external knowledge (…) This first phase is necessarily cross-disciplinary"; (2) the C (Concept) stage

> organizes "oriented" creativity according to precise rules. The innovation team is split into subgroups, each of which is asked to freely explore a concept called a "disruptive searchlight": these are surprising and strongly contrasting propositions (…) each subgroup is also asked to identify new research areas

and (3) the P (Proposal) stage, whose "goal is to build a roadmap that plans a series of actions, which include immediate solutions, research projects and new prototypes" (Elmquist & Segrestin, 2009, p. 235).

3.8.1 Modelling Interventions in Prospective Ergonomics

The literature in Human Factors comprises several examples of models detailing the structure of interventions in corrective and design ergonomics (Guérin et al., 2007) and in user-centered design (Cooper et al., 2014). In contrast, there are very few models of what constitutes an intervention in prospective ergonomics. A notable exception is the stage-based model reported in Robert et al. (2021). Figure 3.4 below reprises and completes the list of activities from that model, and divides them into the

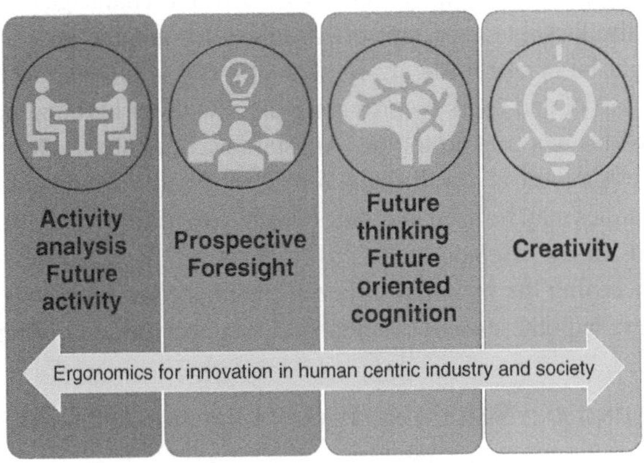

FIGURE 3.4 Main concepts for developing innovation with prospective ergonomics.

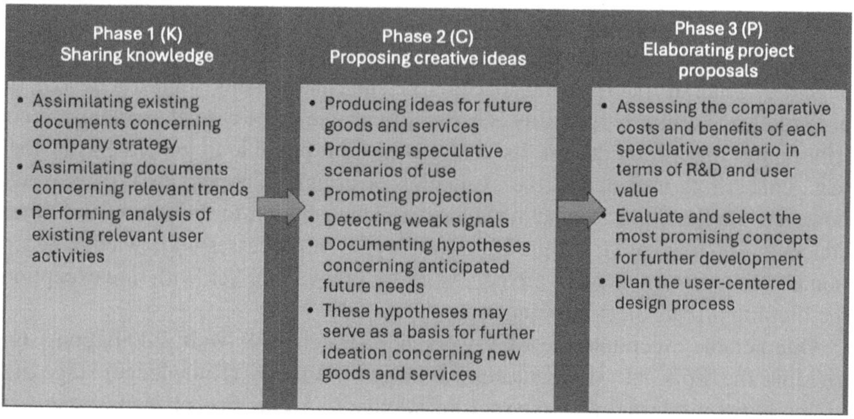

FIGURE 3.5 Typical tasks in a prospective ergonomics intervention viewed from the lens of a KCP workshop.

K, C, and P stages. This division makes it possible to trace a roadmap for prospective ergonomics in particular:

- To produce knowledge related to future human activities;
- To use this knowledge to imagine new concepts of goods and services, as well as descriptions of these concepts in use;
- To formulate proposals – i.e., in this case, to contribute to selecting the most relevant concepts for further development (Figure 3.5)

3.8.2 Identifying the Needs of Future Users

From the point of prospective ergonomics, user needs are hypotheses concerning a future activity. The models and concepts of ergonomics have been introduced concerning "user needs" remain valid. User needs are tied to fundamental psychological drives, and part of ergonomics' role in "user needs analysis" involves ergonomists eliciting feedback from end users concerning their own needs. The challenge in prospective ergonomics lies in the fact that users are often unable to imagine how a very innovative technology might be of use to them.[7] In such cases, it is possible to elicit a reflective attitude on the part of end users by providing them with background knowledge on the technology or with a working prototype (Anastassova et al., 2007; Decotter et al., 2019).

When the design brief is too ill-defined to identify a clear population of end-users, creativity techniques may be used to define possible use frames and, subsequently, speculative scenarios of use (Nelson et al., 2012). However, these scenarios only illustrate hypotheses concerning the potential utility of an emerging technology and underlying user needs. These hypotheses will eventually need to be verified through user research.

3.8.3 Modeling and Simulating Future Uses and Activities

We have noted that prospective ergonomics, like design and corrective ergonomics, made routine use of scenarios to illustrate the benefits of specific design briefs to users (see Section 4.2). In the latter case, scenarios usually serve as a means to illustrate the benefits of a design solution to users, e.g., in the case of a storyboard. The literature emphasizes the role of scenarios as narrative devices, whereby stakeholders in the design process can collectively examine the suitability of a given solution before deciding to accept or reject it (see notably Carroll, 2000). However, a scenario can only ever describe one situation of use at a time,[8] and discussions of design solutions often rely on examining several possible scenarios at one time. Task models can serve as a means to address this variability of possible situations of use (see Section 4.1) while focusing the discussion on higher-order constructs (e.g., user goals, performance, etc.). From this perspective, simulations of future use, whatever device they rely on – e.g., digital modeling, roleplaying, etc. – can be seen as a means to *produce empirical data* on future situations of use during the design process. Indeed, to some authors (e.g., Daniellou, 2007; Guérin et al., 2021), simulations are viewed as the *only* means to produce reliable data on future situations of human activity. In design ergonomics, the primary goal of the scenario is to illustrate the benefits of a potential solution to the end users, even though scenario-based discussions of design solutions can also include other criteria for decision-making (e.g., technical feasibility). Either way, the issues discussed seem to be constrained to the design process or, of course, the subsequent stage of system implementation and use.

In contrast, prospective ergonomics uses scenarios to address different issues – in particular, issues related to the selection of design projects and to planning the user-centered design process. For example:

- Nelson et al. (2009) used scenarios to discuss various concepts in the design of innovative equipments to prevent domestic accidents in infants. Based on initial user research, several different concepts were proposed. These could then serve as a basis for elaborating a strategy foreseeing the launch of multiple products.
- Lauton et al. (2024) are currently tackling the challenges of including emerging technologies in the fields of Artificial Intelligence and robotics in the innovation strategy of a company in the French military sector. This inclusion is itself the result of an ongoing strategy in the French Army, aiming for a stronger inclusion of several key innovative technologies in systems design and development projects. The goal of their work is to help the company anticipate upcoming human factors challenges to help plan strategies for user research.

3.8.4 Creative Stimulation

From the previous section, it should be clear that task models and scenarios are two complementary devices for anticipating future situations of use, as the former can be seen as a generative structure leading to the latter. At the same time, no amount of anticipation will ever reach a level such that one can say that everything has been thought of beforehand. Current work in prospective ergonomics has shown the interest of ideation methods to support the production of speculative scenarios of use, but these works are inspired by laboratory work on the most common methods available (i.e., brainstorming) and limited by the amount of available knowledge. Prospective ergonomics needs to propose new ideation methods that are suited to its specific needs. For example, promising developments are underway with the emergence of data-driven personas (see e.g., Jansen et al., 2021), and several efforts are being made to exploit the potential of generative AIs to support creative work (Lavrič & Škraba, 2023).

3.9 CONCLUSION

Industry 5.0 is the latest in a series of technological evolutions that challenge ergonomics in its methods and ambitions. Although most discourse on innovation focuses on its technical aspects, it is important to realize that innovation is primarily a human phenomenon. It is users who will ultimately decide to accept or reject the goods and services that are provided to them by a particular firm. It is again users that will reappropriate a given piece of technology, giving rise to new forms of interaction and new social practices. And again, it is these evolutions in human activities that lead to the emergence of new user needs, leaving room for the next generation of products. From this point of view, design can be likened to a mutual learning process between the designers and users of technology (Béguin, 2003; Hatchuel et al., 2002).

In user-centered design, designers interact with users to learn more about their needs and requirements. In turn, users learn from designers what are the possibilities afforded by a given piece of technology. This interplay between technology and use is sometimes so intricate that some scholars today question the very goal of design: is it to set the characteristics of an artifact, or of *situations* that are yet to come, in which users interact with a given technology (Preece et al., 2023) or even outcomes in terms of user experience (Hassenzahl, 2010)?

From this perspective, it makes little sense to think of "user needs" in positivistic terms, i.e., as "things" that somehow exist independently from users' relationships with technology. User needs, if anything, are *social* phenomena that cannot exist outside of the relationships that people entertain with the world – in the sense that their interactions with the world are goal-directed – and with each other. The role of ergonomics, insofar as it supports design, is to facilitate these interactions. More specifically, the role of *prospective* ergonomics, addressed here, is to provide organizational stakeholders, designers, and users with the means to discuss – and therefore, to structure – future situations featuring technologies with tantalizing possibilities for creating user value, but which technologists and users alike find difficult to imagine, rendering conventional user-centered design methods difficult to apply. This increased level of expectations relative to ergonomics forces us to refine our treatment of the concept of *future activity* that is so central to all of ergonomics, to include the three complementary facets of probable, possible, and prospective futures.

Because of the potential of emerging technologies for transforming human activity at the level of individuals and social groups alike, prospective ergonomics, which often concerns itself specifically with the development of goods and services that use emerging technologies to produce value for users, is perhaps the form of practice in which ergonomics practitioners are "most" like designers, in the sense that evolutions in user activities lead to questioning the very identity of the artifacts that structure these activities (Le Masson et al., 2010). Let us for example consider how, in recent times, a global pandemic has led to the generalization of remote work, questioning the decades-old concept of the "home office". Or how fruitful encounters between Artificial Intelligence and the automotive industry have opened up such possibilities for the "autonomous vehicle" that the very word "car" seems almost improper to qualify these new artifacts: might they be better described, for example, as "private compartments" or "pods"? And coming full circle, how key technologies such as artificial intelligence and robotics, under the banner of "Industry 5.0", may come to redefine the very meaning of "work".

Prospective ergonomics, whose concepts and methods we have broadly outlined here, aims to usher ergonomics into this new world in a much more active role. The future will, quite literally, be what we make of it.

NOTES

1 Also known as *preventive* ergonomics, since the goal is to prevent flawed designs from ever reaching the implementation stage (de Montmollin, 1967).
2 The shift from usability engineering to user experience design has greatly expanded our view of what constitutes a "user need", notably with the shift from *pragmatic* needs (i.e., task-related needs) to *hedonic* needs, i.e., fundamental psychological drives (Hassenzahl, 2007), but the point still stands that user needs, in this view, can be deduced from the analysis of existing activities.
3 Daniellou (1992) takes care to point out that the purpose of this exercise is not to evaluate probabilities of occurrence of these events as they are described, merely to circumscribe a space of likely forms of future activity.
4 Rosson and Carroll (2002) use the term "problem scenario" in the first case, and the terms "activity scenarios", "information scenarios", and "interaction scenarios" in the second.
5 As Boisadan and Buisine point out, however, these evolutions will not impact the fact that all human activities are bounded a goal of fulfilling the same basic set of fundamental human needs.
6 This is very visible, for example, in the literature on brainstorming. This method is intended to foster idea generation in small groups. Usually, these studies focus only on the ideational stage, ignoring the fact that in actual organizational settings, ideation never occurs in a vacuum but is always preceded by a stage of knowledge-gathering.
7 These needs are sometimes called "undreamed-of requirements" (Robertson, 2001) or "latent user needs" (Loup-Escande et al., 2014).
8 This is perhaps an overstatement: a scenario is a narrative description of one possible sequence of events, but is always an incomplete description. For example, when using a storyboard to represent a scenario, what happens between the images, or outside of the frame? It would probably be better to say that the scenario serves as a means for stakeholders to discuss a *class* of situations of use.

REFERENCES

Adamski, A. J., & Westrum, R. (2003). Requisite imagination: The fine art of anticipating what might go wrong. In E. Hollnagel (Ed.), Handbook of cognitive task design (pp. 193–220). Lawrence Erlbaum Associates.

Akrich, M. (1995). User representations: Practices, methods and sociology. In A. Rip, T. J. Misa, & J. Schot (Eds.), Managing technology in society. The approach of constructive technology assessment (pp. 167–184). Pinter.

Allinc, A., Cahour, B., & Burkhardt, J.-M. (2019). The guided imaginary projection, a new methodology for prospective ergonomics. In S. Bagnara, R. Tartaglia, S. Albolino, T. Alexander, & Y. Fujita (Eds.), Proceedings of the 20th Congress of the International Ergonomics Association (IEA 2018) (Vol. 824, pp. 1340–1347). Springer International Publishing. https://doi.org/10.1007/978-3-319-96071-5_135

Anastassova, M., Mégard, C., & Burkhardt, J. M. (2007). Prototype evaluation and user-needs analysis in the early design of emerging technologies. In J. A. Jacko (Ed.), Proceedings of the 12th International Conference, HCI International 2007, Beijing, China, July 22–27, 2007, Proceedings, Part I (pp. 383–392). Springer. https://doi.org/10.1007/978-3-540-73105-4

Ansoff, H. I. (1975). Managing strategic surprise by response to weak signals. California Management Review, 18(2), 21–33. https://doi.org/10.2307/41164635

Barbé, J., Mollard, R., & Wolff, M. (2014). Ergonomic approach to integrate touch screen in future aircraft cockpits. Journal Européen Des Systèmes Automatisés, 4, 303–318.

Barcellini, F., Détienne, F., & Burkhardt, J.-M. (2014). A situated approach of roles and participation in open source software communities. Human–Computer Interaction, 29(3), 205–255. https://doi.org/10.1080/07370024.2013.812409

Barcellini, F., van Belleghem, L., & Daniellou, F. (2015). Design projects as opportunities for the development of activities. In P. Falzon (Ed.), Constructive ergonomics (pp. 187–203). CRC Press.

Béguin, P. (2003). Design as a mutual learning process between users and designers. Interacting with Computers, 15(5), 709–730. https://doi.org/10.1016/S0953-5438(03)00060-2

Berger, G. (1957). L'attitude prospective [The prospective attitude]. La Revue Des Deux Mondes, 417–426.

Boisadan, A., & Buisine, S. (2025). Prospective ergonomics in the Anthropocene era: Reconsidering human needs. Ergonomics, 1–13. https://doi.org/10.1080/00140139.2025.2474642

Bonnardel, N., & Pichot, N. (2020). Enhancing collaborative creativity with virtual dynamic personas. Applied Ergonomics, 82, 102949. https://doi.org/10.1016/j.apergo.2019.102949

Bornet, C., & Brangier, E. (2016). The effects of personas on creative codesign of work equipment: An exploratory study in a real setting. CoDesign, 12(4), 243–256. https://doi.org/10.1080/15710882.2015.1112814

Brangier, É., Barcenilla, J., & Robert, J.-M. (2017). Prospective ergonomics and innovative ideas in the early stages of design projects. In A. Zunjic (Ed.), Ergonomic design and assessment of products and systems (pp. 47–68). Nova.

Brangier, É., & Bastien, J. M. C. (2006). L'analyse de l'activité est-elle suffisante et/ou pertinente pour innover dans le domaine des nouvelles technologies? [Is activity analysis sufficient and/or relevant to innovate in the field of new technologies?]. In G. Valléry & R. Amalberti (Eds.), L'analyse du travail en perspectives. Influences et évolutions (pp. 143–156). Octarès.

Brangier, É., Brangier, B., Marache-Francisco, C., Kopp, S., & Clausse, J. (2019). An interview process to anticipate future needs. In S. Bagnara, R. Tartaglia, S. Albolino, T. Alexander, & Y. Fujita (Eds.), Proceedings of the 20th Congress of the International Ergonomics Association (Vol. 7, pp. 718–729). Springer. https://doi.org/10.1007/978-3-319-96071-5_76

Brangier, É., & Robert, J.-M. (2014). L'ergonomie prospective: Fondements et enjeux [Prospective ergonomics: Foundations and stakes]. Le Travail Humain, 77(1), 1–20. https://doi.org/10.3917/th.771.0001

Brangier, É., Vivian, R., & Bornet, C. (2019). Méthodes d'ergonomie prospective appliquées à l'identification de besoins pour des systèmes d'énergie à base d'hydrogène: Étude exploratoire [Prospective ergonomic methods applied to the identification of requirements for hydrogen-based energy systems: Exploratory research]. Psychologie Francaise, 64(2), 197–222. https://doi.org/10.1016/j.psfr.2019.02.002

Buisine, S., Guegan, J., Barré, J., Segonds, F., & Aoussat, A. (2016). Using avatars to tailor ideation process to innovation strategy. Cognition, Technology & Work, 18(3), 583–594. https://doi.org/10.1007/s10111-016-0378-y

Buisine, S., Taton, A., & Boisadan, A. (2021). Need-seeking: Creating, discovering, or rediscovering needs? In N. L. Black, W. P. Neumann, & I. Noy (Eds.), Proceedings of IEA 2021 (pp. 624–630). Springer.

Carroll, J. M. (2000). Making use: Scenario-based design of human-computer interactions. MIT Press.

Chapanis, A., Garner, W. R., & Morgan, C. T. (1949). Applied experimental psychology: Human factors in engineering design. Wiley.

Christensen, C. M. (1997). The innovator's dilemma: When new technologies cause great firms to fail. Harvard Business School Press.

Coffman, B. S. (1997). Weak signal research part I. Introduction. *Journal of Transition Management*, 2(1). https://legacy.mgtaylor.com/mgtaylor/jotm/winter97/wsrintro.htm

Colin, C., Martin, A., Bonneviot, F., & Brangier, E. (2021). Unravelling future thinking: A valuable concept for prospective ergonomics. Theoretical Issues in Ergonomics Science, 1–27. https://doi.org/10.1080/1463922X.2021.1943045

Cooper, A. (1999). The inmates are running the asylum (1st ed.). Sams.

Cooper, A., Reimann, R., Cronin, D., & Noessel, C. (2014). About face: The essentials of interaction design (4th ed.). Wiley.

Cropley, D. H., & Cropley, A. J. (2015). The psychology of innovation in organizations. Cambridge University Press.

d'Argembeau, A. (2016). La pensée future épisodique: Entre simulation et contexte autobiographique [Episodic future thinking: Between simulation and autobiographical context]. Revue de Neuropsychologie, 8(1), 55–59. https://doi.org/10.1684/nrp.2016.0367

Daniellou, F. (1992). Le statut de la pratique et des connaissances dans l'intervention ergonomique de conception [The status of practice and knowledge in ergonomics intervention in design] [Habilitation thesis]. Université Toulouse II - Le Mirail.

Daniellou, F. (2004). L'ergonomie dans la conduite de projets de conception de systèmes de travail [Ergonomics in the management of projects for the design of work systems]. In P. Falzon (Ed.), Ergonomie (pp. 359–373). PUF.

Daniellou, F. (2007). Simulating future work activity is not only a way of improving workstation design. Activités, 4(2). https://doi.org/10.4000/activites.1704

Davis, F. D. (1989). Perceived usefulness, perceived ease of use, and user acceptance of Information Technology. MIS Quarterly, 13(3), 319–340. https://doi.org/10.2307/249008

de Montmollin, M. (1967). Les systèmes homme-machine [Man-machine systems]. PUF.

Decotter, D., Burkhardt, J. M., & Lubart, T. (2019). Evaluating users' creativity for service and needs identification in the field of emerging technologies: A comparison of two methods and two production conditions (pp. 983–989). https://doi.org/10.1007/978-3-319-96071-5_100

Dul, J. (2019). The physical environment and creativity: A theoretical framework. In J. C. Kaufman & R. J. Sternberg (Eds.), Cambridge handbook of creativity (2nd ed., pp. 481–510). Cambridge University Press.

Dul, J., Bruder, R., Buckle, P., Carayon, P., Falzon, P., Marras, W. S., Wilson, J. R., & van der Doelen, B. (2012). A strategy for human factors/ergonomics: Developing the discipline and profession. Ergonomics, 55(4), 377–395. https://doi.org/10.1080/00140139.2012.661087

Dul, J., & Neumann, W. P. (2009). Ergonomics contributions to company strategies. Applied Ergonomics, 40, 745–752. https://doi.org/10.1016/j.apergo.2008.07.001

Elmquist, M., & Segrestin, B. (2009). Sustainable development through innovative design: Lessons from the KCP method experimented with an automotive firm. International Journal of Automotive Technology and Management, 9(2), 229–244. https://doi.org/10.1504/IJATM.2009.026399

Flichy, P. (2007). Understanding technological innovation: A socio-technical approach. Edward Elgar.

Gassmann, O., & Schweitzer, F. (2014). Managing the unmanageable: The fuzzy front-end of innovation. In O. Gassmann & F. Schweitzer (Eds.), Management of the fuzzy front end of innovation (pp. 3–14). Springer.

Godet, M. (1986). Introduction to la prospective. Futures, 18(2), 134–157. https://doi.org/10.1016/0016-3287(86)90094-7

Guérin, F., Laville, A., Daniellou, F., Duraffourg, J., & Kerguelen, A. (2007). Understanding and transforming work: The practice of ergonomics. ANACT.

Guérin, F., Pueyo, V., Béguin, P., Garrigou, A., Hubault, F., Maline, J., & Morlet, T. (2021). Concevoir le travail, le défi de l'ergonomie [Designing work: The challenge of ergonomics]. Octarès.

Hassenzahl, M. (2007). The hedonic/pragmatic model of user experience. In E. Law, A. Vermeeren, M. Hassenzahl, & M. Blythe (Eds.), Towards a UX manifesto (pp. 10–14). COST.

Hassenzahl, M. (2010). Experience design: Techology for all the right reasons. Morgan & Claypool.

Hatchuel, A., Le Masson, P., & Weil, B. (2002). From knowledge management to design-oriented organisations. International Social Science Journal, 171, 25–37.

Hatchuel, A., & Weil, B. (2009). C-K design theory: An advanced formulation. Research in Engineering Design, 19(4), 181–192. https://doi.org/10.1007/s00163-008-0043-4

Hendrick, H. W. (1996). The ergonomics of economics is the economics of ergonomics. Proceedings of the Human Factors and Ergonomics Society Annual Meeting, 40(1), 1–10. https://doi.org/10.1177/154193129604000101

Hewing, M. (2014). Collaboration with potential users for discontinuous innovation. Springer Gabler.

Hooge, S. (2010). Performance de la R&D en rupture et des stratégies d'innovation: Organisation, pilotage et modèle d'adhésion [Breakthrough R&D projects performance and innovation strategies: Organisation, management and commitment model] [Doctoral dissertation]. Mines ParisTech.

Huang, S., Wang, B., Li, X., Zheng, P., Mourtzis, D., & Wang, L. (2022). Industry 5.0 and Society 5.0—Comparison, complementation and co-evolution. Journal of Manufacturing Systems, 64, 424–428. https://doi.org/10.1016/j.jmsy.2022.07.010

Jansen, B., Salminen, J., Jung, S., & Guan, K. (2021). Data-driven personas. Springer.

Jaruzelski, B., Staack, V., & Groehle, B. (2014). Proven paths to innovation success. Strategy Business, 77, 2–19.

Kahn, H., & Wiener, A. J. (1967). The year 2000—A framework for speculation on the next thirty three years. Macmillan.

Karwowski, W. (2005). Ergonomics and human factors: The paradigms for science, engineering, design, technology and management of human-compatible systems. Ergonomics, 48(5), 436–463. https://doi.org/10.1080/00140130400029167

Kaufman, J. C., Plucker, J. A., & Baer, J. (2008). Essentials of creativity assessment. John Wiley & Sons.

Lauton, J.-P., Kujawa, L., Loup-Escande, É., & Nelson, J. (2024). Anticiper les retombées de l'intégration de l'IA et de la robotique sur l'activité des opérationnels dans le secteur Défense: Une approche par l'ergonomie prospective [Communication]. Doctoriales ARPEGE.

Lauton, J.-P., Loup-Escande, É., Kujawa, L., & Nelson, J. (submitted-a). Stakeholders' viewpoints concerning innovative project during the fuzzy front-end of innovation: An exploratory study. Applied Ergonomics.

Lauton, J.-P., Loup-Escande, É., Kujawa, L., & Nelson, J. (submitted-b). What to expect from task models in prospective ergonomics? A theoretical review. Theoretical Issues in Ergonomics Science.

Lavrič, F., & Škraba, A. (2023). Brainstorming will never be the same again—A human group supported by artificial intelligence. Machine Learning and Knowledge Extraction, 5(4), 1282–1301. https://doi.org/10.3390/make5040065

Le Masson, P., Weil, B., & Hatchuel, A. (2010). Strategic management of innovation and design. Cambridge University Press.

Liebl, F., & Schwarz, J. O. (2010). Normality of the future: Trend diagnosis for strategic foresight. Futures, 42(4), 313–327. https://doi.org/10.1016/j.futures.2009.11.017

Liem, A. (2017). Prospective ergonomics. John Wiley & Sons.

Loup-Escande, E., Burkhardt, J.-M., Christmann, O., Chaniaud, N., & Richir, S. (2020). Usefulness and needs construction process in innovative artefacts: An exploratory study of designers' viewpoints. International Journal of Virtual Reality, 20(1), 48–71. https://ijvr.eu/article/view/2924

Loup-Escande, É., Burkhardt, J.-M., Christmann, O., & Richir, S. (2014). Needs' elaboration between users, designers and project leaders: Analysis of a design process of a virtual reality-based software. Information and Software Technology, 56(8), 1049–1061. https://doi.org/10.1016/j.infsof.2014.04.008

Loup-Escande, É., & Loup, G. (2021). Designing acceptable emerging technologies: What contribution from ergonomics? Theoretical Issues in Ergonomics Science, 22(5), 581–602. https://doi.org/10.1080/1463922X.2020.1836568

Maline, J. (1994). Simuler le travail: Une aide à la conduite de projet [Simulating work: An aid for project management]. ANACT.

Maniga, E., Adelé, S., & Cahour, B. (2024). How to catch prospective use? A comparative study of virtual environment movie and guided imaginary for projecting future users in a public space. Ergonomics, 1–14. https://doi.org/10.1080/00140139.2024.2418946

Martin, A., Colin, C., Brangier, É., & Robert, J.-M. (2021). Cognition orientée futur [Future-oriented cognition]. In É. Brangier & G. Valléry (Eds.), Ergonomie: 150 notions clés (pp. 171–173). Dunod.

Midler, C. (1995). "Projectification" of the firm: The Renault case. Scandinavian Journal of Management, 11(4), 363–375. https://doi.org/10.1016/0956-5221(95)00035-T

Moon, H., & Han, S. H. (2016). A creative idea generation methodology by future envisioning from the user experience perspective. International Journal of Industrial Ergonomics, 56, 84–96. https://doi.org/10.1016/j.ergon.2016.09.012

Nelson, J., Buisine, S., & Aoussat, A. (2012). A methodological proposal to assist scenario-based design in the early stages of innovation projects. Le Travail Humain, 75(3), 279–305. https://doi.org/10.3917/th.753.0279

Nelson, J., Buisine, S., & Aoussat, A. (2013). Anticipating the use of future things: Towards a framework for prospective use analysis in innovation design projects. Applied Ergonomics, 44(6), 948–956. https://doi.org/10.1016/j.apergo.2013.01.002

Nelson, J., Buisine, S., Aoussat, A., & Duchamp, R. (2009). Elaboration of innovative safety equipment concepts for infants. DS 58-9: Proceedings of ICED 09, the 17th International Conference on Engineering Design, Vol. 9, 103–114. Human Behavior in Design, Palo Alto, CA, USA, 24.-27.08.2009.

Nelson, J., Buisine, S., Aoussat, A., & Gazo, C. (2014). Generating prospective scenarios of use in innovation projects. Le Travail Humain, 77(1), 21–21. https://doi.org/10.3917/th.771.0021

Nielsen, J. (1993). Usability engineering. Morgan Kaufmann.

Norman, D. A. (2010). The way I see it: Technology first, needs last. Interactions, 17(2), 38–38. https://doi.org/10.1145/1699775.1699784

Norman, D. A., & Verganti, R. (2014). Incremental and radical innovation: Design research vs. technology and meaning change. Design Issues, 30(1), 78–96. https://doi.org/10.1162/DESI_a_00250

Perrow, C. (1983). The organizational context of human factors engineering. Administrative Science Quarterly, 28(4), 521–521. https://doi.org/10.2307/2393007

Pinsky, L., & Theureau, J. (1984). Paradoxe de l'ergonomie de conception et logiciel informatique [The paradox of design ergonomics and computer software]. Revue Des Conditions de Travail, 9, 25–31.

Poli, R. (2022). Foresight. In V. Glăveanu (Ed.), The Palgrave encyclopedia of the possible (pp. 577–583). Palgrave Macmillan.

Porter, M. E. (1996). What is strategy? Harvard Business Review, 74(6), 61–78.

Preece, J., Rogers, Y., & Sharp, H. (2023). Interaction design: Beyond human-computer interaction (6th ed.). John Wiley.

Rietzschel, E. F., Nijstad, B. A., & Stroebe, W. (2019). Why great ideas are often overlooked: A review and theoretical analysis of research on idea evaluation and selection. In P. B. Paulus & B. A. Nijstad (Eds.), The Oxford handbook of group creativity and innovation (pp. 179–198). Oxford University Press.

Robert, J.-M., & Brangier, É. (2009). What is prospective ergonomics? A reflection and position on the future of ergonomics. In B.- T. Karsh (Ed.), Proceedings of EHAWC 2009: International Conference on Ergonomics and Health Aspects of Work with Computers (pp. 162–169). Springer Verlag.

Robert, J.-M., & Brangier, É. (2012). Prospective ergonomics: Origin, goal, and prospects. Work, 41, 5235–5242. https://doi.org/10.3233/WOR-2012-0012-5235

Robert, J.-M., & Brangier, E. (2019). Overview of methods for prospective ergonomics. In O. Broberg & R. Seim (Eds.), Proceedings of the 50th Nordic ergonomics and human factors society Conference 2019 (pp. 96–104). Department of Technology, Management and Economics.

Robert, J.-M., & Brangier, É. (2024). Prospective ergonomics for the design of future things. Ergonomics, 1–18. https://doi.org/10.1080/00140139.2024.2409939

Robert, J.-M., Martin, A., Taraghi, M., Colin, C., Maldar, M., Bonneviot, F., & Brangier, É. (2021). Prospective ergonomics in service of technological innovation. In M.- É. Bobillier-Chaumon (Ed.), Digital transformations in the challenge of activity and work (pp. 197–210). Wiley.

Robertson, S. (2001). Requirements trawling: Techniques for discovering requirements. International Journal of Human-Computer Studies, 55(4), 405–421. https://doi.org/10.1006/ijhc.2001.0481

Robertson, S. (2004). Scenarios in requirements discovery. In I. Alexander & N. Maiden (Eds.), Scenarios, stories, use cases (pp. 39–59). John Wiley & Sons.

Robertson, S., & Robertson, J. (2012). Mastering the requirements process: Getting requirements right (3rd ed.). Addison-Wesley.

Rogers, E. M. (1962). Diffusion of innovations (1st ed.). Free Press.

Rosson, M. B., & Carroll, J. M. (2002). Usability engineering: Scenario-based development of Human-Computer Interaction. Academic Press.

Ruault, J.-R., Gardinetti, E., Kolski, C., & Vanderhaegen, F. (2014). Adaptation de l'ergonomie prospective aux systèmes à longue durée de vie [Adaptation of prospective ergonomics to long life systems]. Le Travail Humain, 77(3), 257–257. https://doi.org/10.3917/th.773.0257

Runco, M. A., & Charles, R. E. (1993). Judgments of originality and appropriateness as predictors of creativity. Personality and Individual Differences, 15(5), 537–546. https://doi.org/10.1016/0191-8869(93)90337-3

Runco, M. A., & Jaeger, G. J. (2012). The standard definition of creativity. Creativity Research Journal, 24(1), 92–96. https://doi.org/10.1080/10400419.2012.650092

Schumpeter, J. A. (1934). The theory of economic development. Transaction Publishers.

Schwab, K. (2016). The fourth industrial revolution. Crown business.

Shane, S., & Venkataraman, S. (2000). The Promise of Enterpreneurship as a Field of Research. The Academy of Management Review, 25(1), 217. https://doi.org/10.2307/259271

Simon, H. A. (1996). The sciences of the artificial (3rd ed.). MIT Press.

van Belleghem, L. (2018). La simulation de l'activité en conception ergonomique: Acquis et perspectives [Activity simulation in design: Achievements and perspectives]. Activites, 15(1). https://doi.org/10.4000/activites.3129

van Belleghem, L. (2021). Simulating digital activity in the making: Elements of methodology. In M.- E. Bobillier-Chaumon (Ed.), Digital transformations in the challenge of activity and work (pp. 211–223). Wiley.

Verganti, R. (2011). Radical design and technology epiphanies: A new focus for research on design management. Journal of Product Innovation Management, 28(3), 384–388. https://doi.org/10.1111/j.1540-5885.2011.00807.x

Vermersch, P. (2015). L'entretien d'explicitation [The elicitation interview] (8th ed.). ESF.

Visser, W. (2006). Designing as construction of representations: A dynamic viewpoint in cognitive design research. Human–Computer Interaction, 21(1), 103–152.

Wheelwright, S. C., & Clark, K. B. (1992). Revolutionizing product development: Quantum leaps in speed, efficiency, and quality (p. 364). Free Press.

4 Resilience in Ergonomics 5.0

Christopher Nemeth

4.1 INTRODUCTION

Looking to the future of ergonomics, it can help to consider how what has gone before shapes what is to come. What can we expect? How will life be better? What tools will enable us to create the results we need to develop systems that perform in a resilient way so they are what this text describes as "…efficient, effective, healthy, safe, and sustainable?" The following chapter describes strengths and limitations of human performance and automation, the role of resilience engineering, issues that affect system development and evaluation, and suggests approaches that might best address "5.0" era needs.

4.1.1 BACKGROUND

Terms that have been used in varying ways need to be clearly understood. In this text, the term *ergonomics* includes both cognitive and physical aspects of human performance. It also refers to "human factors," as "human factors and ergonomics (HF/E) refer essentially to a common body of knowledge (Dempsey et al., 2006)." *Ergonomics 5.0* aspires to ensure that future "…interaction between human beings and technology is efficient, effective, healthy, safe, and sustainable." Emery and Trist (1960) coined the term *socio-technical systems* to evoke the integration of both human and mechanical elements to perform work. *Information technology* (IT) is the use of computer systems to manage, process, protect, and share information. *Automation* is mechanization, the use of machines (including computers) that are typically controlled by software, to perform tasks. *Automation surprise* is behavior of the automation that leaves the user asking questions such as "what is it doing?" or "why did it do that?" or "what is it going to do next?" (Wiener, 1989).

Hollnagel (2011) defines *resilience* as "the intrinsic ability of a system to adjust its functioning prior to, during, or following changes and disturbances, so that it can sustain required operations under both expected and unexpected conditions." Beyond its origins in safety, *resilience engineering* (RE) is the practice of developing systems that can adapt. *Resilient performance* is often considered to be the ability to adapt. Woods (2015) described four aspects of resilient performance: how a system *rebounds* from disrupting or traumatic events and returns to previous or normal activities; how some systems are able to *manage* increasing complexity, stressors, and challenge; how a system *extends* performance, or brings extra adaptive capacity

DOI: 10.1201/9781003559399-5

to bear, when surprise events challenge its boundaries; and the ability to *adapt* to future surprises as conditions continue to evolve. To adapt is to make something suitable for a new purpose.

The nature, degree, or timing of adaptation can make a substantial difference in the results it produces. And while *adaptation* to a changing world makes sense, traditional risk and cost management practices such as lean production can make it a challenge to implement by preventing the resources needed to adapt.

4.1.2 The Human Role

Hollnagel and Dekker contend there are two options when trying to improve human-machine performance: alter the human to better fit the machine (e.g., training), or modify the machine to fit the human.

> The Achilles heel of human factors, as a discipline, is that it has adopted the machine analogy as a universal explanatory and analytical principle for what a human is and can and cannot do. Because of this technological bias, humans were from the very beginning seen as inefficient, variable, and unreliable.
>
> *(Hollnagel and Dekker, 2024)*

Automation was embraced as the way to replace "unreliable" humans with reliable technology. However, technology must be suited to real-world needs that continue to increase in both complexity and rate of change.

4.1.3 The Artificial Intelligence Role

Artificial intelligence (AI) is often assumed to be the essence of advanced technology that will characterize "Ergonomics 5.0." Capabilities including machine learning (ML) and AI make rapid adaptation possible. Even with the advantages of modern technology, unexpected outcomes and automation surprises persist. As early as 1987, Bainbridge pointed to IT as an irony: a solution that expands, rather than solves or eliminates, a problem. As a result, it can make things worse rather than better. These issues matter, because while good design might mitigate automation surprises, poor design will spread them. Resource constraints, misunderstanding of work, poor documentation, misunderstanding how automation works has led to unexpected outcomes (Sarter et al., 1997). Systems that change in ways operators do not understand will continue to create unanticipated outcomes.

Choi (2021) advises caution with AI, as it has traits that can impede, rather than foster, a better future. These traits include AI tendencies toward brittleness (vulnerability to misperception), embedded bias (slanted presumptions), catastrophic forgetting (new knowledge overwriting what had already been learned), explainability (how traits are attributed), quantifying uncertainty (assessing confidence in a decision), lack of common sense (ability to reach acceptable, logical conclusions), and math inaccuracy.

AI's tendency toward brittleness is likely to increase in any highly complex system that has emergent traits. Our understanding of the emergent properties of highly complex systems improves when they fail in ways we did not anticipate. This learning requires a deeper analytic ability than the AI/ML that has been developed to automate processes up to now.

4.1.4 Strengths and Limitations

The traits of *HF/E* will shape its ability to create resiliently performing systems. *HF/E* is the practice of making things for human use. This assumes a conception, or design, that results in a useful outcome.

> Design is about shaping something, an artefact, a work process, a task or job, a machine, or a tool, that does not yet exist, in particular shaping how it is going to be used or how it should be used for the joint cognitive system to achieve the goals it intends to achieve.
>
> *(Hollnagel and Dekker, 2024)*

In their comprehensive future view of *HF/E*, Karwowski et al. (2025) consider the field as the "scientific discipline concerned with the understanding of the interactions among humans and other elements of a system and the profession that applies theory, principles, data, and methods to design in order to optimize human well-being and overall system performance." Basing products, processes, and systems on human performance makes it more likely to achieve a world that is more efficient, safer, and healthier. Table 4.1 summarizes the strengths and limitations of these fields of practice.

4.1.5 HFE Strengths

Ergonomics and human factors advance knowledge of human performance in the built environment, from the individual to group to societal levels, from tasks to jobs, from

TABLE 4.1
Strengths and limitations

Field	Strength	Limitation
Human factors/ ergonomics	• Advances knowledge of human performance in the built environment • Reveals actual human experience in the real world	• Difficult to develop generalized rules about human performance • Has been cast as a "soft" practice in comparison to a more quantitative approach • Tends to be based on small and narrow samples
Resilience Engineering	• Can optimize and protect performance by anticipating opportunities and challenges and adapt to meet them • Four abilities (respond, monitor, learn, and anticipate) can be used to assess and develop systems • Leverages "sharp end" operator insights into actual work-as-done (WAD)	• Installed base of product and sunk costs can make change too costly to consider • Regulation creates expectations and procedures that can conflict with each other. • Organization policy can constrain adaptation due to the need to maintain a customer base and protect professional certifications .

tools to systems. While definitions can get complex, Dempsey et al. (2000) describe it more directly: "the design and engineering of human–machine systems for the purpose of enhancing human performance" including health, safety, and productivity.

Methods developed over decades are a valuable resource to learn from and support human performance, particularly by those who work at the "sharp end."

4.1.6 HF/E Limitations

Variability in human behavior has made it difficult to develop generalized rules about human performance. Systems development in recent years has cast human factors as a "soft" practice in comparison to a more quantitative approach. In fact, "Human factors as a field has been preoccupied with the legacy view of humans as a liability and source of error and variability. But again, without this variability, no system would be able to function" (Hollnagel and Dekker, 2024).

Human performance studies have been difficult to generalize beyond the details of particular circumstances. Studies of human performance have tended to be performed on small and narrow samples. For example, "WEIRD" is an acronym Henrich et al. (2010) coined to refer to samples drawn from populations that are White, Educated, Industrialized, Rich, and Democratic and pointed out that 99% of all published studies rely on participants recruited from populations that fit those criteria.

4.1.7 RE Strengths

The ability to anticipate opportunities and challenges and adapt to meet them can optimize and protect performance. Hollnagel's (2011) four abilities (respond, monitor, learn, and anticipate) can be used to assess and develop systems that perform in a resilient manner. Multiple studies have demonstrated how systems can perform resiliently (see Nemeth and Hollnagel, 2021, 2014; Nemeth and Herrera, 2015; Nemeth et al., 2009).

4.1.8 RE Limitations

Despite its conception over 20 years ago, RE is not widely practiced in comparison with methods such as lean production. Installed base of product and sunk costs can make change too costly to consider. Regulation creates expectations and procedures that can conflict with each other. Organization policy can constrain adaptation through the need to maintain a customer base and protect professional certifications necessary to practice.

Future work can look into questions that deserve attention. How well does RE scale? Are some application areas more likely to benefit from RE than others? Small teams such as those in a healthcare emergency department and intensive care unit may enjoy more degrees of freedom than larger-scale industrial enterprises.

4.2 ISSUES

Non-deterministic, complex challenges are typical of evolving post-industrial society. The following issues about socio-technical systems deserve attention.

4.2.1 Variability

Adaptive systems can make it possible for humans to effectively engage dynamic, rapidly evolving problems. To do that, they must be configured to permit rapid change so they constantly minimize the difference between need and solution. System complexity is often underappreciated, and emergent properties discovered over time are not well known. Over the past 20 years, RE has addressed the need for systems that anticipate and adapt to challenges and opportunities.

4.2.2 Inscrutability

Automated systems' distribution across organizations, location, and time makes it diffitule to grasp and accurately understand them as well as what influences them.

> Machines and systems are today non-trivial … in the sense that the transformations that link causes and effects are unknown rather than known. Since systems – and individuals –therefore must perform under conditions that incompletely known, hence partly unpredictable, they must be able to adjust what they do in order to succeed
>
> *(Hollnagel and Nemeth, 2021)*

4.2.3 Fragmentation

There are reasons for a narrow research and development focus. Law, economics, and span of control assign responsibility and compensation for work on portions of systems, not the entire system. Fragmentation can result in gaps that, under certain circumstances, can compromise what is considered safe or reliable. RE has sought to draw attention to how elements and systems rely on each other ("interdependencies").

While interconnections among components and systems can be beneficial, they can also result in cascading failures that a narrower focus cannot anticipate. A failure in one smaller domain can be replicated throughout a larger system.

4.2.4 Human Role in Socio-Technical Systems

Systems that are fully human miss the opportunities that automation affords. Systems that are fully automated miss the context awareness and judgment that humans afford. Blending both effectively varies according to human abilities, system abilities, and circumstances. Early attempts by Fitts (1951) sought to discern whether humans or machines are better at perforning a task, and human supervisory control (Sheridan, 1991) took the concept further to include dynamic allocation. The challenge for Ergonomics 5.0 is to make that blend work at the system level. Leveson (2021) favors a systems safety engineering approach to do that, as "The system must be designed to allow humans to be flexible and resilient and to handle unexpected events," and "If we want resilience in systems, then it needs to be integrated into the design of the entire system." How can systems be designed for resilient performance in the era of Ergonomics 5.0?

4.3 ERGONOMICS, METHODS, AND RESILIENCE

The need in the "5.0" era is to conceive adaptive systems and to accurately understand element interdependence, interaction, and their implications. This is far more than usability. The means that are used to create and evaluate systems also need to evolve.

Traditional socio-technical systems and the methods to develop them are essentially static and deterministic. Systems are stripped down in the interest of efficiency. Responsibility for adaptation is relegated to human operators, either on purpose or by default.

Until now, *HF/E* has relied on methods that decompose systems into their elements in order to develop and evaluate them (see Nemeth, 2004). However, breaking the whole into parts is blind to the complexity, interdependency, and rapid change that are making socio-technical systems far more difficult to understand. This makes traditional methods less than effective. What will replace them?

Methods have been created to support the development of systems that perform resiliently. They include Functional Resonance Analysis Method (FRAM) and the Resilience Assessment Grid (RAG) (Hollnagel, 2011, 2012), and Joint Activity Testing (JAT) (Morey et al., 2022). There is a need for a larger, more dynamic inventory of ways to build and assess adaptive systems suited to human use. Karwowski et al. (2025) endorse the use of "systems HFE methods" including conventional methods such as Cognitive Work Analysis, Event Analysis of System Teamwork, and Accident Mapping. They also favor *computational modeling* to understand "large scale, complex, and multifactorial issues" as well as "emergent properties that arise from interactions between system components."

Learning from simulation can complement other cognitive systems engineering methods (Woods and Roth, 1988). Computational simulation makes it possible to discover success or failure sooner and at less cost than real life. Simulation can: "reveal the need for clearer design specifications, uncover interactions and emergent behavior that may lead to undesirable outcomes, such as coordination surprises, and identify trade-offs between multiple design options" (Paladugu et al., 2025)

Rapid simulation and models offer a potential way to move beyond traditional methods and authentically engage "5.0-era" dynamic challenges and the systems intended to meet them by exploring the implications of decisions at scale and in real time.

Nemeth and Hunte (2021) suggest multiple ways simulation can be used to foster and promote resilient performance, offering the potential to both practice *adaptation* and to cope with *surprise*. Each relates to one or more of Hollnagel's resilient performance abilities.

- *To confirm what goes well (respond, learn)*—Serve as a kind of audit by reconfirming how a system such as a unit or procedure was originally conceived and what about it remains valuable.
- *To explore possibilities (anticipate, learn)*—Test new strategies as experiments in pilot simulations.

- *To understand an undesirable outcome (monitor, learn)*—Reveal what fails, when it fails, and invite consideration of why.
- *As a tool to build resilient performance (anticipate, monitor)*— Improve the ability to foresee what might be possible to better anticipate opportunities, obstacles, and outcomes.

While simulation and modeling are promising, they are not a panacea. Even effective models still need to evolve to reflect real-world changes, and models need to be better when accounting for uncertainty.

4.4 DISCUSSION

Next generation "5.0" ergonomics can be expected to develop socio-technical systems that rely on automation. For Klein et al. (2004), intelligent systems need to have essential traits to participate in the highly adaptive human work domain. These include: engage in "common grounding" to facilitate coordination, work toward shared goals and prevent team coordination breakdowns; coherently manage mutual responsibilities and commitments to facilitate recovery from unanticipated problems; be mutually predictable; be able to deliberately assess and modify others' actions as conditions and priorities change; make targets, states, capacities, intentions, changes, and upcoming actions obvious; be able to signal and form models of teammates, to negotiate, enable a collaboration, participate in managing attention, and help to control the costs of coordinated activity.

Information systems, decision aids, and devices can all suffer from being what Wiener (1989) termed *clumsy automation.* Systems that are clumsy, or poorly designed for human use, do not aid, but rather impede, cognitive work. They add new communication and coordination tasks to an already burdensome workload. In order to adapt to software system shortcomings, practitioners must develop coping skills including work-around procedures. Systems that are hard to operate induce errors. The net effect is to erode effectiveness and performance. In the example of IT, failure flows in part from developers' "...disregard for the ways in which people organize their work coupled with a disdain for the ordinary resources on which they rely..." (Heath and Luff, 2000:3–4). Failure can also stem from underinvestment in technologies such as IT, as well as the failure to take advantage of engineering-based systems design, analysis, and management tools such as human factors research (Reid et al., 2005).

We need to better understand both the opportunities and potential pitfalls when humans and automation are bound together in situations that require resilient performance and develop the future accordingly. Here are suggestions and examples of how that can be done.

4.5 MAKE THE FUTURE EVIDENT

Show all of the decision-making opportunities that are possible and their implications (the "trade space"). The following example from healthcare shows how technology can make the trade space for decision-making evident.

Medical informatics is the application of IT to healthcare, which includes components that range from device control/display interfaces, decision support systems, artificial intelligence, electronic medical records, information retrieval, outcomes assessment, and telemedicine. Complex medical devices have not been obstacles to effective healthcare. The real obstacle has been to understand the emergent, complex operations of health care systems that include sophisticated medical devices. This suggests the best use of complex devices and systems should be to represent the relevant aspects of work domains as the clinician's problem space. The social sciences and engineering, specifically cognitive psychology, sociology, and human factors, provide the means to discover these relevant aspects, or "domain semantics."

Infusion devices are the most widely used IT in healthcare. These pumps provide medications and fluids for patients at a controlled rate. Their displays typically provide information on their current state. However, clinicians need to know what happened in the past (what has been delivered) and what result current settings are likely to yield in the future (what can be expected). Providing past, current, and anticipated states and making connections with related data, such as lab results, makes it easier to recover from unanticipated problems. Showing projected values helps clinicians to *anticipate* and prepare for what will happen. Controls make it possible to explore contingencies before *responding* by committing to a final decision. This enables the clinician to evaluate multiple options and make accurate decisions in real time. Integrating controls with displayed information makes it possible to *monitor* them and deliberately assess and *learn*, modifying programmed infusion actions as patient condition and priorities change.

The combination of infusion device graphic and alphanumeric information in Figure 4.1 makes pertinent aspects of the device's goal state, status, capacities, programming intentions, and upcoming actions obvious to members of the clinical team. These are the kinds of observable and controllable traits that would improve IT support for health care and help to build resilient system performance (Nemeth and Cook, 2010).

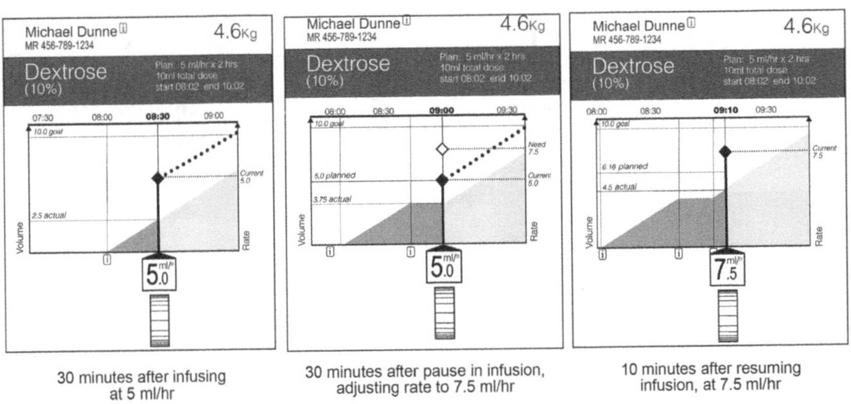

FIGURE 4.1 Infusion device interface displays showing rate changes over three time periods "Infusion" state at 0830, 0900, and 0910. Copyright © 2008 Cognitive Technologies Laboratory. Used by permission.

4.6 MAKE THE IMPLICIT EVIDENT

The decision maker needs to have information that is crucial, or "salient," in order to monitor, respond, learn, and anticipate. The automation and interdependence of socio-technical systems can make that difficult. Even on a small scale, making salient information evident can improve resilient performance. The following example describes how what is important and implicit can be made evident.

It is less likely that casualties will be quickly evacuated to receive medical care during military operations in extreme, austere environments that are remote from the U.S. (e.g., Africa or the Pacific). Future scenarios anticipate the likelihood of "prolonged field care" (PFC), possibly up to three days, by medics whose training and experience can vary dramatically. Under these conditions, casualties are likely to suffer from more severe, complex afflictions such as shock, which is not easily detected by simply observing a patient.

The Trauma Triage Treatment and Training Decision Support (4TDS) System (Figure 4.2) is an application on a smartphone that receives vital signs (e.g., temperature, heart rate, respiratory rate, blood pressure, peripheral oxygen) streamed from a sensor. The system includes an ML model that has been trained and validated on actual patient data and scans vital signs that can identify clinically relevant patterns up to 90 minutes before onset. The model solves two current limitations that have received comparatively little attention: no general shock model exists, and no ability exists to anticipate shock so a medic can avert its occurrence.

Identification and prediction of shock can enable a medic to anticipate and adapt. It can reduce the length of time to differential diagnosis, decrease diagnostic/treatment

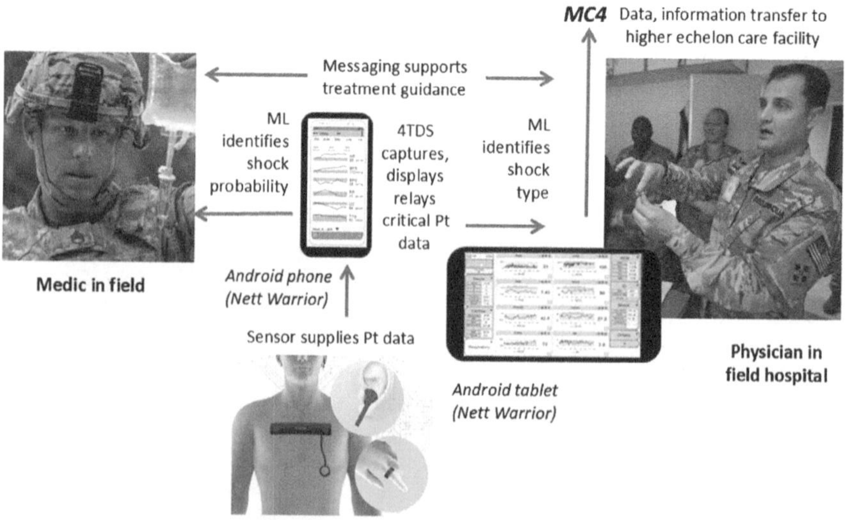

FIGURE 4.2 Diagram showing smartphone and tablet as well as medic and physician that comprise a battlefield trauma remote patient monitoring system—Trauma Triage, Treatment, and Training Decision Support (4TDS) System. Copyright © 2020 Applied Research Associates, Inc. Used by permission.

misadventures and delay, lessen care provider cognitive load, and increase provider efficiency and satisfaction. The model makes medical data collection and analysis easier, improving and speeding the diagnosis of patient conditions for more effective care by making implicit conditions evident.

4.7 MODEL AND SIMULATE WITH INCREASING STRENGTH

Develop more effective ways to anticipate contingent, emergent, and interdependent phenomena through simulation. Simulation can be used in a number of ways to foster and promote resilient performance, offering the ability to routinely practice adaptation and cope with surprise.

Through methodical analysis and design, simulation can be used to carefully study what goes well by cultivating skill and ability to adapt. Simulation can help operators and their systems to improve their adaptive expertise by exposing gaps in operational continuity, presumption, bias, and to realize the implications of trade-off decisions. It can also help to learn about, anticipate and respond to, risks that are part of any socio-technical system.

Nemeth and Hunte (2021) offer these potential uses, along with Hollnagel's abilities shown in parentheses:

- *Confirm what goes well (respond, learn)*—Simulations can serve as a kind of audit method by reconfirming how a system such as a unit or procedure was originally conceived and what about it remains valuable. It can also reveal nuances such as how frontline operators adjust ("tailor") what they do or create informal shortcuts, detecting the difference between what Hollnagel (2017) refers to as "work as imagined (WAI)" and "work as done (WAD)."
- *Explore possibilities (anticipate)*—Develop pilot simulations to test new experimental strategies. Running a simulation can test presumptions about new approaches from policies to procedures. Computer simulation can provide a way to anticipate by showing the need for clearer design specifications, reveal emergent conditions and interactions that may lead to undesired results, and compare different design options to make trade-offs evident. "Insight from simulation can complement other cognitive systems engineering methods to refine and enhance the feasibility and robustness of envisioned operations" (Paladugu et al., 2025).
- *Understand an undesirable outcome (monitor, learn)*—Simulation can reveal what fails, when it fails, and invite consideration of why. Post-event reenactment is one traditional approach to understand how an undesirable result happened. Simulation can also be used to "break the system" by deliberately forcing failure. This tactic can clarify the boundaries of safe performance, help to understand what safety margin may exist, reveal hidden interdependencies, learn in what ways systems are likely to fail, make trade-off decisions evident, and anticipate how future challenges might compromise performance.

- *Build vision (anticipate)*—Having adequate insight to foresee possible future events requires what Adamski and Westrum (2004) term "requisite imagination." Envisioning the future relies on making a deliberate effort to stretch perceptions and expectations to anticipate opportunities and challenges. Simulation makes this kind of exploration possible and can be used to understand and evoke resilient performance.

4.8 MOVE BEYOND PROFESSIONAL LIMITATIONS

Collaboration among disciplines will be essential to produce substantive results. The development and implementation of successful technology requires an understanding of the human context. Broaden study samples to authentically represent the user population. Build effective relations among other development disciplines, as the rest of the engineering community needs data to verify performance improvement. The risk assessment and management community needs assurance that an adaptive system does not increase the potential for loss of property and life. More powerful analytic capabilities can make it possible to overcome some of the traditional inability to generalize by assessing the similarities and differences among diverse applications. This has the potential to identify appropriate ways to "…drive the design of AI support towards solutions for typical operator troubles that can be applied across domains, as well as identifying when domain-specific solutions need to be developed" (Lundberg et al., 2024).

4.9 CONCLUSION

The next generation of human factors and ergonomics, Ergonomics 5.0, is an opportunity to create systems that better serve the complex, dynamic, rapidly evolving problems of post-industrial society. Make both the future evident and the implicit evident, and reveal their implications through an expanded inventory of methods including rapid simulation. Use these processes to develop systems that provide accurate, timely information that enables informed decisions about what and how to anticipate and adapt for more resilient performance. Informed use of advanced IT and AI can support the development of adaptive systems while understanding the human context and triangulating them with cognitive systems engineering methods.

ACKNOWLEDGMENT

The author is grateful to David Mendonça, PhD, and Michael O'Connor, MD for insightful comments during the development of this chapter.

REFERENCES

Adamski, J.A. & Westrum, R. (2004). Requisite imagination: The fine art of anticipating what might go wrong. In E. Hollnagel (Ed.), *Handbook of Cognitive Task Design*, 193–220. Boca Raton, FL: Taylor and Francis/CRC Press.

Bainbridge, L. (1987). Ironies of automation. In J. Rasmussen, K. Duncan, & J. Leplat (Eds.) *New Technology and Human Error*, 271–283. Chichester: Wiley.

Choi, C.Q. (2021). Seven revealing ways AIs fail: Neural networks can be disastrously brittle, forgetful, and surprisingly bad at math. *IEEE Spectrum*, 58(10), 42–47.

Dempsey, P.G., Wogalter, M.S., & Hancock, P.A. (2000). What's in a name? Using terms from definitions to examine the fundamental foundation of human factors and ergonomics science. *Theoretical Issues in Ergonomics Science*, 1(1), 3–10.

Dempsey, P.G., Wogalter, M.S., & Hancock, P.A. (2006). Defining ergonomics/human factors. In W. Karwowski (Ed.), *International Encyclopedia of Ergonomics and Human Factors*, 2nd ed., 32–35. London: Taylor & Francis.

Emery, F.E. & Trist, E.L. (1960). Socio-technical systems. *Management science, models and techniques*, 2, 83–97.

Fitts, P.M. (Ed.) task allocation(1951). Human engineering for an effective air navigation and air traffic control system. Washington, DC: National Research Council.

Heath, C. & Luff, P. (2000). *Technology in Action.* New York: Cambridge University Press.

Henrich, J., Heine, S.J., & Norenzayan, A. (2010). The weirdest people in the world? *Behavioral and Brain Sciences* 33(2–3), 61–83. Epub 2010 Jun 15. doi:10.1017/S0140525X0999152X

Hollnagel, E. (2011). RAG – The resilience analysis grid. In E. Hollnagel, J. Pariès, D. D. Woods, & J. Wreathall (Eds.), *Resilience Engineering in Practice. A Guidebook.* Farnham: Ashgate.

Hollnagel, E. (2012). *FRAM. The Functional Resonance Analysis Method for Modelling Complex Sociotechnical Systems.* Farnham: Ashgate.

Hollnagel, E. (2017). Why is work-as-imagined different from work-as-done? In *Resilient Health Care, Volume 2*, 279–294. Boca Raton, FL: CRC Press.

Hollnagel, E. & Dekker, S. (2024). The ironies of 'human factors.' *Theoretical Issues in Ergonomics Science.* doi:10.1080/1463922X.2024.2443976

Hollnagel, E. & Nemeth, C. (2021). From resilience engineering to resilient performance. In Nemeth, C. & Hollnagel, E. (Eds.). *Advancing Resilient Performance*, 1–9. New York: Springer Nature.

Karwowski, W., Salvendy, G., Endsley, M., Rouse, W., Salmon, P., Stanney, K., et al. (2025). Grand challenges for human factors and ergonomics. *Theoretical Issues in Ergonomics Science*. doi:10.1080/1463922X.2024.2431336.

Klein, G., Woods, D., Bradshaw, J.M., Hoffman, R.R., & Feltovich, P.J. (2004). Ten challenges for making automation a team player in joint human-agent activity. *IEEE Computer*, 91–95.

Leveson, N. (2021). Safety III: A systems approach to safety and resilience. *MIT Engineering Systems Lab, 16.* Available at therm.ward.bay.wiki.org

Lundberg, J., Nylin, M., Praetorius, G., Jansson, A., & Bång, M. (2024). Modelling operator control work across traffic management domains: Implications for interaction design. *Cognition, Technology & Work*, 26, 281–299. doi:10.1007/s10111-024-00754-w

Morey, D. A., Della Vella, D., Rayo, M. F., Zelik, D. J., & Murphy, T. B. (2022, September). Joint activity testing: towards a multi-dimensional, high-resolution evaluation method for human-machine teaming. In *Proceedings of the Human Factors and Ergonomics Society Annual Meeting* (Vol. 66, No. 1, pp. 2214–2219). Los Angeles, CA: SAGE Publications. https://doi.org/10.1177/1071181322661537

Nemeth, C. (2004). *Human Factors Methods for Design.* New York: Taylor and Francis/CRC Press.

Nemeth, C. & Cook, R. (2010, July). Infusing healthcare with resilience. *Proceedings of the 12th Annual International Symposium.* International Council of Systems Engineers, 20 (1), 1073–1087.

Nemeth, C. & Herrera, I. (2015). Preface: Building change- Resilience engineering at ten years. In C. Nemeth & I. Herrera (Eds.), Special issue on Resilience Engineering. *Reliability Engineering and System Safety*, 141. doi:10.1016/j.ress.2015.04.006

Nemeth, C. & Hollnagel, E. (Eds.). (2014). *Becoming Resilient.* Resilience Engineering in Practice. 2. Farnham: Ashgate Publishing. Japanese ed., 2017, Juse Press.

Nemeth, C. & Hollnagel, E. (Eds.). (2021). *Advancing Resilient Performance.* New York: Springer Nature.

Nemeth, C., Hollnagel, E., & Dekker, S. (Eds.). (2009). *Preparation and Restoration.* Resilience Engineering Perspectives. 2. Farnham: Ashgate Publishing.

Nemeth, C. & Hunte, G. (2021). Chapter 2.4 Foresight: How simulation can promote resilient performance. In E. Deutsch, S. Perry, & H. Gurnaney (Eds.), *Comprehensive Healthcare Simulation: Improving Healthcare Systems.* New York: Springer, 37–42.

Paladugu, A., Fernandes, A., & IJtsma, M. (2025). Computational simulation of distributed work as a discovery tool for envisioning future operations. *Journal of Cognitive Engineering and Decision Making*. doi:10.1177/15553434251327698

Reid, P.R., Compton, W.D., Grossman, J.H., & Fanjiang, G. (Eds.). (2005). *Building a Better Delivery System: A New Engineering/Healthcare Partnership.* Washington, DC: The National Academies Press.

Sarter, N., Woods, D., & Billings, C. (1997). Automation surprises. In G. Salvendy (Ed.), *Handbook of Human Factors & Ergonomics*, 2nd ed. New York: Wiley, 553–569.

Sheridan, T. (1999). Human supervisory control. In Sage, A. and Rouse, W. (Eds). *Handbook of Systems Engineering and Management*, 591–628. New York: John Wiley and Sons.

Wiener, E.L. (1989). *Human Factors of Advanced Technology ("Glass Cockpit") Transport Aircraft.* (NASA Contractor Report No. 177528). Moffett Field, CA: NASA Ames Research Center.

Woods, D.D. (2015). Four concepts for resilience and the implications for the future of resilience engineering. *Reliability Engineering & System Safety*, 141, 5–9. ISSN 0951-8320. doi:10.1016/j.ress.2015.03.018

Woods, D. D., & Roth, E. M. (1988). Cognitive engineering: Human problem solving with tools. *Human Factors*, *30*(4), 415–430.

5 Inclusive Design in Ergonomics 5.0 for an Enhanced Workplace Experience

Matteo Zallio

5.1 ERGONOMICS 5.0 AND THE IMPERATIVE OF INCLUSIVE DESIGN

In the era of Ergonomics 5.0, workplace inclusive design moves beyond traditional efficiency metrics to embrace human-centric values and emerging evidence on inclusivity, wellbeing, and sustainability. This paradigm aligns with the Industry 5.0 vision of combining advanced technologies with a human-first approach, ensuring that productivity gains develop alongside improvements in employee experience, including health, work behaviour, inclusion, and diversity.

A cornerstone of this approach is inclusive design in workplace ergonomics, a practice focused on creating adaptive work environments that accommodate the widest possible range of users without the need for later adjustments (Zallio, Chivăran, Capece, Clarkson, and Buono, 2023). Inclusive design promotes innovation and engagement by removing barriers and enabling everyone to participate fully and independently in work activities (Zallio and Clarkson, 2021b).

This chapter outlines the fundamental principles of inclusive design, examines its influence on workplace outcomes, introduces the IDEA (Inclusion, Diversity, Equity, and Accessibility) Toolkit, and discusses the role of emerging technologies in advancing Ergonomics 5.0. Drawing on illustrative case studies and current academic research, it presents inclusive design as both a moral imperative and a strategic driver of organisational performance.

5.2 CORE PRINCIPLES AND METHODOLOGIES OF INCLUSIVE DESIGN

Inclusive design is rooted in the belief that environments, products, experiences, and systems should be inclusive of the variety of people's needs and accessible to all people, regardless of differences in ability, age, or background (Zallio and Clarkson, 2021a). This approach goes beyond the minimum requirements of accessibility legislation to embrace universal design principles, creating spaces that are not only

DOI: 10.1201/9781003559399-6

functional but also empowering (Story, Mueller, and Mace, 1998). Inclusive design also aligns with the legal imperatives outlined by Eleanor Kirk (2021), ensuring that design processes meet and, where possible, exceed statutory expectations. As a result, inclusive workplace design is guided by several interrelated principles:

- **Equitable use:** Work environments should be useful and usable for people with diverse abilities without excluding any user group (Story et al., 1998).
- **Flexibility in use:** Work environments must adapt to a wide range of individual preferences and abilities, such as adjustable workstations and intuitive digital interfaces (Coleman, Clarkson, and Cassim, 2008).
- **Simple and intuitive use:** Work environments should be easy to understand regardless of the user's experience or cognitive capabilities (Clarkson, Coleman, Keates, and Lebbon, 2019).
- **Perceptible information:** Information in work environments should be communicated effectively through multiple modalities, such as visual, auditory, and tactile formats, to support diverse sensory needs. (Story et al., 1998).
- **Tolerance for error and low physical effort:** Work environments should minimise the risk of adverse consequences from accidental actions and reduce physical strain (Story et al., 1998).

The implementation of inclusive design relies primarily on user-centred and participatory design methodologies. Engaging diverse stakeholders, ranging from employees with different abilities to neurodiverse individuals with varied cognitive and sensory profiles, at every stage of workplace design and operation is essential. Methods such as ethnographic studies, co-design workshops, and simulation exercises have proven effective in uncovering hidden barriers and fostering empathy among designers (Zallio and Clarkson, 2023). These approaches translate theoretical principles into practice and ensure that innovative solutions are tailored to actual user needs (Dul, Ceylan, and Jaspers, 2011). In addition to participatory methods, data-informed strategies that draw on behavioural and attitudinal insights, such as post-occupancy evaluations, sensor-based environmental data, and comfort or wellbeing surveys, provide an evidence base to monitor workplace performance and continuously refine design decisions based on real patterns of use (Zallio and Clarkson, 2021b).

5.3 CASE STUDIES

5.3.1 The Impact of Inclusive Design on Workplace Experience

Inclusive design directly influences workplace experience by creating environments where all employees can work comfortably and effectively. Research indicates that workplaces designed with inclusivity in mind see significant improvements in employee engagement, creativity, and overall productivity (Schriefer, 2005). For example, ergonomic interventions such as height-adjustable desks and noise-reducing partitions have been shown to reduce physical strain and minimise cognitive overload, thereby reducing error rates and boosting performance (Margaritis and Marmaras, 2007).

Consider a case study from a multinational corporation that reconfigured its open-plan offices based on inclusive design principles. The introduction of quiet zones, adjustable lighting, and sensory-friendly layouts resulted in a 25% reduction in reported employee stress levels and a marked increase in collaborative innovation among neurodiverse teams (Narenthiran, Torero, and Woodrow, 2022). This example illustrates how investments in inclusive design not only meet legal standards but also drive substantial improvements in workplace efficiency and satisfaction.

Diverse teams consistently generate richer and more innovative solutions, a finding well established in social psychology and organisational research. Inclusive design contributes directly to these outcomes by creating environments in which all employees have the resources, support, and conditions they need to thrive. In doing so, organisations are able to draw on a broader range of perspectives and capabilities, strengthening their capacity for innovation, adaptability, and long-term resilience.

5.3.2 The IDEA Toolkit: A Strategic Framework for Inclusion, Diversity, Equity, and Accessibility for Built Environments

To evaluate and implement inclusive design in workplace contexts, researchers at the University of Cambridge developed the IDEA Toolkit. Its purpose is to provide a systematic and evidence-informed approach for improving employee experience, productivity, satisfaction, and the overall perception of workplace environments (Fernandez, Zallio, Berry, and McGrory, 2021). The toolkit comprises two primary components: the IDEA Audit and the Inclusive Design Canvas. Together, these instruments offer a structured method for assessing existing work environments and generating innovative, inclusively designed workplace solutions. Beyond diagnosis and ideation, the toolkit also supports organisations in prioritising interventions, tracking progress over time, and embedding inclusive design principles into ongoing planning cycles, thereby enabling a continuous improvement process grounded in behavioural and attitudinal evidence.

The IDEA Audit is an evidence-informed evaluation tool that collects quantitative and qualitative data on how effectively a workplace aligns with inclusion, diversity, equity, and accessibility standards. It examines environmental, organisational, technological, and behavioural factors, providing a multidimensional profile of the strengths and limitations of a given workspace (Zallio and Clarkson, 2021b). Pilot studies have demonstrated that organisations using the IDEA Audit can identify and address specific deficits in their physical and digital environments, leading to measurable improvements in accessibility, employee satisfaction, and overall workplace experience. The audit's structured indicators also enable longitudinal tracking, allowing organisations to monitor progress over time and evaluate the impact of interventions with greater accuracy. (Zallio, Chivăran, and Clarkson, 2024).

The Inclusive Design Canvas is a participatory design tool that supports practitioners in systematically exploring user diversity and translating insights into actionable design strategies. It provides a structured template that guides cross-disciplinary teams through key stages of the inclusive design process, including identifying user personas, mapping capability needs, defining accessibility challenges, and generating design requirements. By visually organising complex information related to physical, sensory, cognitive, and cultural factors, the canvas helps teams surface assumptions, uncover hidden barriers, and ensure that design decisions are grounded in real user experiences. The tool is particularly effective in co-design contexts, as it enables users, designers, and stakeholders to collaborate equitably and articulate priorities in a shared, comprehensible format.

By mapping the journeys and lived experiences of diverse users, from individuals with mobility or sensory impairments to neurodiverse employees, the Inclusive Design Canvas fosters innovative ideas that might otherwise be overlooked in conventional design workflows. It also supports iterative refinement, allowing teams to revisit and adjust their ideas as new evidence emerges from prototyping or stakeholder feedback. Users have reported increased team engagement, improved communication among different professional groups, and a more rigorous integration of user feedback into design decisions (Zallio and Clarkson, 2022a). As such, the Inclusive Design Canvas functions not only as a creative tool but also as a methodological bridge that connects inclusive design principles with practical implementation in workplace and built environment contexts.

5.3.3 Emerging Technologies: AI, Big Data, and Digital Twins as Enablers of Inclusive Ergonomics

Recent technological advancements are transforming the landscape of inclusive workplace design, enabling environments to respond more precisely to the diverse capabilities, preferences, and behaviours of building occupants. Advanced, pioneering wearable systems, including robotic exoskeletons, are redefining physical inclusivity in physically demanding work environments. The adoption of exoskeletons across several industry segments illustrates how assistive wearables can reduce musculoskeletal load, support ageing or mobility-limited workers, and expand the pool of employees able to participate in manual tasks (Badri, Nadeau, and Gbodossou, 2018). Such technologies shift inclusivity from accommodation to augmentation, enabling workers of varied physical capacities to contribute safely and meaningfully.

At the environmental scale, the integration of IoT sensors within workplaces enables continuous, data-informed adaptation of spatial conditions. Real-time inputs on temperature, noise levels, lighting, air quality, space utilisation, and movement patterns can be used to automatically adjust environmental parameters to better meet individual needs (Coleman et al., 2008). When complemented by behavioural and attitudinal data drawn from building occupant surveys, post-occupancy evaluations, and in-situ feedback mechanisms, organisations gain a multidimensional understanding of inclusion barriers. These combined datasets support dynamic optimisation, identifying mismatches between environmental conditions and user requirements and enabling proactive interventions.

Furthermore, digital twin technologies further enhance these capabilities by creating virtual replicas of workplaces that integrate sensor data, user flows, ergonomic constraints, and accessibility requirements. Digital twins can simulate how individuals with different capabilities navigate and experience a space, allowing designers and facility managers to test inclusive interventions prior to physical implementation.

This modelling capability strengthens predictive inclusive design, reducing the need for costly retrospective adjustments.

Extended reality technologies such as virtual reality (VR) and augmented reality (AR) are also reshaping inclusive design and training. VR allows designers and decision makers to inhabit the perspective of users with disabilities or sensory differences, generating empathy and improving the accuracy of design reviews. AR can overlay accessibility information, hazards, or navigational cues directly onto the physical environment, supporting users with cognitive or sensory impairments. These technologies have been shown to enhance both design validation and organisational awareness of inclusion through immersive, experiential learning (Zhang and Huang, 2024).

Artificial intelligence extends these capabilities by analysing complex multimodal datasets and generating design or operational recommendations that accommodate extreme user variability. AI-enabled generative design tools can explore thousands of spatial or ergonomic configurations, optimising environments to meet diverse capability profiles and preferences (Clarkson et al., 2019). AI can also support personalised building experiences, adjusting workstation ergonomics, acoustic profiles, or digital interfaces in response to individual behavioural patterns, thereby increasing comfort, wellbeing, and inclusion.

Together, these technologies signal a shift towards adaptive, evidence-informed, and personalised inclusive design. When integrated thoughtfully, they create workplaces that not only meet legal accessibility requirements but also proactively enhance human experience and enable participation across the full spectrum of human diversity (Chivaran, Zallio, Waller, and Clarkson, 2021).

5.3.4 Inclusive Design in Ergonomics 5.0: Towards a Predictive, Human-Centred Workplace

The evolution of ergonomics reflects a gradual but profound shift from a narrow focus on physical efficiency towards a holistic understanding of human diversity in increasingly complex work environments. Ergonomics 5.0 represents the next frontier in this progression, integrating inclusive design principles with behavioural science, data-informed methodologies, and emerging technologies to create workplaces that are equitable, adaptive, and fundamentally human-centred.

Earlier stages laid the foundation for this paradigm. Ergonomics 1.0 concentrated on physical safety and biomechanical optimisation; Ergonomics 2.0 incorporated cognitive ergonomics, emphasising mental workload and decision making; and Ergonomics 3.0 expanded to organisational ergonomics, highlighting team dynamics, social factors, and systems thinking (Dul et al., 2011). Ergonomics 4.0 aligned with Industry 4.0, engaging smart technologies, human-machine partnerships, and digitally mediated work processes.

Ergonomics 5.0 builds upon these developments by placing inclusive design at its core. It recognises that effective workplaces must accommodate the full spectrum of human abilities, characteristics, experiences, and cultural backgrounds, moving beyond compliance-based accessibility towards proactive, predictive, and experience driven approaches.

This perspective emphasises physical, cognitive, sensory, social, and emotional dimensions of human diversity and positions inclusive design as central to wellbeing, participation, and equitable performance in the contemporary workplace.

Despite its demonstrated value, the adoption of inclusive design remains uneven. Persistent challenges include limited organisational awareness, misconceptions about cost, lack of technical capability, and cultural resistance within established corporate environments. In many organisations, accessibility is still perceived as a regulatory obligation rather than a catalyst for innovation and improved employee experience (Winance, 2014). These barriers hinder the development of workplaces that fully support diverse populations, particularly as workforces age and the prevalence of neurodiversity becomes more visible.

At the same time, significant opportunities are emerging. Societal attention to diversity, equity, and inclusion has created a receptive context for inclusive design initiatives. The rising availability of behavioural and attitudinal data from post-occupancy evaluations, environmental sensors, digital twins, and workplace feedback systems enables more nuanced insights into user needs and environmental performance. Tools such as the IDEA Toolkit lower methodological barriers by offering systematic, evidence-informed frameworks for assessment, intervention, and continuous improvement. Advances in AI, extended reality, and generative design further expand the possibilities for designing environments that respond dynamically to individual capabilities and preferences.

To fully realise the potential of Ergonomics 5.0, organisations should take a strategic and integrated approach:

- Embed inclusive design from the earliest stages of workplace planning, rather than relying on retrofitting.
- Build multidisciplinary collaborations among designers, engineers, human resources, health and safety professionals, and end users.
- Develop sustained training and capacity building to strengthen organisational competence in inclusive design.
- Leverage behavioural, attitudinal, and sensor-based data to guide decision making and monitor progress over time.
- Implement emerging technologies ethically, ensuring that AI, digital twins, and extended reality tools augment human capabilities without compromising privacy, autonomy, or psychological safety.
- Contribute to the advancement of inclusive design standards, guidelines, and policy frameworks at national and international levels.

These actions not only advance compliance and risk reduction but also help organisations cultivate environments where innovation, creativity, and wellbeing can flourish. Ergonomics 5.0 positions inclusive design as a strategic imperative, enabling organisations to create adaptable, equitable workplaces that reflect the full spectrum of human diversity. As sensory, cognitive, and behavioural differences become more widely recognised, the future of work will depend on design approaches that seamlessly integrate accessibility, performance, and human experience.

Inclusive design is therefore not only a foundation for social equity but a critical pathway towards resilient, innovative, and future-ready work environments.

REFERENCES

Badri, A., Nadeau, S., & Gbodossou, A. (2018). Occupational safety in the Industry 4.0 era: A cause for major concern? *Safety Science,* 109, 403–411.

Chivaran, C., Zallio, M., Waller, S., & Clarkson, P. J. (2021). Visual accessibility and inclusion: An exploratory study to understand visual accessibility in the built environment. In *Proceedings of Smart Accessibility 2021* (pp. 1–7). IARIA. ISBN: 978-1-61208-876-1.

Clarkson, J., Coleman, R., Keates, S., & Lebbon, C. (2019). *Inclusive design: Design for the whole population.* Springer.

Coleman, R., Clarkson, J., & Cassim, J. (2008). *Design for inclusivity: A practical guide to accessible, innovative and user centred design.* Routledge.

Dul, J., Ceylan, C., & Jaspers, F. (2011). Knowledge workers' creativity and the role of the physical work environment. *Human Resource Management, 50*(6), 715–734. https://doi.org/10.1002/hrm.20454

Fernandez, C., Zallio, M., Berry, D., & McGrory, J. (2021). Towards a people-first engineering design approach. A comprehensive ontology for designing inclusive environments. In *Proceedings of the International Conference on Engineering Design (ICED21),* Gothenburg, Sweden, 16–20 August 2021. https://doi.org/10.1017/pds.2021.579

Kirk, E. (2021). Law and legalities at work: HR practitioners as Quasi-legal professionals, *Industrial Law Journal, 50*(4), 583–609. https://doi.org/10.1093/indlaw/dwab025

Margaritis, S., & Marmaras, N. (2007). Supporting the design of office layout meeting ergonomics requirements. *Applied Ergonomics, 38*(6), 781–790. https://doi.org/10.1016/j.apergo.2006.10.003

Narenthiran, O. P., Torero, J., & Woodrow, M. (2022).Inclusive design of workspaces: Mixed methods approach to understanding users. *Sustainability, 14,* 3337. https://doi.org/10.3390/su14063337

Schriefer, A. E. (2005). Workplace strategy: What it is and why you should care. *Journal of Corporate Real Estate,* 7(3), 222–233.

Story, M. F., Mueller, J. L., & Mace, R. L. (1998). *The principles of universal design.* NC State University.

Winance, M. (2014). Universal design and the challenge of diversity: reflections on the principles of UD, based on empirical research of people's mobility. *Disability and Rehabilitation, 36*(16), 1334–1343. https://doi.org/10.3109/09638288.2014.936564

Zallio, M., & Clarkson, P. J. (2021a). The IDEA Toolkit: A framework to foster inclusion, diversity, equity, and accessibility in the built environment. *Design Studies,* 73(2), 102–118.

Zallio, M., & Clarkson, P. J. (2021b). On inclusion, diversity, equity, and accessibility in civil engineering and architectural design. A review of assessment tools. In *Proceedings of the International Conference on Engineering Design (ICED21),* Gothenburg, Sweden, 16–20 August 2021. https://doi.org/10.1017/pds.2021.491

Zallio, M., & Clarkson, P. J. (2022a). The inclusive design canvas: A strategic design template for architectural design professionals. In *Proceedings of the Design Society, 2* (pp. 81–90). https://doi.org/10.1017/pds.2022.9

Zallio, M., & Clarkson, P. J. (2022b). The inclusion, diversity, equity, and accessibility audit: A post-occupancy evaluation method to help design the buildings of tomorrow. *Building and Environment,* 109058. https://doi.org/10.1016/j.buildenv.2022.109058

Zallio, M., & Clarkson, P. J. (2023). A study to depict challenges and opportunities building industry professionals face when designing inclusive and accessible buildings. *Architectural Science Review.* https://doi.org/10.1080/00038628.2023.2270983

Zallio, M., Chivăran, C., & Clarkson, P. J. (2024). Exploring inclusion, diversity, equity, and accessibility in the built environment: a case study. *Buildings, 14*, 3018. https://doi.org/10.3390/buildings14093018

Zallio, M., Chivăran, C., Capece, S., Clarkson. P. J., Buono, M. (2023). Inclusive spatial learning experience: An exploratory framework to deliver human-environment interactions. *Strategic Design Research Journal. 15*(3), September - December 2022. 262-276. DOI: 10.4013/sdrj.2022.153.04

Zhang, Y., & Huang, X. (2024). Integrating extended reality (XR) in architectural design education: a systematic review and case study at Southeast University (China). *Buildings, 14*, 3954. https://doi.org/10.3390/buildings14123954

6 Artificial Intelligence and the Future of Neo-Human Workers

An Analysis from the Perspective of Activity Ergonomics

Fabiana Raulino

6.1 INTRODUCTION

The book *Social Acceleration: A New Theory of Modernity* by Hartmut Rosa (2019) begins with a text by Goethe (Johann Wolfgang Von Goethe, taken from his work *Maxims and Reflections*) about the importance of "symmetry" in a person's life to achieve fulfillment and realize their potential. The concept of "symmetry" here does not merely refer to external harmony but to an internal balance and the ability to work within one's own limitations, whether natural or acquired. Goethe observes that even great talents can be hindered and even destroyed when a person fails to find this balance or does not know how to work within their capabilities. The lack of "symmetry" can lead to a waste of potential.

This reflection is contextualized by Hartmut Rosa (2019) within the modern era, with the acceleration of changes and the high standards demanded by society, where it becomes increasingly challenging for people to meet expectations, especially in the workplace. Given the intensity and speed at which modern society moves, it becomes even more difficult for someone to find this balance and fully develop themselves.

Work continues to be our expression of usefulness in the world. It is through work that we build our identity and connect with the challenges we face in our society, delivering results under conditions that are often inadequate. In this context, Ergonomics emerges as an increasingly necessary transdisciplinary science that safeguards and nurtures the pursuit of symmetry—or, in a more ergonomic context: a healthier, more sustainable, safer approach that ensures efficient performance by workers.

According to Guérin et al. (2001, p. 11), Ergonomics takes work as its object of study. However, it is essential to recognize that the term "work" encompasses a variety of realities depending on the context. For example, it can refer to working

DOI: 10.1201/9781003559399-7

conditions (such as in the case of unhealthy work environments), to the outcome of work (such as work that needs to be redone to meet different quality standards), or to the labor activity itself (such as performing one's work and how it affects the psychophysiological characteristics of human beings).

With the advent of generative AI and AI agents, we are not only witnessing a major disruption of humanity but also an unprecedented acceleration of processes.

Acceleration. Of time. Of processes. Of everything. And human beings have biological barriers that prevent them from processing and memorizing data at the same scale as machines. This is the context of the *neo-human*, an anti-dualist concept that considers humans, machines, and nature from a cyborg perspective, as translated by Donna Haraway (1985).

The report presented by Amy Webb at the SXSW 2025 innovation event, developed by the Future Today Strategy Group, points out that *living intelligence*—resulting from the convergence of AI, advanced sensors, and biotechnology—is on the verge of creating intelligent systems capable of perceiving, learning, and evolving beyond the capabilities of human programming. This convergence is an emerging trend and marks the rise of systems that intrinsically interact with the physical world, adapting to it dynamically and autonomously.

The integration of these technologies is creating integrated feedback cycles between digital and biological systems, allowing adaptation and evolution that would be impossible to achieve with any isolated technology. These systems do not merely process data but are capable of perceiving, interpreting, and modifying the environment in real time, implying a profound capacity for adaptation to contextual needs and changes (FTSG, 2025, p. 13).

According to the same report, when multiple AI agents work together, each specializing in different tasks while coordinating toward common goals, networks of AI systems emerge that are capable of handling complex and interconnected challenges that would overwhelm isolated agents. According to a study by Stanford HAI (2024), 72% of companies already using AI agents report efficiency gains in business processes.

This affects the entire human experience and raises many questions and reflections.

As Hartmut Rosa (2019) notes, the temporal structures of a given society are inevitably encountered by its members as normative expectations, and this development of standardized times—through clocks, calendars, and increasingly regularized schedules (by trains, school days, stock markets, factory work, and so on)—forces individuals to conform their behavior and plan their lives around the socially dominant time schemes.

6.2 THE FUTURE OF WORK FOR NEO-HUMANS: UNDERSTANDING THE IMPACT OF ARTIFICIAL INTELLIGENCE

Technological acceleration is driven by the economic engine encapsulated in the proverbial expression "time is money." To maximize the production of marketable goods or services per unit of time and to be the first to introduce innovations reflect the categorical demands of capitalism. This logic of temporal efficiency implies that, especially in large-scale operations, small temporal advantages over competitors can result in substantial profits for market leaders and catastrophic losses for laggards (Rosa, 2019).

As ergonomists, we need to understand these new technologies, their different types, and how they integrate into the world of work so that we can design solutions that maintain organizational coherence and sustainable advances for the future (Breque et al., 2021). Social change acceleration is fueled by the socio-structural engine of functional differentiation, which manifests in our daily lives through professional specialization and the division of labor. This phenomenon, characteristic of modernization, allows for remarkable gains in processing speed and productivity (Rosa, 2019). However, as these gains materialize, they tend to accelerate social changes, as they increase the number of social arenas of action. Consequently, there arises an urgent need for coordination and synchronization between these spheres, initially demanding a rigorous discipline in time management, with an increasingly structured temporal regimen.

Although organizations are highly focused on the development of AI, many of them fail to recognize the potential impact of biotechnology and advanced sensors in amplifying these systems' capabilities. This narrow vision prevents companies from fully understanding the scope of this transformation, which is not limited to automation or data analysis but involves systems that can autonomously and continuously modify their environment. This technological convergence will bring about a new wave of innovation, enabling systems to learn, evolve, and react in highly dynamic ways. The future of work will be shaped by this transformation, as interactions between humans and machines become increasingly complex, adaptive, and efficient (FTSG, 2025, p. 13).

6.2.1 Artificial Intelligence

AI, as a field of study and practice, refers to the simulation of human cognitive processes in machines, with the goal of creating systems capable of learning, reasoning, perceiving, and interacting adaptively with the environment. Since its conception in the 20th century, AI has evolved from systems based on rigid rules to more sophisticated predictive models driven by machine learning and deep learning (DL).

In the early studies of AI, researchers adopted an approach based on explicit logical rules, known as symbolic or classical AI (as we will discuss later). These systems, such as expert systems, were designed to perform specific tasks by following instructions predefined by human programmers (Franco and Primi, 2022).

The intention was to simulate human reasoning through a logical sequence of steps, but without any form of real learning or adaptation by the machine. Although these systems proved effective in limited and controlled contexts, the approach soon proved insufficient to cope with the complexity of the real world, as encoding a rule for every possible scenario became impractical, especially for more complex and dynamic tasks (Franco and Primi, 2022).

In the 1980s, with the development of machine learning algorithms, AI took a new direction. These algorithms no longer depended on explicit rules but on large volumes of data to train models capable of recognizing patterns and making decisions based on those patterns. Over the following decades, machine learning evolved, eventually giving rise to DL—a technique that uses deep neural networks to learn complex data representations, driven by significant advances in computational power and data availability.

It was within this context of DL evolution that generative AI emerged—a revolutionary subcategory that, instead of merely recognizing patterns, is capable of creating new content from the learned data. Tools such as generative adversarial networks (GANs) and transformers have brought this content generation capability to new heights, enabling machines to produce texts, images, music, and videos that resemble, in quality and innovation, what would be created by a human.

Generative AI thus represents a stage of continuity and innovation in the long history of computational development, highlighting the transformative role of technology in extending human thought and expanding the frontiers of knowledge and creation.

6.2.1.1 Types of Artificial Intelligence

Historically, classical AI was known as symbolic AI and had roots in formal logic and human cognitive intelligence, where ideas and information were treated as discrete entities (symbols), and reasoning processes followed clear logical rules to manipulate these symbols. Classic examples of symbolic AI include expert systems and predicate logic, which attempt to represent and manipulate knowledge in a structured and explanatory manner.

There are several ways to classify and explain the different types of AI. For didactic purposes and to simplify, we chose to divide them into the following categories: capacity, functionality, techniques or approaches, application areas (business), and cognitive and emotional aspects.

6.2.1.2 AI According to Capacity

- *Weak AI (Narrow AI):* Designed to perform specific tasks, such as movie recommendations, voice recognition, or medical diagnosis.
- *Strong AI (General AI):* AI with general cognitive abilities, capable of performing any human cognitive task. It is still an area in development and a future goal of AI.
- *Superintelligent AI (Superintelligent AI):* A hypothetical form of AI that would surpass human intelligence in every aspect.

In *The Conscious Mind* (1996, p. 320), Chalmers proposes that consciousness is a fundamental and irreducible phenomenon of nature, suggesting that a Strong AI could eventually become conscious. However, much more would need to be understood about human consciousness to replicate it in a machine. He also suggests that, in the future, machines could be created with mental capacities that are not only symbolic or behavioral but also endowed with subjective experiences—a form of true "mind."

6.2.1.3 AI According to Functionality

- *Supervised Learning:* AI is trained with a labeled dataset, where each input has a known output. It is used for tasks like classification and regression.

- *Unsupervised Learning:* AI is fed with unlabeled data and must find patterns or structures within the data on its own. Used in clustering and dimensionality reduction.
- *Reinforcement Learning:* AI learns through trial and error, receiving rewards or penalties according to its actions in a dynamic environment.
- *Semi-Supervised Learning:* A combination of supervised and unsupervised methods, where a small amount of labeled data and a large amount of unlabeled data are used to train the AI.
- *Transfer Learning:* AI applies what it has learned in one context to solve problems in a similar context, saving time and resources during training.

6.2.1.3.1 AI According to Technique or Approach

- *Artificial Neural Networks:* Computational models inspired by the human brain, used for pattern recognition and classification.
- *Support Vector Machines:* Supervised learning techniques primarily used for classification and regression, focusing on finding the best margin of separation between data classes.
- *Decision Trees:* Algorithms that create tree-shaped decision models, where decisions are made based on a series of rules.
- *Genetic Algorithms:* Based on the process of natural selection, used for optimization and search problems.
- *Fuzzy Logic:* Used to model uncertainties and make decisions under ambiguous conditions, commonly applied in control systems and automation.
- *Natural Language Processing (NLP):* Techniques that enable machines to understand, interpret, and generate human language. Used in chatbots, translators, and virtual assistants.
- *Computer Vision:* Algorithms designed to interpret and process images and videos, applied in object recognition, facial recognition, and autonomous vehicles.

6.2.1.4 Application-Based AI

Some examples include:

- *NLP AI:* Tools such as chatbots and automatic translators.
- *Image Recognition AI:* Used to classify and identify objects in images and videos.
- *Medical Diagnostic AI:* Algorithms used to predict diseases, suggest treatments, and interpret medical exams.
- *Autonomous Vehicle AI:* AI systems that enable vehicles to navigate without human intervention.
- *Recommendation AI:* AI algorithms used in streaming services and e-commerce to suggest content or products to users.

6.2.1.4.1 Cognitive and Emotional AI

- *Cognitive AI:* Systems that attempt to replicate the functioning of the human brain to solve complex problems.

- *Emotional AI (Affective Computing):* AI that aims to understand, interpret, and respond to human emotions. Used in virtual assistants and customer support.

Beyond this classification, it is important to consider other divisions of AI based on how the system learns or processes data:

- *Machine Learning (ML):* Machine learning is a subfield of AI that focuses on developing algorithms capable of identifying patterns in data and improving performance as they are exposed to more data. The main premise of ML is that, rather than being explicitly programmed to perform tasks, ML algorithms "learn" from the data.
- *Deep Learning:* DL is a subfield of ML based on deep neural networks, computational structures inspired by the human brain. Deep neural networks consist of multiple processing layers, allowing the model to learn data representations at progressively more abstract levels. Deep neural networks are responsible for remarkable advances in areas such as image recognition, automatic translation, and natural language generation.
- *Generative AI:* Generative AI is a type of AI capable of creating new data from an existing dataset. Unlike discriminative models, which classify or predict based on given inputs, generative models are trained to learn the data distribution and generate new examples that resemble the training data. Popular examples of generative AI include models like GANs and, of course, the GPT model (Generative Pre-trained Transformer), which is a form of generative AI applied to natural language.

6.2.1.5 Spotlight on Generative AI

In 2017, a group of researchers from Google Brain, including Ashish Vaswani, Noam Shazeer, Niki Parmar, Jakob Uszkoreit, Llion Jones, Aidan N. Gomez, Łukasz Kaiser, and Illia Polosukhin, published the influential paper "Attention Is All You Need." This work marked a revolutionary breakthrough in the field of NLP by proposing a new neural network model known as the Transformer, which addressed the limitations of recurrent networks by introducing the attention mechanism.

The transformer architecture enabled more efficient processing of sequential data, allowing models to capture long-range dependencies within texts. This innovation laid the groundwork for advanced generative models like GPT (Generative Pre-trained Transformer), which leverage attention mechanisms to produce highly coherent and contextually relevant text. Transformative in scope, this model revolutionized the way machines understand and generate human language, giving rise to a new era of AI applications in text generation, machine translation, and conversational agents.

In other words, the order of words matters for its correct functioning. The central difference between traditional predictive models and modern architectures like the Transformer, launched in 2017 by Vaswani et al., lies in the approach to handling large volumes of data and identifying complex patterns. While traditional predictive models rely on linear equations or classification algorithms, the Transformer introduced attention as a central mechanism, enabling parallel and efficient processing of information distributed across large corpora of data.

The Transformer model is composed of two main blocks: the **encoder** and the **decoder**. The encoder is responsible for processing the input, while the decoder generates the output. Each of these blocks contains multiple layers of attention and feed-forward neural networks. The primary advantage of the Transformer is its ability to process information in parallel, unlike recurrent neural networks (RNNs), which process sequences sequentially and, therefore, are slower. This makes the Transformer particularly efficient for tasks such as language translation, text summarization, and even text generation tasks using models like GPT.

Distributional semantics is an approach to understanding the meaning of words based on their context. Instead of viewing words as isolated entities, distributional semantics treats word meaning as something that can be inferred from patterns of word co-occurrence in large text corpora. This concept is fundamental for models like Word2Vec and GloVe, which map words into vector embeddings where semantically similar words are closer in the vector space.

The Transformer brought a paradigmatic shift to the field of AI by overcoming the limitations of recurrent architectures, such as RNNs and Long Short-Term Memory (LSTM) networks, in modeling temporal dependencies in sequential data. Through the self-attention mechanism, the Transformer assigns relative weights to words or textual elements within a context, capturing semantic nuances and structural relationships more efficiently. This innovation propelled the development of massive language models such as GPT and BERT, which now dominate applications in NLP, automatic translation, and text generation.

The improvements in speed, scalability, and utilization of massive amounts of zetabytes have impacts on previously unimaginable scales. Today, it is possible to automate tasks that once required human creativity and 100% manual operation, including images, videos, music, and text. For example, videos can be transcribed in seconds, and even new videos can be created by choosing the camera positioning (Santaella, 2024).

However, caution is needed regarding cybersecurity and the lack of transparency about which data were used to train the machines and what biases they may generate. Generative AIs operate within what semiotics calls "enunciative opacity," referring to the invisibility of the "machine's hand" due to our familiarity with the reference styles upon which the machine works. This opacity is a fundamental part of the semiotic manipulation process, hiding the technical mechanisms that produce the final effect (Dondero et al., 2024).

The accessibility of tools, no matter how intuitive they may seem, often conceals a dependence on technical knowledge. Documentation and the need to understand underlying concepts become a competitive advantage for those who dedicate time to learning and practice. For instance, understanding how to build a command that specifies camera angles in a video or the color palette of an image allows specialists to produce results that far exceed the standards offered by basic interactions (Santaella, 2023).

In this scenario, mastering prompts—command phrases that instruct AIs to perform tasks—highlights how semantic awareness is essential to extract the maximum potential from these technologies.

In the context of generative AI, understanding distributional semantics is vital since models like the Transformer and GPT need to capture complex relationships between words to generate coherent and meaningful text. These models generate responses or content based not only on the probability of isolated words but also on a deep understanding of how words relate within a context (Santaella et al., 2025; Arboleda, 2025).

General AI, still theoretical, proposes the ability to fully replicate human cognition. In contrast, AI agents are autonomous units that perceive the environment, make decisions, and act according to specific objectives. These agents can be integrated into complex networks, interacting with each other and with humans, as advocated by Bruno Latour's Actor-Network Theory (1996). In this perspective, both humans and technological agents are considered actors with action capabilities, collaborating to co-create hybrid realities.

AI agents are autonomous or semi-autonomous systems designed to perform specific tasks, react to external stimuli, and make decisions based on environmental information. They are essential components of more complex systems, such as autonomous robots, virtual assistants, and recommendation systems. AI agents can range from simple ones, like a voice assistant that recognizes commands and answers questions, to more complex ones, such as agents that manage traffic in a smart city or recommend movies based on user behavior (Calvaresi et al., 2023).

The technical explanation for each level is as follows:

- *Level 0*—LLMs (Large Language Models): This level involves agents that operate with language models based on AI, such as GPT. The main characteristic of agents at this level is the absence of memory. They are capable of generating responses and performing tasks based on single inputs, without retaining any knowledge or context from previous interactions.
- *Level 1*—Memory: Agents that incorporate a simple memory system. This means that, besides interacting with input data, they are capable of retaining information throughout interactions, allowing for greater personalization in responses based on the history of conversations or previous interactions.
- *Level 2*—RAG (Retrieval-Augmented Generation): Agents with personalized data and the ability to access external information through search systems or database queries. This level allows the agent to search for and utilize specific data to enhance its responses or perform more complex tasks.
- *Level 3*—Tools: Agents capable of accessing and utilizing external tools to execute actions. These agents not only process data but can also interact with other systems or devices, performing practical tasks and offering more robust solutions, such as data manipulation, executing commands on external platforms, and more.
- *Level 4*—Auto Feedback (Self-Assessment): Agents at this level possess the ability to evaluate the results of their actions and feed back into their process to improve task execution. This self-assessment cycle enables the agent to enhance its performance by adapting as needed to achieve better results.
- *Level 5*—Multi-Agents: At this level, multiple agents, each specialized in different areas, work in a coordinated way to solve a complex task. Each

agent brings a unique skill or specialization, and the joint effort is necessary to reach a common goal. This collaboration among agents offers a higher level of complexity, allowing more sophisticated solutions to multifaceted problems.

6.3 THE IMPACT ON THE WORLD OF WORK AND THE ROLE OF ERGONOMICS

Ergonomists, particularly those who follow the activity ergonomics methodology, now face the challenge of understanding the knowledge engineering that structures and processes this data. Additionally, they must grasp how work organization addresses microtasks, cognitive overload, and psychosocial risks.

The impact of AI on human cognition is profound, reflecting how we perceive and interact with the digital world. From the perspective of digital semiotics, as explored by Lucia Santaella (2025), AI systems transform meaning-making by mediating and amplifying our interactions with digital signs. AI interprets and produces signs independently, challenging the concept of authorship and human agency.

This dynamic leads to the concept of the post-human, where cognition is distributed between humans and technical systems, overcoming the subject-object dichotomy and expanding the boundaries of intelligence.

In the future of work, AI redefines the skills and practices needed in an increasingly digital and interconnected economy. The automation of routine tasks and the amplification of human cognitive capabilities suggest a transition toward symbolic and creative work, requiring new forms of technological literacy and adaptability.

However, AI also exacerbates existing inequalities, increasing precariousness and the risk of unemployment. Studies in neuroscience and labor sociology highlight the importance of inclusive policies and the promotion of digital literacy to prevent the marginalization of vulnerable populations (Vainzof, 2024; Fagundes et al., 2024).

The challenge lies in leveraging AI to enhance human potential while mitigating the risks associated with rapid technological change, ensuring that the future of work remains equitable and sustainable.

AI emerges at a time when society's cognition needs to evolve beyond our biological, geographical, temporal, and contextual barriers. As pointed out by Santaella (2007, p. 136),

> the human species is indeed taking on unpredictable contours today. To refer to the heterogeneity of the body hybridized with technologies, I have been using, since 1998, the term 'biocybernetic' in a broader sense than 'prosthetic' or 'cyborg,' though it includes both.

The intersection of cognition, semiotics, and social practice highlights the need for critical and reflective engagement to navigate the challenges and potentials of this new technological era.

We live in a scenario of displacement and decentering of the human, where the agents shaping our reality are not only human beings but also what Bruno Latour termed the "parliament of things" (Latour, 1996). This concept, presented in alignment with Actor-Network Theory (ANT), challenges the classic dichotomies that have dominated Western thought for centuries: human versus non-human, subject versus object, materiality versus sociality. Instead, Latour invites us to envision a world where everything—people, devices, objects, institutions—can act as agents, playing active roles in dynamic and interconnected networks rather than merely serving as symbolic projections (Santaella, 2023; Latour, 1996).

As we transition into the era of Industry 5.0 and Society 5.0, the boundaries between technology and human experience become increasingly blurred. This transformation goes beyond technological advancements—it represents a profound shift in how we interact with the digital world and with each other in the workplace. At the core of this evolution lies the fusion of Generative AI, Digital Intelligence, and the future of work—a convergence point that is giving rise to a new vision of ergonomics designed for the digital age.

Generative AI represents a paradigm shift in how we understand AI systems. Unlike traditional AI, which identifies patterns, Generative AI pushes boundaries by creating innovative solutions, simulating possible futures, and even anticipating needs before they arise. In the workplace, Generative AI will not only automate routine tasks but also generate creative insights, foster innovation, and provide intelligent and personalized support systems. By dynamically adjusting to tasks, preferences, and individual environments, these technologies will enable more adaptable workspaces, enhancing both productivity and well-being.

Digital Intelligence, closely related to Generative AI, expands this concept by allowing systems to reason, learn, and adapt using digital tools and information. Digital Intelligence emphasizes human-machine collaboration, merging the potential of advanced technologies with the richness of human experience. Together, Generative AI and Digital Intelligence are revolutionizing ergonomics, creating environments where human potential is maximized through intuitive and responsive systems.

As technological advancements accelerate, a new form of resilience is emerging—one that blends human adaptability with the predictive capabilities of AI. In the workplace, resilience goes beyond merely surviving in a world of constant change. It involves thriving in volatile, uncertain, complex, and ambiguous environments. Generative AI and Digital Intelligence provide tools that help individuals and organizations navigate these challenges effectively while ensuring that well-being remains a priority.

Moreover, resilience today means integrating technology with a human touch, ensuring that as digital solutions boost productivity, they also support emotional and psychological health. Ergonomics 5.0 emphasizes this delicate balance, designing systems that help workers thrive during periods of stress or transformation, thereby reinforcing the connection between well-being and performance at work.

Sustainability is crucial to the future of work and is embedded in the development of Generative AI and Digital Intelligence. It is no longer sufficient for technologies to be efficient; they must also contribute to a broader vision of ethical, inclusive, and socially responsible progress. In the context of Ergonomics 5.0, sustainability means developing AI systems that are not only energy-efficient but also promote equity, reduce biases, and support long-term social health.

Sustainable digital systems also promote collaboration, innovation, and education, while respecting the planet and human dignity. The goal is to create an ecosystem where technology benefits everyone, ensuring that advancements contribute positively to both the environment and society as a whole. This human-centered approach ensures that technology remains a tool for collective well-being rather than a force that exacerbates inequalities.

Looking to the future, human-centered design remains central to the development of digital systems and AI (Soares, 2021). An ethical by design perspective ensures that technology is accessible, intuitive, and empathetic, placing human needs and behaviors at the core of design. In this context, Generative AI becomes a crucial enabler of human-centered design, dynamically responding to human inputs and adapting to individual preferences and demands.

Ergonomists need to develop skills to understand AI so that their analyses become more comprehensive. It is not necessary to be a great specialist in AI architecture, but rather to know how these technologies impact work organization and the cognitive, psychosocial, physical, and environmental aspects through work redesigns that we have yet to fully comprehend. More than ever, applied lifelong learning is essential.

Integrating Generative AI and Digital Intelligence into the workplace reaffirms the central role of humans in shaping their own future. While AI can enhance our capabilities and improve decision-making, it is human creativity, empathy, and judgment that remain irreplaceable. This partnership between humans and machines will define the future of work—a future where technology amplifies human potential rather than replacing it.

Ergonomics 5.0 represents a vision where humans and technology coexist harmoniously, creating environments that are not only more productive but also more supportive and sustainable. By embracing Generative AI, Digital Intelligence, and design principles focused on human-centered approaches within the neo-human context, we can build work environments that prioritize resilience, inclusion, and sustainability. This is the future of work—a future where technology enhances well-being, fosters personal growth, respects our relationship with non-human agents, and drives social transformation.

REFERENCES

Breque, Maija; De Nul, Lars; Petridis, Athanasios. *Industry 5.0: towards a sustainable, human-centric and resilient European industry*. Luxemburgo: Publicações da União Europeia, 2021. Available at: https://op.europa.eu/en/publication-detail/-/publication/468a892a-5097-11eb-b59f-01aa75ed71a1. Accessed on: March 25, 2025.

Calvaresi, Davide; Eggenschwiler, Stefan; Mualla, Yazan; Schumacher, Michael; Calbimonte, Jean-Paul. Exploring agent-based chatbots: a systematic literature review. *Journal of Ambient Intelligence and Humanized Computing*, v. 14, p. 11207–11226, 2023. DOI:10.1007/s12652-023-04626-5. Available at: https://link.springer.com/article/10.1007/s12652-023-04626-5. Accessed on: March 25, 2025

Chalmers, David J. *The conscious mind: in search of a fundamental theory*. New York: Oxford University Press, 1996.

Dondero, Maria Giulia; Castro, Gustavo Henrique Rodrigues de; Schwartzmann, Matheus Nogueira; Pereira, Daniervelin. Semiótica da inteligência artificial: análise computacional de grandes bases de dados e geração automática de imagens. *MATRIZes: Revista do Programa de Pós-Graduação em Comunicação da Universidade de São Paulo*, v. 18, n. 3, p. 29–54, set./dez, 2024. DOI:10.11606/issn.1982-8160.v18i3p29-54. Available at: https://www.revistas.usp.br/matrizes/article/view/232626?utm_source=chatgpt.com. Accessed on: March 25, 2025.

Fagundes, Alan Cordeiro; Chaves, Rômulo Oliveira; Miranda, Hellen Patriny Soares; Calunga, Walter Paulo de Oliveira Celestino. Precarização do trabalho docente: plataformas digitais e inteligência artificial no ensino superior. *RevistaFT*, v. 29, n. 140, nov. 2024. https://revistaft.com.br/precarizacao-do-trabalho-docente-plataformas-digitais-e-inteligencia-artificial-no-ensino-superior/. Accessed on: March 25, 2025.

Franco, Víthor Rosa; Primi, Ricardo. El Futuro de las Evaluaciones Mediante las Redes Sociales. Avaliação Psicológica [online]. 2022, vol.21, n.4, pp.437–445. ISSN 1677-0471. Available at: https://pepsic.bvsalud.org/scielo.php?pid=S1677-04712022000400008&script=sci_abstract&tlng=es. Accessed on: March 25, 2025.

FTSG (Future Today Strategy Group). *2025 tech trend report*. 18h ed. Available at: https://ftsg.com/wp-content/uploads/2025/03/FTSG_2025_TR_FINAL_LINKED.pdf. Accessed on: March 25, 2025.

Guérin, François; Bardy, Benoît; Cavaillé, A.; Lavoie, C.; Lemaire, E. *Ergonomie: concepts et méthodes*. Paris: Masson, 2001.

Haraway, Donna. *Simians, cyborgs, and women: the reinvention of nature*. New York: Routledge, 1985.

Latour, B. On actor-network theory: a few clarifications. *Soziale Welt*, v. 47, n. 4, p. 369–381, 1996.

Martínez Arboleda, Antonio. Language sustainability in the age of artificial intelligence. *Alfinge. Revista De Filología*, v. 36, p. 1–37, 2025. DOI:10.21071/arf.v36i.17761. Accessed on: March 25, 2025.

Rosa, Hartmut. *Aceleração: a transformação das estruturas temporais na modernidade*. São Paulo: Editora Unesp, 2019.

Santaella, Lucia. A expansão artificial da inteligência humana. *Revista da UFMG*, Belo Horizonte, v. 30, n. fluxo contínuo, 2024. DOI:10.35699/2965-6931.2023.48125. Available at: https://periodicos.ufmg.br/index.php/revistadaufmg/article/view/48125. Accessed on: March 25, 2025.

Santaella, Lucia. Por que é imprescindível um manual ético para a Inteligência Artificial Generativa? *TECCOGS: Revista Digital de Tecnologias Cognitivas*, São Paulo, v. 28, p. 7–24, 2023. Available at: https://revistas.pucsp.br/index.php/teccogs/article/view/67064?utm_source=chatgpt.com. Accessed on: March 25, 2025.

Santaella, Lucia. Pós-humano: por quê? *Revista USP*, n. 74, p. 126–137, jun./ago, 2007. DOI:10.11606/issn.2316-9036.v0i74p126-137. Available at: https://www.researchgate.net/publication/313778241_Pos-humano_por_que. Accessed on: March 25, 2025.

Santaella, Lucia; Raulino, Fabiana; Cruz, Kalynka. DeepSeek e a transformação da inteligência artificial. *Instituto Humanitas Unisinos – IHU*, 03 fev. 2025. Available at: https://www.ihu.unisinos.br/648258-deepseek-e-a-transformacao-da-inteligencia-artificial-artigo-de-lucia-santaella-fabiana-raulino-e-kalynka-cruz. Accessed on: March 25, 2025.

Soares, Marcelo M. *Ergodesign methodology for product design: A human-centered approach*. 1st ed. Boca Raton: CRC Press, 2021. DOI:10.1201/9781003214793. Accessed on: March 25, 2025.

Stanford Institute for Human-Centered Artificial Intelligence (HAI). *Artificial Intelligence Index Report 2024*. Available at: https://hai.stanford.edu/ai-index/2024-ai-index-report. Accessed on: March 25, 2025.

Vainzof, Rony; Gutierrez, Andriei; Godinho, Gustavo; Krastins, Alexandra. Inteligência Artificial Generativa: um novo patamar no desenvolvimento da humanidade e os desafios da Caixa de Pandora. In: Vainzof, Rony; Gutierrez, Andriei; Godinho, Gustavo; Krastins, Alexandra (Orgs.). *Comentários ao EU AI Act - Ed. 2024*. São Paulo: Editora Revista dos Tribunais, 2024. Available at: https://www.jusbrasil.com.

br/doutrina/secao/3082143668/1-ia-e-suas-aplicacoes-25-inteligencia-artificial-generativa-um-novo-patamar-no-desenvolvimento-da-humanidade-e-os-desafios-da-caixa-de-pandora. Accessed on: March 25, 2025.

Vaswani, Ashish; Shazeer, Noam; Parmar, Niki, Uszkoreit, Jakob; Jones, Llion; Gomez, Aidan N.; Kaiser, Łukasz; Polosukhin, Illia. Attention is all you need. *Advances in Neural Information Processing Systems*, v. 30, 2017. Available at: https://arxiv.org/abs/1706.03762?utm_source=chatgpt.com. Accessed on: March 25, 2025.

7 Privacy Interfaces
A Human Factors and Ergonomics Approach

Fabiana França and Cláudia Mont'Alvão

7.1 INTRODUCTION

7.1.1 Information Age

In recent decades, the way people browse the Internet has been completely transformed. If previously access had logistical limitations such as equipment size and connection quality, today the ubiquity of the service is a reality. In theory, people can access whatever they want, whenever they want, anywhere.

The mobility of information, the arrival of smartphones, and social networks were directly responsible for this. This, of course, changed the dynamics of Internet users' interaction with the network and information. User behavior has also changed due to the fascination brought by the possibility of uninterrupted and increasingly efficient access.

In other words, companies and online services aligned themselves with users' preferences, and the wishes of these users gained value in guiding which paths businesses should follow. Knowing the wants, desires, and needs of these users is possible thanks to the data and information they produce on the internet, when simply browsing or choosing a product. This can be condensed into data that can be collected by different institutions in order to guide their purposes.

The truth is that, with the consolidation of the Information Age, data collection and a data-based economy dominated the internet. The preferences that users leave behind in their online data serve many purposes. They serve to meet people's needs and guide business, but also to manipulate and influence people based on what is known about them.

This scenario is known as data capitalism. It was quickly developed by exploiting evolutionary lapses, seeking profit at the expense of public understanding and improving data protection laws (Zuboff, 2018). This happened because the way data collection scaled too quickly to keep up with the consolidation of the Information Age. It was faster than the problematizations about the collection of internet users' personal data, and its use.

The commercial advantages identified by the collecting, storage and processing of personal data have eclipsed the protection of a fundamental right of the individual, the right to privacy. Data is useful for companies. Services collect more and more

DOI: 10.1201/9781003559399-8

data and what is personal and intrinsic to being is subject to this chain of activities. Therefore, the risks to people's privacy are real today.

7.1.2 Interaction with Privacy

In this scenario, the way society thinks about privacy is important, since choices related to the subject influence the direction of political campaigns, the intentions of large corporations, and government power (Véliz, 2022). As a result, public institutions need to establish and strengthen legal mechanisms that ensure the privacy of individuals in the face of the reality of data capitalism.

From this need emerge laws such as the General Data Protection Regulation (GDPR) in Europe and the Lei Geral de Proteção de Dados (LGPD) in Brazil. These regulations are essential to guarantee citizens' right to privacy in the digital environment. In Brazil, the LGPD determines that people need to be informed in advance about the collection, storage, and processing of data when using an online service.

The way companies choose to communicate to people about LGPD provisions involves interfaces that contain the following elements:

- hyperlinks that lead to texts containing privacy policies or consent terms;
- assent checkboxes for displayed content;
- dialog boxes communicating the use of cookies;
- consent or accept all buttons; and others.

Considering the importance of protecting privacy and the use of user data, it is important that what the law determines is clearly communicated to the user. However, this does not appear to be the case in Brazil. What the LGPD determines is confusing for ordinary citizens.

7.1.3 Difficulty for Users

Digital transformations and the inclusion of information and communication technologies in people's lives have brought about a reality in which collected data is important. This data provides information about individuals and provokes economic movement. To protect users' personal information and privacy, regulations have been created.

However, as is the case in Brazil, understanding and enforcing these laws proves to be an exhausting task for users. According to the 2021 Privacy and Personal Data Protection survey (cgi.br, 2023) by the Internet Steering Committee in Brazil, on reading privacy policies on the screens of websites and applications in the country, 81% of internet users do not read them in full because they are too long, and 69% consider them difficult to understand.

Until 2023, 94.3% of the Brazilian population had been using the internet every day (IBGE, 2024). Of this population, 98.8% access mostly via cell phone. 87.6% of people over 10 years of age have their own cell phone in hand. These people use it for the most diverse interests, from accessing social networks to banking services and financial institutions. And all these users provide data that is collected either by the public administration or by private companies.

The elements contained in the interaction interfaces that contain the LGPD determinations are repeated on different websites and apps. These elements end up not considering the specificities of the different types of users who access the pages, but need to preserve their privacy. Of internet users in Brazil, 66% are over 60 years old (IBGE, 2024). These people access the internet due to its presence in everyday society and the need to adapt to new standards. In the country, for example, 44.4% of users are uneducated people who have not completed formal basic education (IBGE, 2024). The table below shows the context of Brazilian users with the kind of difficulty they reported in interaction with privacy policies in privacy interfaces.

The interaction with privacy interfaces for Brazilian users		
Context of the Brazilian users		
94.3% of the Brazilian population has been using the internet every day.	98.8% access mostly via cell phone, and 66% are over 60 years old.	44.4% of users are people who have not completed formal basic education.
Privacy interface difficulties reported		
81% of internet users do not read privacy policies in full because they are too long	69% of internet users consider privacy policies difficult to understand	

These two audiences may face difficulties when interacting with privacy-related interfaces. In other words, it is necessary to consider the satisfaction needs of users when performing online tasks, as well as the protection of their privacy. Especially, consider the heterogeneity and difficulties of different types of users.

7.2 PURPOSE OF THE STUDY

After identifying this problem, we sought to systematically review the existing literature to map the state of research on ergonomics or human factors in data protection. This approach seeks to optimize system usability and improve the user experience.

Human Factors/Ergonomics is thought to be fundamental to reducing user effort, discomfort in digital interactions, and privacy issues. They can address the identified problems related to how difficult it is for Brazilian users to interact with privacy interfaces.

The main objective was to identify the key concepts for studying human interaction with privacy interfaces in previous research and to identify research gaps to support future studies.

The hypothesis that guided this search was that how data protection is presented to users requires a significant cognitive effort. Therefore, this work addresses the following question: What are the key concepts for studying human interaction with privacy interfaces in previous research?

Next, we present the methodology used to create this study. The following topics present the key concepts identified, their relationship with ergonomics, mapped challenges, and research gaps. Finally, the study's conclusions are presented.

7.3 METHODOLOGY

7.3.1 Systematic Literature Review

This literature review chose to search different databases for the most up-to-date work developed on the topic. The first choice was the CAPES journal portal of the Brazilian Coordination for the Improvement of Higher Education Personnel (CAPES), which brings together several Brazilian and international academic scientific productions in Portuguese and English. Secondly, the Brazilian Open Access Publications and Scientific Data Portal (Oasisbr) of the Brazilian Institute of Information in Science and Technology (Ibict). The third was the Scientific Electronic Library (SciELO).

Therefore, a protocol was defined to guide the search, specifying the inclusion and exclusion criteria. The first parameter was studies produced from 2016 to 2023. The starting year was the publication of the GDPR of the European Union, which inspired the Brazilian data privacy law. The search considered peer-reviewed publications, both articles and proceedings. The domain area was applied social sciences, ergonomics, and public innovation.

The keywords defined were LGPD, GDPR, privacy by design, data protection, ergonomics, human factors, consent, and simple Portuguese and English. The main inclusion criteria were combining two terms using the Boolean operator "AND" and alternating between the Portuguese and English versions.

The search found 122 results across the three databases, with 33 works selected for initial consideration after previously reading the abstract and keywords. After reading the abstracts and keywords, duplicate studies and those with limited download access were excluded to narrow the focus to studies directly addressing human factors/ergonomics in privacy. The final result, combined across all databases, was 20 searches available for filtering in the final consideration set, as illustrated in the table below.

The SciELO database did not provide any results for the combinations of terms tested. Nineteen results originated from the CAPES database, and one result originated from the Oasisbr database. The filtering criteria for these 20 selected papers were the complete reading of their introduction and conclusion. After that step, we obtained six papers for full reading. Five papers were ultimately included in this work due to the main objective of the search.

After analyzing the articles selected from the systematic review, the findings were grouped into five thematic groups: user context, transparency and trust, induction of consent, and user rights. This grouping allowed a better understanding of these five topics as key concepts that should guide studies.

A previous work reported part of these results. Here, they are analyzed and discussed in depth.

7.4 RESULTS

7.4.1 Key Concepts

The five papers included in the final filtering phase of this research were read in full. This action made it possible to map the recurring themes between each one—the recurring idea raised by the authors in different papers. The most frequently

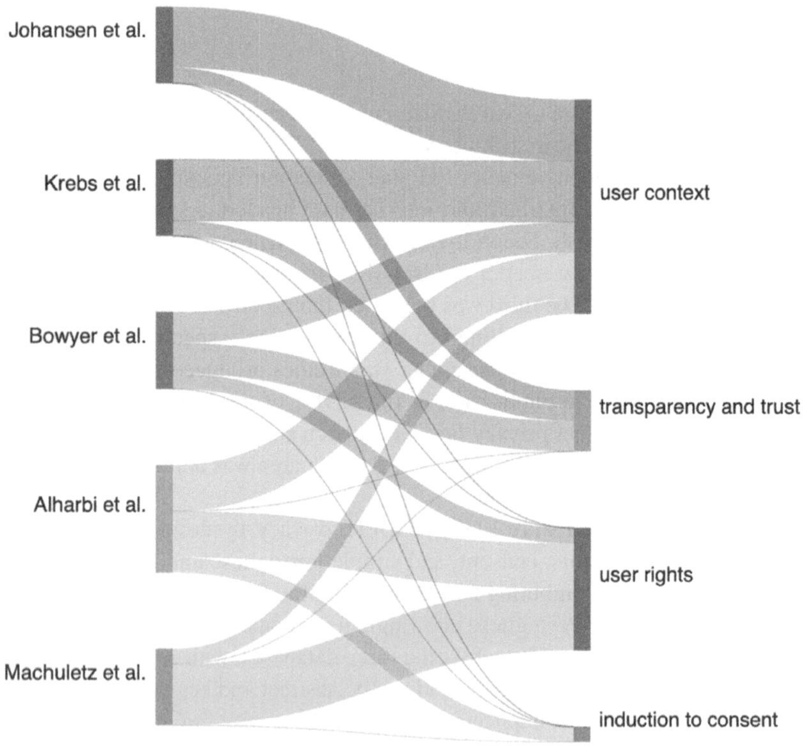

FIGURE 7.1 Diagram of the authors' relationship with the identified topic.

identified ideas were user context, transparency and trust, user rights, and induction to consent.

The repetition of these themes allowed us to identify them as key concepts for research on user interaction with privacy interfaces. Furthermore, other approaches were identified within the concepts. The Diagram in Figure 7.1 explains the distribution of the key concept according to its approach by the author of the included paper.

A Diagram with two columns connected by colorful bars. On the left are the authors' names, and on the right are the identified topics, ordered from top to bottom. The colorful bars extended from the left, crossing over one another until they reached the right, linking each author's name to the different topics.

The concept of **user context** addresses the need to consider users' specificities when studying user interaction with interface privacy. All included papers address this concept.

Transparency and trust are concepts that address the need for privacy interfaces that inspire trustworthiness and convey trust in users.

User rights address how user rights are a non-negotiable part of privacy interfaces and need to be adapted and fully understood depending on the user's context.

Inducing consent presents the main challenge for adequate interaction with privacy interfaces: the application of dark patterns that harm the system's usability and trick users into consenting to data collection.

It is also possible to list a series of topics that should be highlighted to clarify what the reviewed research reveals about human interaction-privacy policies, and for the positive interaction of users with privacy policies in digital interfaces:

- Use practices focused on the context of users;
- Raise awareness among users through creative solutions;
- Evaluation with all types of users;
- Ensure clear information and communication of privacy policies;
- Improve the usability of privacy policies;
- Respect users' real choices;
- Guarantee user rights;
- Data controller transparency.

7.5 DISCUSSION

This part discusses the main ideas behind the key concepts identified during the systematic review. It also adds new works that dialogue with human factors/ergonomics and privacy interfaces.

7.5.1 Interacting with the Privacy Interfaces

Information security and privacy are still abstract and ambiguous topics for most people (Das *et al.*, 2021). Therefore, the objective of laws is to make citizens aware of their choices and their power to consent to have control over them (Bowyer *et al.*, 2022). Awareness about personal data allows people to make informed decisions.

Under the law, users can refuse data collection, request personal data back, or agree to collection. A right that must be protected (Alharbi *et al.*, 2023). However, users must be informed about this. However, this information should not be difficult to understand and should not increase the digital divide or impact users' relationships with the services.

In other words, interacting with the privacy interface should be a protective, informative, and satisfying experience. According to Peterson (2025), ergonomics can optimize interaction with systems through digital interfaces that promote comfort, efficiency, and safety, just as human-centered design (HCD) prioritizes the individual's needs over a satisfactorily usable and accessible experience.

Therefore, the user can only preserve their privacy when the privacy interface is user-friendly. For Chong *et al.* (2018), researchers and designers should adopt a human factors/ergonomics approach by evaluating the following factors:

- how users think about their security;
- motivation for care regarding the security of one's information;
- level of user trust in devices integrated with the internet;
- how these factors will affect how users interact with the devices themselves;
- awareness of the amount and types of risk to which they may be subject.

Das *et al.* (2021) state that Human Factors is critical to computing security and data privacy. According to Peterson (2025), their intersection with HCD is important in improving the user experience as they are fundamental concepts for designing user-friendly systems.

Controllers can trick the user into providing more data than necessary (Johansen *et al.*, 2020). Therefore, data transparency is important to increase loyalty and trust and improve the relationship between the service and the customer or user because people's data represents themselves, their experiences, and their lives (Bowyer *et al.*, 2022).

Therefore, the screens on which the interaction takes place must be designed to meet these needs and guarantee your right to data protection (Krebs *et al.*, 2019). One discovery among the works is how users are persuaded to consent. For example, cookie interfaces, including government websites, are being designed using dark patterns to minimize user choices (Alharbi *et al.*, 2023).

"Accept all" buttons are consciously highlighted to trick users into agreeing to more data collection than necessary. Alternatively, even accept all options when users do not actively select them (Machuletz; Böhem, 2020; Bowyer *et al.*, 2022; Alharbi *et al.*, 2023).

This calls into question the morality and legitimacy of these actions due to the regret reported by users when they realize the implications surrounding their choices to accept everything (Machuletz; Böhem, 2020). The deceptive design of the consent window misinforms users and results in users regretting it when they realize it (Machuletz; Böhem, 2020).

Users tend to consent to fewer options in the consent window when they know their privacy concerns. The design must respect users' interests rather than inducing them to collect their data (Machuletz; Böhem, 2020). Users should be able to change their preferences easily (Alharbi *et al.*, 2023).

Furthermore, it is difficult for users to understand what companies do with their data. For example, requesting the return of one's data is a requirement under existing laws. However, suppose users manage to obtain this information from data collection services (which is not guaranteed to happen despite being required by law). In that case, they will be unlikely to understand the material received due to the way companies process the data (Bowyer *et al.*, 2022).

In other words, a complex system involves communicating information about personal data. What should be clear becomes increasingly complex, especially when lay users interact with it. This means that the users' perspective is not being prioritized.

However, it is essential to consider users' contexts (Krebs *et al.*, 2019; Alharbi *et al.*, 2023). Especially since people are not aware of data tracking during their interactions with digital interfaces, they want to complete their online tasks effortlessly (Johansen *et al.*, 2020; Alharbi *et al.*, 2023).

Much of people's lack of interest in data collection is partly because many of its factual purposes are hidden from users (Krebs *et al.*, 2019; Machuletz; Böhem, 2020; Bowyer *et al.*, 2022). It is up to designers and developers to think of creative solutions that contain all the privacy information offered by the service or product. Solutions that are useful and easy for communicating with lay users (Johansen *et al.*, 2020), which is important because privacy is complex to understand (Krebs *et al.*, 2019; Alharbi *et al.*, 2023).

Ergonomics and HCD approaches must converge to design perfect, intuitive, and safe user interfaces, in which cognitive demand is not excessive, but reliable and

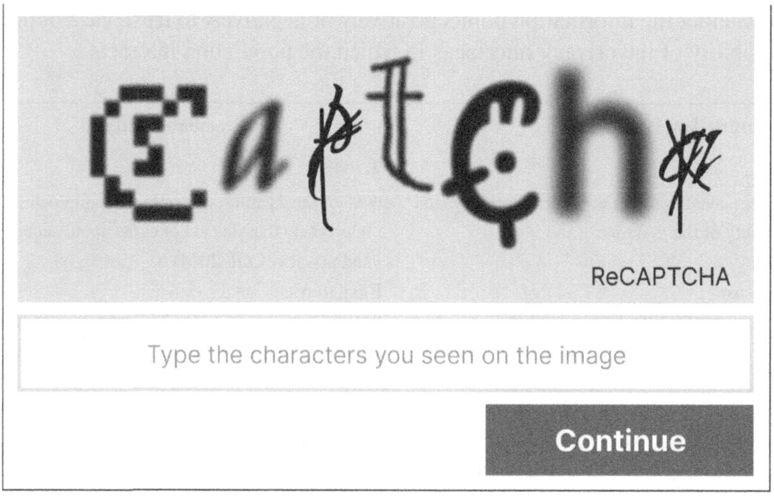

FIGURE 7.2 Using CAPTCHA security confirmation to validate identity.

accessible (Peterson, 2025). In other words, user-friendly privacy interfaces that meet user expectations.

Thus, even if people are on autopilot, they can be directed correctly regardless of the intentions of the controller or data collector (Johansen *et al.*, 2020). It is recommended to adopt practices related to human-computer interaction and user-centered design approaches (Krebs *et al.*, 2019; Johansen *et al.*, 2020).

Ergonomic interfaces have improved efficiency by overcoming usability issues such as complex navigation, task performance, and error prevention (Peterson, 2025). An inclusive privacy and security design seeks to reach the most significant number of users, with different skills, characteristics, and needs. Diversity is an essential requirement for the success of such research and must be integrated into security policies (Das *et al.*, 2021).

For example, people with low vision or reduced manual mobility are excluded from security measures that employ visual aids and passwords typed on minimal virtual keyboards without haptic feedback (Das *et al.*, 2021). Figure 7.2 illustrates a dialog box in which the CAPTCHA identity security measure is adopted. It usually adopts a security to identify the human user which is visually difficult to understand for some people.

Elderly people and LGBTQIAPN+ people can be specific and differentiated targets of cyberattacks and face more serious consequences. Colors and sounds can cause overstimulation. Care must be taken to design an adequate interface to prevent this. Therefore, creative solutions are needed for individuals with different learning or attention difficulties (Das *et al.*, 2021).

When these designs do not consider the particularities of users, there are designs whose use of tools, techniques, and mechanisms is problematic for users. Therefore, the difficulty of use makes vulnerable users even more vulnerable to privacy risks (Das *et al.*, 2021).

Peterson (2025) proposes key factors for user-centric platforms that meet users' expectations. This is contrasted with the data collected on Brazilian internet users to infer whether the information points positively or negatively to trust, ease of use, and accessibility of the privacy interfaces in which the population interacts.

User-centric key factors	Brazilian reality
Trust	**Trust**
The user is confident in the security, privacy, and reliability of the system	40% of the 21 million people who stop using the internet do that due to concerns about security and privacy (CGI, 2024)
Easy to use	**Easy to use**
Prevents users from feeling overwhelmed	81% of internet users do not read privacy policies or consent terms in full because they are too long, and 69% of them find it difficult to understand (CGI, 2022)
Accessible	**Accessible**
Inclusive for people with disabilities or limited digital literacy.	66% of internet users are over 60 years old, and 44.4% are uneducated people who have access to the same type of privacy policies or terms of consent and other elements (IBGE, 2024)

Crossing Peterson's (2025) definition with data on Brazilian internet users suggests that the context of Brazilian users is not considered. No data suggests that different profiles, such as people over 60 and people without formal education, have accessible interaction with privacy interfaces.

The privacy interfaces also don't seem user-friendly for understanding data collection. More than 80% of users do not read the documents that inform them about this. The claim about extension and complexity demonstrates how cognitively demanding the privacy issues exposed to users are. This can also impact these people's trust in the systems.

More than 8 million people have trust issues regarding their data privacy. They stop using the internet because the systems do not seem reliable enough for them. This may even suggest a scenario of exclusion, considering that everyday tasks and public services are increasingly carried out online due to the Information Age.

7.6 CHALLENGES

Users must be assured that their information will be stored appropriately and that they can revoke access to their data at any time. This means empowering them and making them aware of risks to improve their confidence (Chong *et al.*, 2018).

The main challenge is developing methods to facilitate this awareness. Prevention and education methods that ignore individual aspects of human identity and experience are useless (Chong *et al.*, 2018). Public initiatives, dissemination of recent discoveries, production of educational materials, training, or formal education are necessary (Das *et al.*, 2021).

It is important that learning is available to everyone. However, it is challenging to assimilate the challenges faced by different populations. What we have today regarding training and education does not consider the individual's different skills, characteristics, needs, and values. Inclusion should be considered for all forms of subject awareness (Das *et al.*, 2021).

7.7 RESEARCH GAPS

Few studies explore the technological perceptions of users from different cultural, technical, and physical backgrounds. Understanding the security and privacy of vulnerable people is a critical point to be explored and a research gap. This includes people with disabilities, children, the elderly, and those from economically emerging countries (Das *et al.*, 2021).

This gap demonstrates the presence of non-inclusive designs in privacy and security due to many biased preconceptions. Tools and technologies must be sought that empower users to protect their privacy. (Das *et al.*, 2021).

A study gap found among the research analyzed was the participation of everyday and real users. In all studies analyzed, the sample group comprised people linked to academia. People with little literacy in the digital environment, whether elderly or individuals with low education, were not considered. In other words, those subject to privacy and consent issues, whether due to inclusion in the e-government system, data governance policies, or the imperative progress of the Information Age through their mobile devices, such as smartphones, were not subjects of investigation.

As reported by the authors, there is a need to emphasize the importance of evaluations with real users (Johansen *et al.*, 2020). Only through it will its actual context be considered. There is also a lack of consideration of e-government in other studies (Alharbi *et al.*, 2023). There is also a gap related to the location of this research, most of which is restricted to European countries. Although one of the search terms used in this literature review was GDPR, the Portuguese search did not find significant results that would contribute to a Brazilian context.

One of the articles has a Brazilian researcher on the research team, while the others are linked to institutions in Belgium and the Netherlands. One of the five selected articles includes researchers from an institution in Saudi Arabia, even when researching interfaces around the globe, only Chilean, Argentinean, and Uruguayan sites were considered. This demonstrates a gap and research opportunities in regions beyond the global north.

Although research discusses the topic, there are gaps in what should be addressed to empower users. There is a consensus regarding the need to consider users' perceptions of different characteristics, skills, and values. However, where is the research that provides these people's perceptions?

Scientific production and public policies seem to have already realized that adequate data privacy protection will only be effective when vulnerable people are also heard. We have also designed technologies and features that meet your needs. Nevertheless, why does the existing research reviewed not include this population in the studies developed when pointing out this gap?

7.8 FINAL CONSIDERATIONS

In the digital age, where a data-based economy permeates every facet of society, privacy interfaces are the frontline defense in safeguarding individuals' personal information.

This study reveals that, while regulatory frameworks like Brazil's LGPD and Europe's GDPR provide essential legal protection, there remains a critical disconnect between these mandates and how privacy information is communicated to users. Many current interfaces employ confusing language and dark patterns that undermine users' capacity to make informed decisions about their data.

Our review underscores that the challenges associated with privacy interfaces are not uniform. Users with limited digital literacy, including older populations and those with lesser formal education, face unique hurdles when navigating lengthy privacy policies and consent forms. By integrating ergonomics and HCD principles, future developments should aim to reduce cognitive load and present critical privacy information in a clear, accessible manner.

Moreover, the research highlights significant gaps in existing studies, notably the limited participation of non-academic users and an overrepresentation of European-centered research. These gaps require a broader methodological approach, including real-world user testing across diverse socio-cultural contexts. Future investigations should seek to incorporate empirical evaluations that reflect the lived experiences of a broader user base, ensuring that the design of privacy interfaces is both inclusive and equitable.

Collaboration among interface designers, developers, legal experts, and policymakers is crucial. Practical steps include standardizing formats for privacy notices, investing in user education, and developing design guidelines prioritizing transparency and genuine user consent. Such multi-stakeholder efforts will improve user trust and bridge the gap between regulatory intent and practical implementation.

Although the number of available studies was limited in this review, its findings provide a foundation for future research. Expanding this work through interdisciplinary studies, especially those that involve direct user engagement, will be key to evolving privacy interfaces into tools that empower all users. Addressing these challenges head-on is essential for transforming privacy legislation from a regulatory requirement into a meaningful component of the digital experience.

REFERENCES

Alharbi, J. A.; Albesher, A. S.; Wahsheh, H. A. An empirical analysis of e-governments' cookie interfaces in 50 countries. *Sustainability* 15, 1231 (2023).

Bowyer, A.; Holt, J.; Jefferies, J. G.; Wilson, R.; Kirk, D.; Smeddinck, J. D. Human-GDPR interaction: practical experiences of accessing personal data. In: CHI Conference on Human Factors in Computing Systems (CHI'22) Proceedings (2022).

Chong, I; Xiong, A; Proctor, R. W. Human factors in the privacy and security of the internet of things. Ergonomics in Design 27(3), 5–10 (July 2019). 2018 *by Human Factors and Ergonomics Society; Article Reuse Guidelines.* https://doi.org/10.1177/1064804617750321

CGI Comitê Gestor da Internet no Brasil - Pesquisa sobre o uso das tecnologias de informação e comunicação nos domicílios brasileiros: TIC Domicílios 2023. *Comitê Gestor da Internet no Brasil, São Paulo* (2024).

CGI Comitê Gestor da Internet no Brasil - Privacidade e proteção de dados pessoais 2021: perspectivas de indivíduos, empresas e organizações públicas no Brasil. *Comitê Gestor da Internet no Brasil, São Paulo* (2022).

Das, S.; Gutzwiller, R. S.; Roscoe, R. D.; Rajivan, P.; Wang, Y.; Jean Camp, L.; Hoyle, R. Humans and technology for inclusive privacy and security. *Proceedings of the Human Factors and Ergonomics Society Annual Meeting*, 64(1), 461–464 (2021). https://doi.org/10.1177/1071181320641104 (Original work published 2020)

Instituto Brasileiro de Geografia e Estatística (IBGE). Em 2023, 88,0% das pessoas com 10 anos ou mais utilizaram Internet. Available: https://agenciadenoticias.ibge.gov.br/agencia-noticias/2012-agencia-de-noticias/noticias/41026-em-2023-87-2-das-pessoas-com-10-anos-ou-mais-utilizaram-internet. Acessed: 18 dez. 2024.

Johansen, J.; Pedersen, T.; Fischer-Hübner, S.; Johansen, C.; Schneider, G. Roosendaal, A.; Zwingelberg, H.; Sivesind, J.; Noll, J. A multidisciplinary definition of privacy labels: the story of princess privacy and the seven helpers. *Information and Computer Security*, 30, 452–469 (2020).

Krebs, L. M.; Dewitte, P.; Geerts, D.; Verbert, K.; Rodriguez, O. L. A.; Ausloos, J.; Naudts, L. Tell me what you know: GDPR implications on designing transparency and accountability for news recommender systems. In: CHI'19 Extended Abstracts (2019).

Machuletz, D.; Böhem, R. Multiple purposes, multiple problems: a user study of consent dialogs after GDPR. *Proceedings on Privacy Enhancing Technologies* (2020):2: 481–498 (2020).

Peterson, B. (2025). Human-centered Design and Ergonomics: Enhancing Educational Environments and User Experience in Digital Banking. Available: https://www.researchgate.net/publication/390108720_Human-Centered_Design_and_Ergonomics_Enhancing_Educational_Environments_and_User_Experience_in_Digital_Banking/citations. Accessed in: 28 apr. 2025.

Véliz, C. Privacidade é poder. Editora Contracorrente, São Paulo (2022).

Zuboff, S. Big other: capitalismo de vigilância e perspectivas para uma civilização de informação. *In*: Bruno, F.; Cardoso, B.; Kanashiro, M.; Guilhon, L.; Melgaço, L. *Tecnopolíticas da vigilância: perspectivas da margem*. Boitempo, São Paulo. 17–68 (2018).

8 Modular Approaches for an Efficient UX Measurement in Ergonomics 5.0

Martin Schrepp

8.1 INTRODUCTION

The rapid technological advancements of the past two decades, for example, the massive growth in smartphones, e-commerce, and communication platforms, have had a massive impact on the way people live, communicate, and work. It is likely that technology will continue to develop at an increasing pace. Emerging technologies like virtual and augmented reality, artificial intelligence, Internet of Things, and big data will become increasingly integrated into daily life, and new technologies may continue to evolve. As a result, we can anticipate even greater changes in our living and working environments in the future.

Industry 5.0 (Nahavandi, 2019; Adel, 2022) describes a new phase of industrial evolution that is characterized by an increased collaboration of workers and smart technologies as well as increased possibilities to produce products adapted to the personal preferences of customers. Correspondingly, Society 5.0 (Fukuyama, 2018) envisions a transformation of society using new technologies and by merging cyberspace and physical space (Deguchi et al., 2020). Both paradigms highlight the increasing impact of new technologies to improve productivity, quality of life, inclusion, or sustainability. Therefore, the importance of UX will massively increase, as this quality is essential to enable as many individuals as possible to interact effectively and efficiently with a growing number of digitalized services built in an increasing variety of technologies.

On the other hand, more digitalized services must be designed, built, and maintained. And due to the fast technological change, their lifecycle will become shorter. Thus, there will be a greater need for professional UX designers who can transfer user requirements and expectations into user-friendly solutions. But design resources are of course not unlimited. Thus, it is necessary to increase the efficiency of design and development processes. New AI capabilities already help developers to massively increase their productivity (Coutinho et al., 2024; Sauvola et al., 2024). It is important to get a clearer picture of how these capabilities, especially Large Language

DOI: 10.1201/9781003559399-9

Models, can help UX designers and researchers to be more efficient, for example by generating persona descriptions, task descriptions for usability tests, or the analysis of customer comments and what the borders and risks of AI usage are in automating UX tasks (Graser et al., 2024).

A well-known quote attributed to Lord Kelvin is *If you cannot measure it, you cannot improve it.* This underlines the importance of reliable quantitative data as a basis for informed decisions. This is of course also true for UX quality. Several standardized UX questionnaires have been developed in the past (for a broader overview, see Schrepp, 2021) to measure UX based on user feedback. However, the development of such standardized UX questionnaires requires a lot of effort. Several empirical studies are required to select good scales and items, and the objectivity, reliability, and validity of a new questionnaire must be investigated in subsequent studies (Schrepp, 2024). Each UX questionnaire covers, by its scales and items, only a part of the UX spectrum. Many questionnaires are restricted to special use cases and technologies. There are, for example, a lot of specialized questionnaires that measure different UX aspects of classical GUI applications. For other interaction paradigms, for example virtual or augmented reality, voice interaction or gesture interaction, the number of suitable UX questionnaires is currently quite limited, and for some relevant research questions, no fitting UX questionnaire exists.

We will discuss in this chapter two approaches that try to increase the efficiency of UX questionnaires. Given an increasing demand to measure the UX of newly developed scenarios, these approaches will hopefully play a role in Ergonomics 5.0. We will discuss the advantages but also limitations of these approaches.

8.2 UX MEASUREMENT WITH STANDARDIZED QUESTIONNAIRES

We explain in this section the basic concepts that are required to understand the advantages and limitations of modular approaches to UX measurement.

8.2.1 What Does UX Mean?

According to the standard ISO 9241-210 (ISO, 2019), UX is defined as *a person's perceptions and responses resulting from the use or anticipated use of a product, system, or service*, i.e., as a purely subjective perception.

Since the subjective perception of the user is at the center of this definition, we must ask users about their opinions to measure UX. There is no objective way to do this. Assume, for example, that we track the navigation of users in a user interface and that we detect that sometimes long loading times occur when the user navigates to another screen. Is this a UX problem? The answer to this question can be *Yes*, if users are dissatisfied with these delays. But the answer can also be *No*, if they do not care about that, for example, since they perform the corresponding task quite rarely, or if their attention is captured during the waiting time by decisions about next steps, and they do not even recognize the delay. And, of course, due to personality or experience of the users with the product or differences in the typical tasks they perform some users may be annoyed by the response times, while others are not.

8.2.2 Standardized UX Questionnaires

Standardized UX questionnaires allow to measure UX quantitatively. From the answers of users to the items of the questionnaire, one or several UX scale scores are calculated. Each of these scores corresponds to a dedicated UX aspect, for example, efficiency, learnability, controllability, fun of use, aesthetic appeal, or trust. This allows us to compare different products or product versions concerning these UX aspects. If, for example, a new product version is compared to the previous one, the researcher can check if the new version has a higher score and if the difference is statistically significant.

Another frequent use case is associated with benchmarks. If only a single measurement with a UX questionnaire is available, the question arises how the obtained score can be interpreted. Does a score of 67 on a scale from 0 to 100 represent good, adequate, or bad UX? Of course, this question can only be answered if the measured score can be compared to the scores of other products. A benchmark is simply a collection of a large set of product evaluations with a given UX questionnaire. If a comparison to the scores in the benchmark shows, for example, that the score is in the 10% best results, this indicates a pretty good quality. If the comparison shows that 70% of the products in the benchmark data set showed a better score, this does not speak for a good UX of the product. Many established UX questionnaires offer larger benchmark data sets, for example, the SUS (Brooke, 1996; 2013), the UEQ (Laugwitz et al., 2008), or VISAWI (Moshagen & Thielsch, 2010).

To ensure reliable results and avoid sampling biases, it is necessary to collect UX data from larger samples. This is why standardized UX questionnaires are so popular in this field. These questionnaires can be integrated into a product through a feedback button that initiates an online survey, or they can be distributed through email campaigns, links on websites, or social media platforms. Therefore, it is cheap and easy to collect data with this method.

It is essential that a UX questionnaire covers those UX aspects that are important for the overall UX impression of an application. These aspects differ (Schrepp et al., 2023a) between applications. For example, for a social network, other UX aspects are important than for an enterprise software used for professional work. In addition, the complexity of the items must fit the cognitive and language skills of the participants. It is, for example, problematic to apply a questionnaire developed for adults to children (Hinderks et al., 2014). Thus, each standardized UX questionnaire can only be used for special classes of products or user groups. Let us explain the points above on the example of two well-known standardized questionnaires.

8.2.3 System Usability Scale

The System Usability Scale (SUS) (Brooke, 1996) is a short questionnaire that measures the usability of a product. It consists of ten statements that describe aspects of a product that are related to good or bad impressions concerning usability. Participants can express their disagreement or agreement to a statement on a 5-point response scale.

Example of a SUS item:

I think that I would like to use this system frequently
Strongly disagree ○·○·○·○·○ *Strongly agree*

Other examples of such statements are *I thought that the system was easy to use*, *I felt confident using the system*, or *I found the system unnecessarily complex*.

The SUS provides an overall usability score ranging from 0 (worst) to 100 (best). To calculate the score the answer alternatives of the positive items (agreement represents positive UX, for example, *I thought that the system was easy to use*) are scored from 0 to 4 from left to right. The negative items (disagreement represents positive UX, for example, *I found the system unnecessarily complex*) are scored from 0 to 4 from right to left. Per participant the scores of all ten items are added (resulting in a sum score between 0 and 40) and then transferred to a 0–100 scale by dividing by 40 and multiplying the result by 100. The mean of these calculated scores over all participants is then the resulting SUS score for the product.

The SUS is one of the most popular usability questionnaires. There is a large body of research dealing with the psychometric properties of the SUS, so it is quite well understood. The SUS benchmark contains data from 241 industrial usability studies (Lewis & Sauro, 2018) and is shown in Table 8.1. Alternative approaches to provide interpretations of SUS scores can be found in Bangor et al. (2009).

If you measure, for example, a SUS score of 82 for your product you are in category A and your result is better than the results of 90% of the products in the benchmark data set. Therefore, the usability of your product is pretty good compared to an average product. If your score is 66, then your product is in category C, and the usability of your product is of average quality.

TABLE 8.1
The curved grading scale for interpreting SUS scores from Lewis and Sauro (2018)

Category	Score interval		Percentile
A+	84.10	100.00	96–100
A	80.80	84.00	90–95
A-	78.90	80.70	85–89
B+	77.20	78.80	80–84
B	74.10	77.10	70–79
B-	72.60	74.00	65–69
C+	71.10	72.50	60–64
C	65.00	71.00	41–59
C-	62.70	64.90	35–40
D	51.70	62.60	15–34
F	0.00	51.60	0–14

The SUS measures usability, with a strong focus on learnability and efficiency. Thus, it is a good choice to evaluate products that are mainly used for professional work or to achieve clearly defined goals. The SUS does not consider hedonic UX aspects (non-task-related qualities, for example, aesthetic impression or fun of use). Therefore, it is not optimal to use it to evaluate a social network or an app that is used mainly for fun or leisure activities.

8.2.4 User Experience Questionnaire

The goal of the User Experience Questionnaire (UEQ) (Laugwitz et al., 2008) is to allow a quick product evaluation concerning a broader set of different task-related (pragmatic) and non-task-related (hedonic) UX aspects. It contains 26 items which are grouped into the six scales *Attractiveness*, *Efficiency*, *Perspicuity*, *Dependability*, *Stimulation*, and *Novelty*. A detailed semantic interpretation can be found in Laugwitz et al. (2008) or on the UEQ homepage https://ueq-online.org.

The UEQ items are semantic differentials with a 7-point answer scale, where the positive term is in half of the items placed in the right and in the other half in the left position.

Example of a UEQ item:

attractive ○·○·○·○·○ *unattractive*

Examples of other items (corresponding scale in parenthesis): unpleasant/pleasant, good/bad (*Attractiveness*); efficient/inefficient, fast/slow (*Efficiency*); easy to learn/difficult to learn, clear/confusing (*Perspicuity*); unpredictable/predictable, obstructive/supportive (*Dependability*); boring/exciting, not interesting/interesting (*Stimulation*); and usual/leading edge, creative/dull (*Novelty*).

If the positive term in an item is on the right side, then the responses are scored from −3 to +3 from left to right. If the positive term is in the left position, then responses are scored from −3 to +3 from right to left. The score for a scale is then simply the average over all responses of all participants to all items in a scale.

The UEQ can be filled in approximately 3–5 minutes and is thus short enough to be used as an online questionnaire but can also be used at the end of a usability test. Because of the item format and the balanced choice of pragmatic and hedonic scales, it can be applied to a wide range of products.

The UEQ offers a benchmark (Schrepp et al., 2017a) that contains data from more than 400 studies with different products and is included in the data analysis sheet that can be downloaded from the UEQ homepage (https://ueq-online.org). The benchmark provides per scale a grouping into five categories (*Excellent*, *Good*, *Above Average*, *Below Average*, and *Bad*). Similar to the SUS benchmark each category is linked to a percentile of products from the benchmark data set. If a product falls, for example, for a scale into category *Excellent* this means it is amongst the 10% of the best products. If it falls into category *Above Average*, then 25% of the products from the benchmark data set have better results, and 50% have worse results concerning this scale (Figure 8.1).

The UEQ scales describe quite general UX aspects; thus, the UEQ can be applied in many scenarios. But for special applications, for example, voice interfaces or

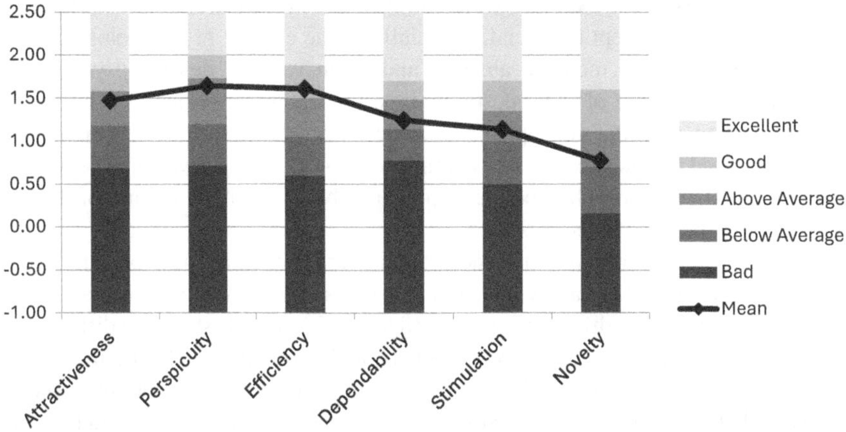

FIGURE 8.1 Example for a comparison of a product evaluation to the UEQ benchmark. The black line shows the measured UEQ scale scores for the product. The colored bars show the benchmark categories per scale. Scores range from –3 to +3. To increase visibility the range of the y-axis was restricted to –1 to +2.5 in the figure above.

household appliances, naturally some special aspects of the interaction are important that are not fully covered by the UEQ scales.

8.2.5 Quality Criteria: Objectivity, Reliability, and Validity

To deliver valuable and comprehensible outcomes, UX questionnaires must meet specific quality criteria. These criteria, similar to those used in psychological test development, include objectivity, reliability, and validity. We will briefly explore each of these key aspects.

Objectivity refers to the independence of the results of a measurement tool, such as a UX questionnaire, from the person administering it. Generally, objectivity is not problematic for UX questionnaires due to the use of pre-written instructions that guide participants on how to complete the questionnaire. Additionally, the analysis and interpretation of the data are usually outlined in the scientific publications that describe the construction of the questionnaire, handbooks, or data-analysis tools that are provided by the authors. Consequently, this standardized approach leaves little room for subjective interpretation by the UX researcher who applies the questionnaire.

Reliability refers to the consistency of scores obtained from repeated measurements under identical conditions (such as the same target group and product). This is a crucial criterion for a UX questionnaire. For example, assume you measure the UX of the same product twice, recruiting a large number of users from nearly identical target groups (similar in age, gender, experience with the product, frequency of use, etc.). Obviously, you would expect the results to be highly similar. While some random fluctuations are inevitable, significant discrepancies between the two sets of results would undermine our trust in the accuracy of the scale. Reliability of UX questionnaires is typically assessed using the Alpha Coefficient (Cronbach, 1951; Schrepp, 2020).

Validity refers to the extent to which an instrument measures what it is intended to measure. This concept is crucial, especially in the context of psychological tests, which often intend to measure abstract and often vaguely defined psychological constructs. In the case of UX questionnaires, the items are typically semantically closely aligned with the concept that a scale within the questionnaire represents. Consequently, UX questionnaires generally exhibit a strong level of face validity (Holden, 2010), ensuring that the items appear to be a plausible measure of the concept they are intended to assess.

However, validity remains a critical quality criterion that must be guaranteed. Consider a scenario where a UX scale that measures efficiency is used to evaluate two products known to have different levels of efficiency. If the scale shows no difference in the efficiency scores between the two products, it clearly fails to deliver a meaningful result. This indicates that the scale does not measure what it claims to measure, making it unreliable to base important decisions on the results. This aspect of validity, which assesses whether a tool measures the intended construct accurately, is often referred to as construct validity or overall validity (Cronbach & Meehl, 1955).

When a new UX questionnaire is developed, the three quality criteria mentioned above are usually checked in subsequent studies.

8.3 LIMITATIONS OF CURRENT APPROACHES

UX is quite a heterogeneous concept. Depending on the concrete use case of an application different qualities of a user interface play a role in perceived UX (Schrepp, 2021). In addition, many applications are used by different user groups, each with unique needs. A questionnaire that intends to measure UX perception must consider this complexity.

This is the reason for the huge variety of existing UX questionnaires. Some are applicable to many usage scenarios since they measure some high-level concepts, for example, SUS (Brooke, 1996), UMUX (Finstadt, 2010), or UEQ (Laugwitz et al., 2008). Others are highly specialized to measure isolated UX aspects in detail, for example, VISAWI (Moshagen & Thielsch, 2010) for the aesthetic impression of a graphical user interface, QUESI (Hurtienne & Naumann, 2010) for intuitive use, ICTQ (Ozok & Salvendy, 2001) for consistency of web page design. Other questionnaires are restricted to certain types of applications, for example, WEBQUAL (Loiacono et al., 2007) for websites, SUISQ (Polkosky & Lewis, 2003) for speech systems, SUPR-Q (Sauro, 2015) for commercial websites, or HARUS (Santos et al., 2015) for augmented reality applications on handheld devices. For a more comprehensive overview of commonly used UX questionnaires, refer to Schrepp (2021).

Developing a UX questionnaire that accurately measures the important UX qualities for a specific type of application is a labor-intensive process. It typically begins with experts constructing a comprehensive set of potential items, often derived from a thorough literature review. This initial item set is then refined through one or more empirical studies. The preliminary version of the questionnaire produced from this process is further evaluated in multiple subsequent studies to ensure that it meets the common quality criteria. Typically, several years are required to collect enough data for a high-quality benchmark.

The methodology used to develop UX questionnaires is adapted from the field of psychological test development. Psychological tests are typically stable over time and have broad applicability, which justifies the substantial effort required to develop high-quality measurement instruments. However, if we consider the need for methods in Ergonomics 5.0 that allow for rapid adaptation to new use cases, application types, or target groups, the traditional approach to developing UX questionnaires may prove to be problematic. In the following section, we will outline two initial strategies to address these challenges.

8.3.1 Modular User Experience Questionnaire (UEQ+)

The UEQ+ (Schrepp & Thomaschewski, 2019) is a modular extension of the UEQ. It tries to address the problem that different UX studies require different combinations of UX scales to answer the underlying research questions. It extends the UEQ in the sense that the six UEQ scales are available together with several new scales (available scales and materials including translations and a data analysis tool can be found on https://ueqplus.ueq-research.org/).

The UEQ+ is thus not a UX questionnaire, but a modular framework to build concrete questionnaires with a scale structure adapted to the underlying research questions. The UEQ+ contains at the moment 27 UX scales that share the same format. This common scale format allows to combine them in any order for a concrete study (see Figure 8.2 for an example).

As an example, we show the UEQ+ scale *Efficiency*:

To achieve my goals, I consider the product as								
slow	o	o	o	o	o	o	o	fast
inefficient	o	o	o	o	o	o	o	efficient
impractical	o	o	o	o	o	o	o	practical
cluttered	o	o	o	o	o	o	o	organized
I consider the product property described by these terms as								
Completely irrelevant	o	o	o	o	o	o	o	Very important

Each scale starts with a short sentence that sets a common context for the four items of the scale. These items have (like the items in the original UEQ) the form of a semantic differential. Thus, each item consists of two adjectives that represent the extreme ends of a semantic dimension. The ratings correspond to values –3 (most negative) to +3 (most positive). In contrast to the original UEQ the negative term is always positioned left, and the positive term is always positioned right. The scale score is simply the average over all participants and all items in a scale and therefore also scaled from –3 to +3. Due to the change in the item format scores obtained with the corresponding UEQ and UEQ+ cannot be directly compared (Schrepp et al., 2023b). Short forms of UEQ+ questionnaires with just two items per scale are also available (Schrepp et al., 2021).

In some use cases, it is beneficial to measure not only the means for the different scales, but also to provide a single number (a key performance indicator, or KPI) that can

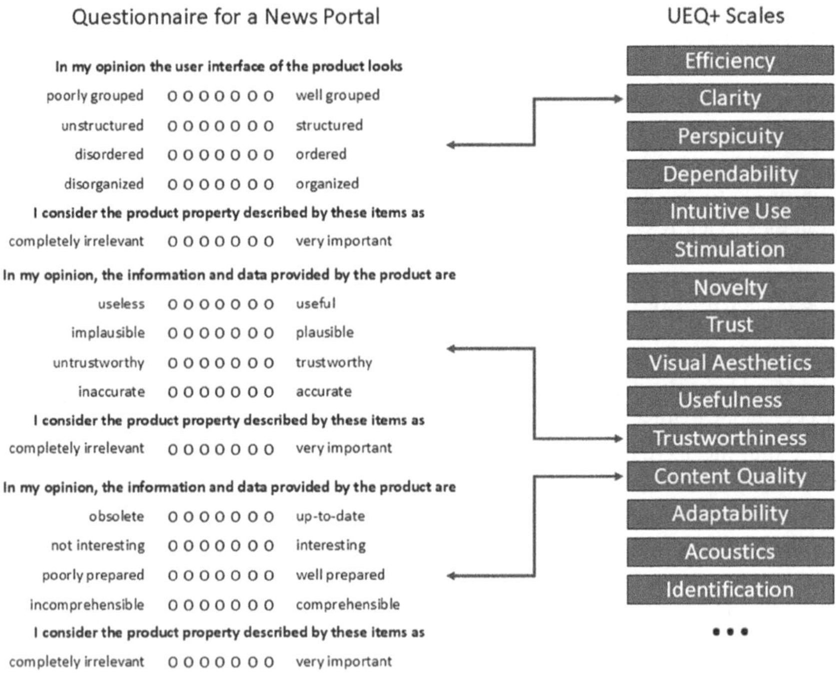

FIGURE 8.2 Depending on the type of application different scales are important and should therefore be selected for the UX measurement.

be interpreted as a measure for the overall UX impression. To allow such a calculation, a question concerning the importance of the UX aspect measured by a scale is displayed below the items. This rating is used to weight the scale values by their importance. If this is not needed for the research, this part of the scale can be dropped. In detail, the calculation of the KPI works as follows. Assume we have created a concrete questionnaire by selecting m scales S_l to S_m from the UEQ+ framework and that n participants of a study evaluated a product with these m scales. The mean rating of the items in scale S_j by participant i is denoted by v_{ij} and the importance rating of this scale is called w_{ij}. We calculate now per scale S_j the relative importance r_{ij} for participant i by:

$$r_{ij} = \frac{w_{ij}}{\sum_{j=1}^{m} w_{ij}}$$

With these relative importance ratings, we can now calculate the KPI by:

$$\text{KPI} = \sum_{i=1}^{n} \sum_{j=1}^{m} r_{ij} v_{ij}$$

Thus, we simply weight the ratings on the scale per participant by the importance for the participant. This procedure is equivalent with the KPI calculation for the extended UEQ (Schrepp et al., 2017b; or Hinderks et al., 2019). If a measurement of the overall UX impression is not required for a research project, then the importance question can be dropped to reduce the length of the questionnaire.

Some UEQ+ scales are applicable to a wide range of products, such as efficiency, intuitive use, usefulness, and trust (Hinderks, 2016). Other scales are specific to certain types of products. For example, sound created by the operation of a device, especially designed to evaluate household appliances (Boos & Brau, 2017), *Response Quality* of a voice assistant (Klein et al., 2020), *Trustworthiness of Content* of a web pages (Schrepp & Thomaschewski, 2019), or *Hardware Security* (Lindemann & Weber, 2023) of large medical devices. Another type of scale is designed to measure the impact of using or owning a product on social connections or status, for example, *Identification* or *Social Acceptance* (Mortazavi et al., 2023).

One significant advantage of the UEQ+ framework is that scales can be independently constructed and tested for reliability and validity, enabling the framework to be easily expanded to cover new application scenarios. Thus, there is no single owner for the UEQ+ scales. After the initial launch of the UEQ+, which already described 16 scales, 11 new scales were contributed by various authors. So, the UEQ+ is a community approach, and the UEQ+ home page tries to offer an overview of existing scales and translations to simplify the usage and distribution of new scales in the community.

This makes the adoption of the UEQ+ to new application scenarios highly efficient. Existing scales can be combined with new ones, and the framework can be adapted step-by-step to new application domains by different researchers. The only restriction is that new scales must be delivered in the common format.

We have seen above that the original UEQ contained a benchmark that allows to interpret the scores of single scales. This is difficult to obtain with a modular approach. First, of course in all UX evaluations with the UEQ usually all six scales are contained. Thus, each evaluation can contribute in principle to the benchmark (if the study authors share their data). For the UEQ+ this is different. In an evaluation with the UEQ+ only some selected scales are used. There are of course huge differences concerning the frequencies with which scales are selected. Thus, it is unlikely that it will be possible to develop a reliable benchmark for all scales. A similar argument applies for a benchmark concerning the KPI. The KPI will for different studies be calculated from different scales. Thus, a direct comparison and therefore also an aggregation into a benchmark does not make much sense.

Therefore, a first approach to allow an interpretation is provided by publishing a set of UEQ+ KPI results for common products (Meiners et al., 2023a). For a visual representation of the results, see Figure 8.3. This is of course far away from the quality of the UEQ benchmark but provides at least some guidance on the interpretation of UEQ+ scores.

The basic idea of this simple benchmark is to allow a comparison to the KPI score of well-known products that support a similar use case. Thus, if, for example, a social network is evaluated, it provides some additional insight to compare the score to the result for Facebook shown in the UEQ+ benchmark.

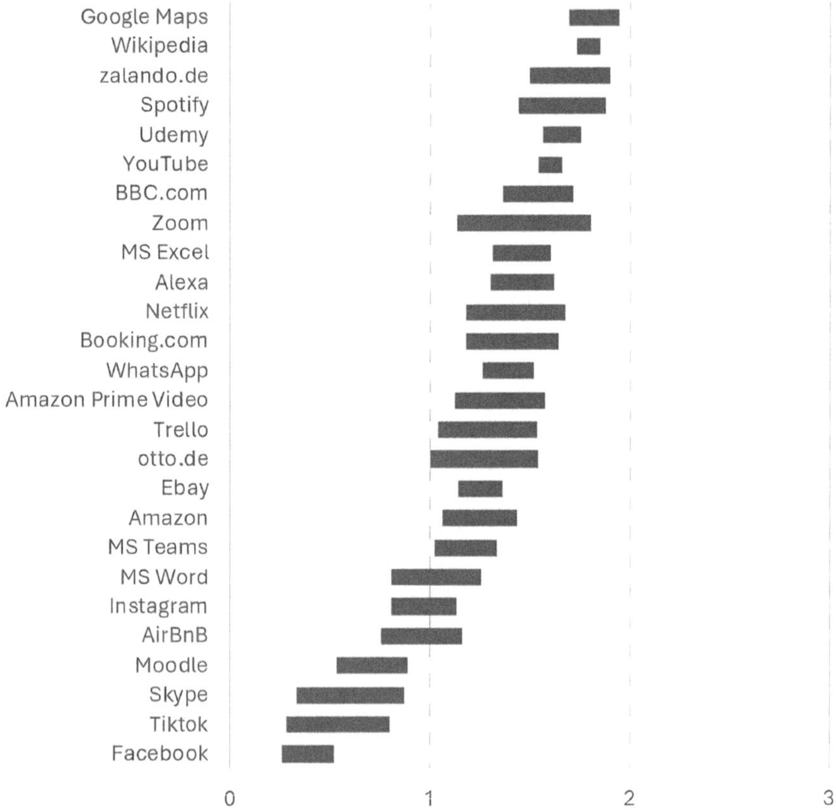

FIGURE 8.3 Measures UEQ+ KPI's for several well-known products (Meiners et al., 2023a). The width of a bar corresponds to the 95% confidence interval of the score.

8.4 USER EXPERIENCE EXPECTATION AND EVALUATION

As a second example, we describe a method (called UEEE, for *User Experience Expectation and Evaluation*) that was originally developed to capture user expectations toward a new product (Oelfin & Schrepp, 2021). While UEQ+ is modular on the level of scales, UEEE is modular on the level of items. The method contains a list of predefined "established" items (items are adjectives, for example, *efficient*, *useful*, *interesting*, that describe positive UX aspects), and the researcher can simply pick from this list. But it is also possible to add new items. Thus, the UX researcher is free to define the items as long as the common item format (adjective describing a positive UX aspect) is used.

The original method uses two phases. In the first phase, a sufficiently large number of participants judge the importance of the UX aspects represented by the adjectives for their overall UX assessment of the product. The participants for this first rating can be users of the product, but it is also possible that participants rate expectations concerning a product they have not yet used, i.e., the anticipated use.

In the second phase, the participants rate how well the product fulfills the UX quality described by these adjectives. Participants in this phase must have usage experience with the product; thus, this rating refers to the actual use.

Thus, this method defines an item format and a standardized way to collect data but allows full flexibility concerning the used items.

The first phase collects information concerning the importance of pre-selected attributes for the product under investigation. In the original paper (Oelfin & Schrepp, 2021), a web application (see Figure 8.4) was used. Participants get all selected adjectives in a container on top and could assign them by drag & drop to the categories *unimportant*, *rather unimportant*, *rather important*, and *important*.

But this can also be realized without any application support. It is sufficient to print the adjectives on simple cards and let participants sort these cards into the

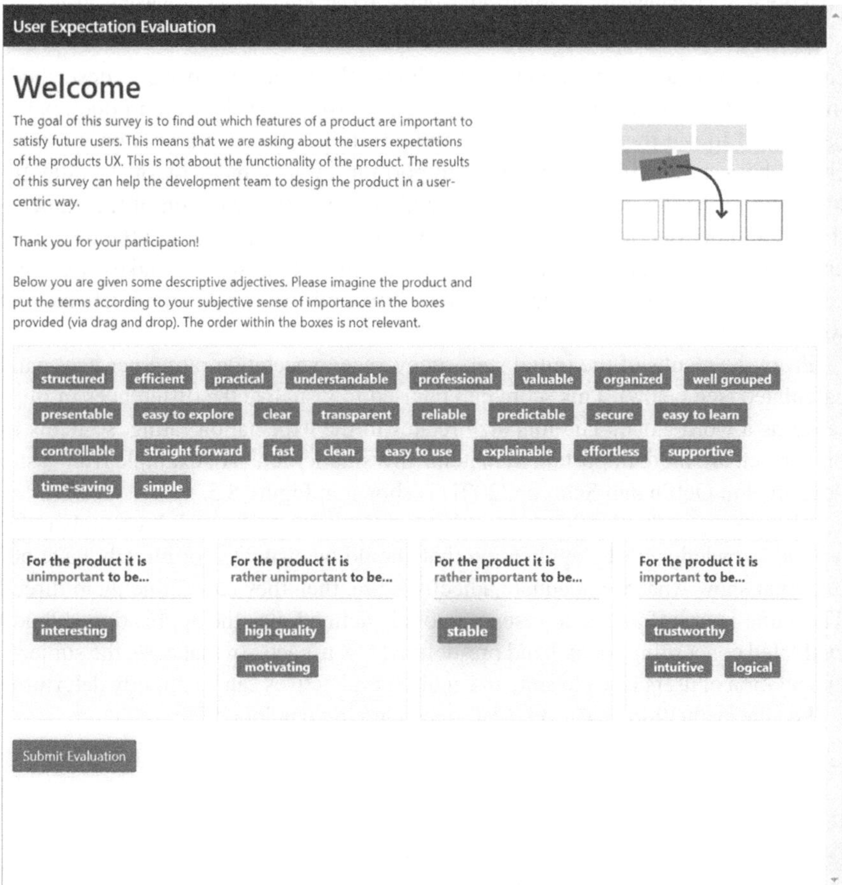

FIGURE 8.4 Part of the screen of a web tool to classify the adjectives as described in (Oelfin & Schrepp, 2021).

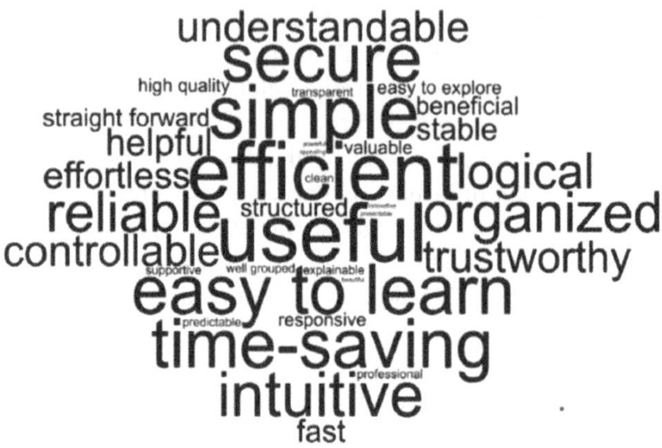

FIGURE 8.5 Example for a visualization of the expectation ratings as a word cloud.

categories. It is also not mandatory to choose the same categories as described in this example. Since all adjectives describe positive properties of a product, participants tend to see them all as somehow important. Thus, if there is only a binary choice between "unimportant" and "important," a huge majority of the adjectives will land in the "important" category, which makes the selection of the adjectives that should be used in a study difficult. The four categories used in the tool should encourage participants to differentiate a bit more. This simply makes it easier to select the most important ones. But, of course, other categorization schemes may work as well.

From the results of the initial step, an average expectation rating per item can be calculated (see below). This score can be used to visualize the differences in importance as a word cloud. The font size relates to the expectation rating, so items in a bigger font are more important than items in a smaller font. An example from a study described in Oelfin and Schrepp (2021) is shown in Figure 8.5.

There are possible variations to this procedure. First, if the UX researchers have a good knowledge of the application that should be evaluated or already have other data that show what the adequate adjectives are, then they can define them directly. The same is true if a special research goal is defined, i.e., the application should be evaluated concerning special and pre-defined UX aspects. In that case, the subjective expectation of users is irrelevant, and again the adjectives can be directly determined.

For the evaluation phase, a UEEE questionnaire can look like:

The product is…
valuable
does not apply at all o o o o o o o fully applies
predictable
does not apply at all o o o o o o o fully applies

…

Thus, the selected items are simply shown below each other in a classical UX questionnaire which can be distributed, for example, as an online survey. The item format is similar to the Product Reaction Cards (Benedek & Miner, 2002), a set of 118 adjectives that describe positive UX aspects.

If possible, all items from the first part of the study should also appear in the questionnaire. But if the questionnaire gets too long and there is a danger that participants will not fill it out because of a too high effort, it is possible to define a cut-point concerning the importance rating and include only the most important adjectives in the questionnaire. In fact, researchers are mainly interested in finding out if a product fulfils the expectations concerning the UX properties most important for the target group. Thus, by dropping the least important items not much information is lost.

The results of the importance ratings and the questionnaire can be presented in an easy-to-understand graphic (see Oelfin & Schrepp, 2021 for details). We just describe the calculation briefly. Assume that A is one of the adjectives used in the study and that n participants have provided expectation ratings. Let a_k be the rating of participant k. If the participant classified A as important (see Figure 8.4) then $a_k = 1$, otherwise $a_k = -1$. Then we define the expectation score of A by:

$$\text{Expectation}(A) = \frac{1}{n}\sum_{k=1}^{n} a_k$$

Thus, the expectation score ranges between -1 and 1. Similarly we calculate an evaluation score from the results of the questionnaire. The ratings in the questionnaire are scaled from −3 (does not apply at all) to +3 (fully applies). Assume that m participants rated the corresponding item (the participants who rate their expectations are usually not the same as the participants who rate how well the adjectives describe the product) and that b_k is the rating of participant k. Then we calculate the evaluation score by:

$$\text{Evaluation}(A) = \frac{1}{m * 3}\sum_{k=1}^{m} b_k$$

Again, this is a score between −1 and +1. Thus, the results can be shown in a simple 2 × 2 grid. Figure 8.6 shows such a comparison from a study with a real product (Oelfin & Schrepp, 2021).

Such a comparison clearly shows which expectations are met and where there is a mismatch between user's expectations and the actual experience with the product. For example, users have a high expectation concerning security and stability and these expectations are obviously met. For efficiency (adjectives *efficient*, *fast*, *time-saving*) or understandability (adjectives *simple*, *easy to learn*, *intuitive*, *logical*) a high expectation conflicts with a low experience. This display allows relatively easy to detect potential problem areas in a design and to decide which areas should be investigated with other methods, for example by expert reviews or a dedicated usability test.

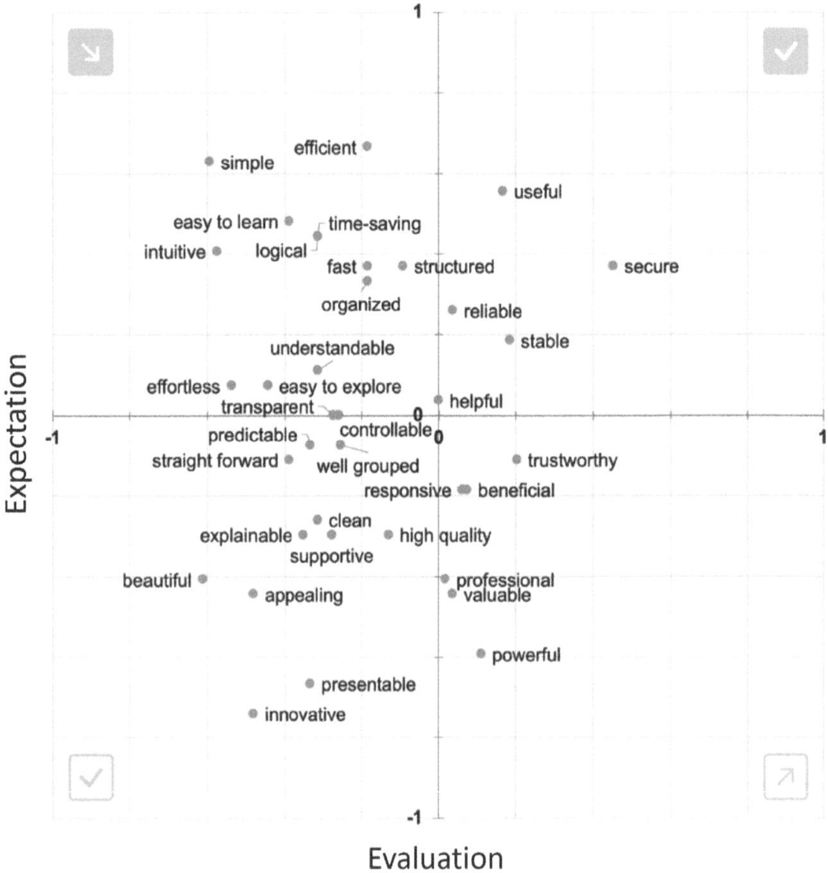

FIGURE 8.6 Comparison of expected and perceived UX qualities for a product.

8.5 ADVANTAGES AND CHALLENGES FOR MODULAR APPROACHES

Both UEQ+ and UEEE are modular approaches that allow to create UX questionnaires tailored for special usage scenarios. Thus, the main advantage of both methods is their flexibility. They can be used to define questionnaires that measure exactly those aspects of UX that are most relevant for the underlying research question.

To determine the scales (UEQ+) or items (UEEE) that should be used in a study both methods require knowledge about the importance of certain UX aspects for the overall UX impression of users. This information can be easily generated by a simple pre-study. But in fact this is not even required in most cases, since this type of knowledge is usually generated as part of user research activities that involve potential end users.

Typical UX research projects collect user requirements and use cases. Potential users are interviewed about their needs, typical tasks, pain points with a previous version or competitor products and their expectations toward the final product. Implicitly or explicitly these data always contain information about which UX aspects are most important for the users. In addition, if users are involved it is not much additional effort to use card sorting or another method to determine such information more formally. A detailed information about the importance of UX quality aspects is usually also quite helpful to handle conflicting requirements in the design phase. Assume, for example, that the design team discusses to realize a use case by a guided procedure (a user interface that guide the user step by step through a process). If it is known that intuitive use is quite important while efficiency does not really play a role (e.g., since the task is done quite infrequently), this would clearly speak for such a solution. If, on the other hand, efficiency is the clearly most important aspect (e.g., if the application is used many times during a typical workday), then using such a slow and inflexible design would obviously be a bad idea.

The UEQ+ is still relatively close to classical UX questionnaires. It contains a number of scales that measure distinct UX quality aspects. The single scales are carefully constructed, and their quality (reliability and validity) is checked. In addition, the correct interpretation of these scales is described in accompanying material, for example the UEQ+ handbook or on the UEQ+ web page (https://ueqplus.ueq-research.org).

The UEQ+ framework can be easily enhanced by adding new scales in the predefined scale format. In fact, the number of scales has grown since the first publication of the UEQ+, and many independent researchers have contributed scales for new usage scenarios that can now be used by other researchers. Thus, the UEQ+ is a community approach, and UX researchers are invited to share their results with others. Given the expected growth of different new application scenarios and new UI technologies, this is a promising method to improve the efficiency of UX methods by independently developing and sharing UX scales.

However, this flexibility also comes with a price. A questionnaire derived from a selection of UEQ+ scales can be used to compare different products or product versions. For this use case, there are no restrictions. Of course, products that are compared must be measured by the same scales. A comparison of results obtained from different combinations of scales makes semantically not much sense, since simply different things are measured in this case.

But concerning the interpretation of a single measurement, the UEQ+ is not as powerful as a standardized questionnaire, for example the UEQ. The reason is simply that the provided UEQ+ benchmark is not of the same quality as for example the UEQ or SUS benchmark. Usually, benchmarks contain results from a huge number of products. Thus, a comparison of a result to the benchmark can be used to determine how many of the reference products score better or worse per scale or in general. For a modular approach like the UEQ+, it will not be possible to provide such large benchmarks for all scales. In each study, different scales are used, and they are not all used with the same frequency. There are more general scales, for example, *Efficiency*, that are important for many application scenarios. But many of the UEQ+ scales are either limited to certain types of products, for example, *Haptic* (the perception of the user when touching the product) is more or less only used for household appliances. Thus, the benchmark clearly can be improved over time, but it will never reach the quality of a benchmark for a questionnaire with a fixed scale structure.

A similar problem exists for the materials that come together with the UEQ+. It is important that users can fill out a UX questionnaire in their natural language. Thus, translations are important. But since the number of available UEQ+ scales expands over time, existing translations must be updated when new scales appear. This makes it nearly impossible to offer at each point in time translations for all scales.

The UEEE method leaves the area of classical UX questionnaires. For maximum flexibility, several established principles of UX measurement are sacrificed. The method allows to create a UX questionnaire that is completely tailored to the use case. UX researchers simply select the items that seem to be most important for their users or that are of special interest to their project situation. Thus, a first principal question is what such a questionnaire measures. In fact, it simply measures what the researcher has defined as the scope. This makes it difficult for scientific use cases where it must be clearly communicated what the concepts underlying the data mean. However, science is not the intended use case for this method.

The UEEE is developed to be used in industrial design projects. It is more a structured way of user research than a scientific method to measure UX. The artefacts created, for example the word cloud to visualize the importance of adjectives, or the 2 × 2 matrix that visualizes the importance compared to the perception, can be easily communicated to stakeholders in projects and allow to justify design decisions based on the collected data. In addition, the method works with limited amounts of data. It is not required to have perfectly accurate measurements, but the results help to make conclusions if the design goals are reached. The adjectives in the right upper corner indicate UX qualities that are seen as important and that are sufficiently realized in the solution. The adjectives in the upper left corner indicate UX qualities that are important but are not realized in the expected quality in the perception of the participants.

In practical design projects, there is always a tradeoff between available time and resources and the quality of the information that can be collected and then used as a basis for the design or redesign of an application. There is always the danger that data-driven methods are ignored, since they are seen as too much effort. UEEE tries to reduce the required effort and also the required amount of data that must be collected to a minimum to allow data-driven UX evaluation in project that have few resources.

Both UEQ+ and UEEE allow UX researchers to set up UX evaluations fitting their research questions with low effort. Of course, they also have some limitations compared to classical UX questionnaires but are flexible and efficient to use. Thus, these methods or further developments in this direction will most likely be a part of the tool set of UX professionals in Ergonomics 5.0.

REFERENCES

Adel, A. (2022). Future of industry 5.0 in society: Human-centric solutions, challenges and prospective research areas. *Journal of Cloud Computing: Advances, Systems and Applications*, 11(1), p. 40. https://doi.org/10.1186/s13677-022-00314-5.

Bangor, A.; Kortum, P. T. & Miller, J. T. (2009). Determining what individual SUS scores mean: Adding an adjective rating scale. *Journal of Usability Studies*, 4(3), pp. 114–123.

Benedek, J. & Miner, T. (2002). Measuring desirability: New methods for evaluating desirability in a usability lab setting. Usability Professionals Association, 2002 Conference Proceedings.

Boos, B. & Brau, H. (2017). Erweiterung des UEQ um die Dimensionen Akustik und Haptik. In: Hess, H. F. (Ed.), *Mensch und Computer 2017– Usability Professionals*, 10–13 September 2017, Regensburg. https://dl.gi.de/items/c4ffa520-2718-444d-9117-3ae26200ce5f.

Brooke, J. (1996). SUS - A quick and dirty usability scale. *Usability Evaluation in Industry*, 189(194), pp. 4–7.

Brooke, J. (2013). SUS - A retrospective. *Journal of Usability Studies*, 8(2), pp. 29–40.

Coutinho, M.; Marques, L.; Santos, A.; Dahia, M.; França, C. & de Souza Santos, R. (2024). The role of generative AI in software development productivity: A pilot case study. In Proceedings of the 1st ACM International Conference on AI-Powered Software (pp. 131–138).

Cronbach, L. J. (1951). Coefficient alpha and the internal structure of tests. *Psychometrika*, 16, pp. 297–334.

Cronbach, L. J. & Meehl, P.E. (1955). Construct validity in psychological tests. *Psychological Bulletin*, 52 (4), pp. 281–302.

Deguchi, A.; Hirai, C.; Matsuoka, H.; Nakano, T.; Oshima, K.; Tai, M. & Tani, S. (2020). What is society 5.0? Society, 5, pp.1–24.

Finstadt, K. (2010). The usability metric for user experience. *Interacting with Computers*, 22(5), pp. 323–327.

Fukuyama, M. (2018). Society 5.0: Aiming for a new human-centered society. *Japan Spotlight*, 27(5), pp.47–50.

Graser, S.; Snimshchikova, A.; Böhm, S. & Schrepp, M. (2024). Enhancing UX research activities using GenAI – Potential applications and challenges. In: Böhm, S. & Berntzen, L. (Eds.), *The Seventeenth International Conference on Advances in Human oriented and Personalized Mechanisms, Technologies, and Services*, CENTRIC 2024, 29.9.2024–3. 10.2024, Venice, Italy (pp. 1–13).

Hinderks, A. (2016). Modifikation des User Experience Questionnaire (UEQ) zur Verbesserung der Reliabilität und Validität. https://doi.org/10.13140/RG.2.2.31619.50722.

Hinderks, A.; Schrepp, M. & Thomaschewski, J. (2014). Evaluation des UEQ in einfacher Sprache mit Schülern. In: Koch, M.; Butz, A. & Schlichter, J. (Eds.), *Mensch & Computer* (pp. 323–326). Oldenbourg: De Gruyter.

Hinderks, A., Schrepp, M., Mayo, F. J. D., Escalona, M. J. & Thomaschewski, J. (2019). Developing a UX KPI based on the user experience questionnaire. *Computer Standards & Interfaces*, 65, pp. 38–44.

Holden, R. B. (2010). Face validity. In: Weiner, I. B. & Craighead, W. E. (Eds.), *The Corsini Encyclopedia of Psychology* (4th ed., pp. 637–638). Hoboken, NJ: Wiley.

Hurtienne, J. & Naumann, A. (2010). QUESI—A questionnaire for measuring the subjective consequences of intuitive use. *Interdisciplinary College*, 2010, p. 536.

International Organization for Standardization. (2019). Ergonomics of human-system interaction -- Part 210: Human-centred design for interactive systems (ISO Standard No. 9241-210).

Klein, A.; Hinderks, A.; Schrepp, M. & Thomaschewski, J. (2020). Construction of UEQ+ scales for voice quality: Measuring user experience quality of voice interaction. *Mensch und Computer 2020- Tagungsband*. New York: ACM, pp. 1–5. MCI: Short Paper (Poster). Magdeburg. 6.–9. September 2020. https://doi.org/10.1145/3404983.3410003.

Laugwitz, B.; Schrepp, M. & Held, T. (2008). Construction and evaluation of a user experience questionnaire. In: Holzinger, A. (Ed.), *USAB 2008, LNCS 5298*, pp. 63–76. Springer-Verlag Berlin Heidelberg.

Lewis, J.R. & Sauro, J. (2018). Item benchmarks for the System Usability Scale. *Journal of Usability Studies*, 13(3), pp. 158–167.

Lindemann, M. & Weber, H. (2023). UX in Zahlen: Der User Experience Questionnaire for Medical Devices. *Usability Professionals 23*. Gesellschaft für Informatik e.V. Rapperswil. 3–6. September 2023. https://dl.gi.de/items/b4b4ba26-a69a-4b99-9bcb-ea16e622d637.

Loiacono, E. T.; Watson, R. T. & Goodhue, D. L. (2007). WebQual: An instrument for consumer evaluation of web sites. *International Journal of Electronic Commerce*, 11(3), pp. 51–87.

Meiners, A. L.; Schrepp, M.; Hinderks, A. & Thomaschewski, J. (2023). A benchmark for the UEQ+ framework: Construction of a simple tool to quickly interpret UEQ+ KPIs. *International Journal of Interactive Multimedia and Artificial Intelligence*. https://doi.org/10.9781/ijimai.2023.05.003.

Mortazavi, E.; Doyon-Poulin, P.; Imbeau, D. & Robert, J. M. (2023). Development and validation of four social scales for the UX evaluation of interactive products. *International Journal of Human–Computer Interaction*, 40(20), pp. 6608–6621.

Moshagen, M. & Thielsch, M. T. (2010). Facets of visual aesthetics. *International Journal of Human-Computer Studies*, 68(10), pp. 689–709.

Nahavandi, S. (2019). Industry 5.0—A human-centric solution. *Sustainability* 2019, 11, 4371. https://doi.org/doi:10.3390/su11164371.

Oelfin, D. & Schrepp, M. (2021). Evaluation of user expectations and anticipated UX. In: Mensch und Computer 2021, Workshopband, Workshop on "Quantitative Messung von User Experience." https://dl.gi.de/items/b4b4ba26-a69a-4b99-9bcb-ea16e622d637.

Ozok, A. A. & Salvendy, G. (2001). How consistent is your web design? *Behaviour & Information Technology*, 20(6), pp. 433–447.

Polkosky, M. D. & Lewis, J. R. (2003). Expanding the MOS: Development and psychometric evaluation of the MOS-R and MOS-X. *International Journal of Speech Technology*, 6(2), pp. 161–182.

Santos, M. E. C.; Polvi, J.; Taketomi, T.; Yamamoto, G.; Sandor, C. & Kato, H. (2015). Toward standard usability questionnaires for handheld augmented reality. *IEEE Computer Graphics and Applications*, 35(5), pp. 66–75.

Sauro, J. (2015). SUPR-Q: A comprehensive measure of the quality of the website user experience. *Journal of Usability Studies*, 10(2), pp. 68–86.

Sauvola, J.; Tarkoma, S.; Klemettinen, M.; Riekki, J. & Doermann, D. (2024). Future of software development with generative AI. *Automated Software Engineering*, 31(1), 26.

Schrepp, M. (2020). On the usage of Cronbach's Alpha to measure reliability of UX scales. *Journal of Usability Studies*, 15(4), pp. 247–258.

Schrepp, M. (2021). User Experience Questionnaires: How to use questionnaires to measure the user experience of your products? *KDP*, ISBN-13: 979–8736459766.

Schrepp, M. (2024). Designing and analyzing questionnaires and surveys. In: Stephanidis, C. & Salvendy, G. (Eds.), *Human Computer Interaction: User Experience Methods and Tools*, pp. 121-169. Boca Raton, FL: CRC Press.

Schrepp, M.; Hinderks, A. & Thomaschewski, J. (2017a). Construction of a benchmark for the User Experience Questionnaire (UEQ). *International Journal of Interactive Multimedia and Artificial Intelligence*, 4(4), pp. 40–44.

Schrepp, M.; Hinderks, A. & Thomaschewski, J. (2017b). Die UX KPI - Wunsch und Wirklichkeit. In: Hess, S. & Fischer, H. (Hrsg.), *Mensch und Computer 2017- Usability Professionals. Regensburg: Gesellschaft für Informatik e.V.* (pp. 117–125). Magdeburg: German UPA e.V.

Schrepp, M., Kollmorgen, J., Meiners, A. L., Hinderks, A., Winter, D., Santoso, H. B. & Thomaschewski, J. (2023a). On the importance of UX quality aspects for different product categories. *International Journal of Interactive Multimedia and Artificial Intelligence*, 8(2), pp. 232–246. https://doi.org/10.9781/ijimai.2023.03.001.

Schrepp, M., Kollmorgen, J. & Thomaschewski, J. (2023b). Impact of item polarity on the scales of the User Experience Questionnaire (UEQ). In: *Proceedings of the 19th International Conference on Web Information Systems and Technologies - WEBIST*; ISBN 978-989-758-672-9; ISSN 2184-3252, SciTePress (pp. 15–25). https://doi.org/10.5220/0012159900003584.

Schrepp, M.; Sandkühler, H. & Thomaschewski, J. (2021). How to create short forms of UEQ+ based questionnaires? In: Wienrich, C., Wintersberger, P. & Weyers, B. (Hrsg.), *Mensch und Computer 2021- Workshopband.* Bonn: Gesellscha.t für Informatik e.V. https://doi.org/10.18420/muc2021-mci-ws01-230.

Schrepp, M. & Thomaschewski, J. (2019). Design and validation of a framework for the creation of user experience questionnaires. *International Journal of Interactive Multimedia and Artificial Intelligence*, 5(7), pp. 88–95.

Part II

Implementations

While the first part of this book provided the theoretical and conceptual scaffolding of Ergonomics 5.0, the second part—Implementations—moves into the concrete application of these principles across a variety of real-world contexts and industries. The chapters collected here illustrate how the core pillars of digital intelligence, sustainability, and resilience are operationalized in diverse scenarios, reflecting both the challenges and transformative potential of ergonomics in the 5.0 era.

Collectively, the chapters in this section serve as a testament to the versatility, relevance, and urgency of Ergonomics 5.0. They demonstrate how theoretical principles translate into actionable solutions that not only improve productivity and safety but also support the development of equitable, adaptive, and human-centered work environments. These implementations underscore the discipline's expanded scope and the need for ergonomists to be proactive agents of innovation in a rapidly changing world.

DOI: 10.1201/9781003559399-10

9 Understanding Accidents and Sick Leave from Work in Light of Technological Developments in the Automotive Industry

Maria Victoria Cabrera Aguilera, Nilo Antonio de Souza Sampaio, Robson José de Moura Barbosa and Bernardo Bastos da Fonseca

9.1 INTRODUCTION

The automotive sector is recognized for its role as a manufacturing industry that impacts the economy through job creation, the country's gross domestic product with investments in innovation, and production chains with suppliers of parts and services (Beier et al., 2022). Its companies are recognized for their importance as transformation industries for carrying projects through from the planning to the sales of their products, besides their contributions to technological innovations, linkages with the segment of inputs, components, parts, and services (Barbosa et al., 2020; Doner, Noble and Ravenhill, 2021). Furthermore, the sector has economic relevance, with the generation of income, taxes, exports, and skilled jobs, and each of these elements is fundamental to the sustainable development plans of organizations in the sector (Araujo et al., 2021; Silva et al., 2021).

In Brazil, the sector has approximately 1.2 million direct and indirect contracts, as it has 26 manufacturers and 491 parts manufacturers. In 2022, the country was ranked as the eighth largest global producer of vehicles and the sixth largest domestic market, according to the 2024 Brazilian Automotive Industry Yearbook (ANFAVEA, 2024). In 2019, it represented 2.5% of the Gross Domestic Product (GDP) and 20% of the GDP of the manufacturing industry. In 2024, companies of the sector announced investments on the order of R$95 billion until 2032 in Brazil for new factories and expansion with new products.

 DOI: 10.1201/9781003559399-11

The sector's production activities are workforce-intensive, despite the fact that they carry out automated production stages. The automotive industry requires workers to perform repetitive assembly tasks and use excessive force (Lee and Jung, 2020; Sales et al., 2022). There are manual tasks that can be automated; however, many others present difficulties because they require human precision, skills, decision-making capacity, flexibility, and movement capacity (De Looze et al., 2015). These working conditions affect the health and occupational safety of workers (Silva et al., 2023).

In assembly lines, cognitive and physical factors often impose high demands on human performance, which results in errors, quality deficiencies, and other failures related to vehicle assembly (Falck et al., 2014; Stoyneva et al., 2015; Cheng, Li and Chen, 2019). Conti (1996) points out quality problems caused by workers who replaced those who had a sick leave and did not learn the task previously, because there is a demand for the job to be filled by a specialized worker (Mateo et al., 2014). In addition, it affects the activities of managers in management decisions (Blumenfeld and Inman, 2009) and results in a loss of productivity in local work, with financial repercussions for the company (Ramaswamy and Mosher, 2017).

These assembly lines and their working characteristics, combined with organizational factors, result in working conditions that allow for cases of sick leave and work accidents in the automotive industry. Thus, ergonomics 5.0 seeks to ensure that work on assembly lines considers the principles of human-centered manufacturing (Trstenjak et al., 2025) in the interaction between workers and the various elements of the work system in an efficient, healthy, safe, and sustainable way (Santos, 2024).

Official data from government agencies of sick leave and accidents at work show that between 2012 and 2021, Brazil spent R$23.4 billion on sick leave benefits, recorded a total of 6.2 million accidents, and 461.5 million workdays lost in this period, according to the National Institute of Social Security (Brasil, 2023a) and the Ministry of Labor and Employment (Brasil, 2023b). The data is related to workers with formal employment contracts and does not include information on informal workers.

On the one hand, the automotive sector is recognized as a processing industry with a decisive role in the local, regional, and national economy due to technological development, innovation, and job creation. On the other hand, it is directly related to sick leave, which entails complex relationships related to health, safety, economic, and sustainable development, as they imply human and economic losses that are difficult to translate into monetary terms for the country. These losses distance us from decent work and economic growth, sustainable development goals (SDG-8), the capacity to reduce, prevent, and manage health risks (SDG-3) and inclusive industrialization with a quality, reliable, sustainable, and resilient infrastructure to support economic development and human well-being (SDG-9), objectives of 2030 Agenda from the United Nations.

9.2 OBJECTIVE

In this regard, Ergonomics contributes to achieving high levels of productivity that consider the various risks from line balancing, organizational aspects, human factors, and work environments combined with technological modernization and innovation to achieve a wide diversity of full employment and decent work. Reduce workforce-intensive activities that demand human efforts with physical and mental overload, promoting the well-being of citizens in safe and protected work environments, as well as contributing to economic and social development at regional and national levels. This chapter presents the accident rates in the automotive sector regarding Occupational Health and Safety, focusing on the automotive hub of Rio de Janeiro, which comprises four cities: Resende, Porto Real, Quatis, and Itatiaia, considered the second largest in number of companies, the sixth in employees and the fourth in production capacity. In this sense, this chapter points out the need to understand how ergonomics and new technologies contribute positively to reducing accident and sick leave rates in the automotive industry toward achieving the targets of the Sustainable Development Goals.

9.3 METHODOLOGY

This chapter presents official data from the government agencies responsible for recording social security leave, type B91, according to Brazilian legislation (Brasil, 1991), and Accident rates in the automotive sector. The data was obtained from the National Social Security Institute (INSS) and the Ministry of Labor and Employment (MTE), focusing on the automotive hub of Rio de Janeiro. This data is gathered and available on SmartLab – Digital Observatory of Occupatonal Health and Safety, an official tool reinforced by the International Labor Organization (ILO) and the Brazilian Public Prosecutor's Office, which gathers information from the INSS Reports and Yearbooks and the Ministry of Labor and Employment's Annual Social Information Report (RAIS), providing statistics and indicators in an integrated and synergistic way (Brasil, 2023c).

The study was based on official data made available in the databases of the aforementioned bodies, between the years 2012 and 2022, to obtain the historical series with absolute numbers of sick leaves of the population with regular employment, duration of sick leaves in days, social security expenditure related to Social Security Benefits – Sick Pay for accidents at work granted each year, sick leaves according to the chapters of the International Classification of Diseases (ICD); sick leaves according to ICD codes; and occupations with the highest number of sick leaves according to the Brazilian Classification of Occupations (Brasil, 2002).

About accidents at work, the study used data on accidents with Working Accident Communication (WAC), communication and official registration by Decree No. 3048 (Civil House, 1999). The data were historical series, absolute numbers of WACs, most frequent occupations cited in WACs, causative agents most cited in WACs, most frequent injuries cited in WACs, and most affected body parts according to WACs.

The economic activities related to the automotive industry, focusing on the automotive hub of Rio de Janeiro, portrayed in the chapter are the manufacture of cars, vans, and utility vehicles, and the manufacture of parts and accessories for motor vehicles. These activities are the main ones in the sector in terms of final product, economic relevance, and number of jobs generated in the automotive hub of Rio de Janeiro.

9.4 RESULTS AND DISCUSSION – TIME OFF WORK

Figure 9.1 shows the time series in absolute numbers of sick leave where social security benefits were granted for the formally employed population in Brazil and in the automotive hub of Rio de Janeiro, where all economic activities are considered, between 2000 and 2022. The figures show similarities between Brazil and the automotive hub in the period from 2004 to 2014 when there were higher rates of sick leave among the working population. In the following years, there was an increase of more than 60%, possibly related to the number of people formally employed, despite the high number of workers who left their jobs in the goods and services transformation sector. The Brazilian Institute of Geography and Statistics' National Household Sample Survey Panel (IBGE, 2022) shows that in 2014 there was a drop in the formal hiring of workers, with stabilization until 2019. Sterud and Johannessen (2018) cite studies in which socioeconomic characteristics and labor market conditions are determining factors in the variation of workers' illness rates over time.

The decrease in rates in 2020, both in Brazil and in the automotive hub, occurred at the beginning of the COVID-19 pandemic, with the paralysis of several economic sectors for months and social isolation. That year, the INSS brought forward the sickness benefit for applicants, as provided for in Law No. 13,982, of April 2, 2020, in its Article 4 and regulated by Decree No. 10,413, of July 2, 2020 (Brasil, 2020).

Data was also examined on the duration of sick leave in days and the Social Security Expenditure related to Social Security Benefits – Sick Pay for Accidents at Work granted each year, as well as the accumulated in the period and the rate per inhabitant. The sum of the duration of each benefit granted makes it possible to identify the accumulated number of working days lost, those in which the worker did not show up at the company to carry out their work activities. The following formula

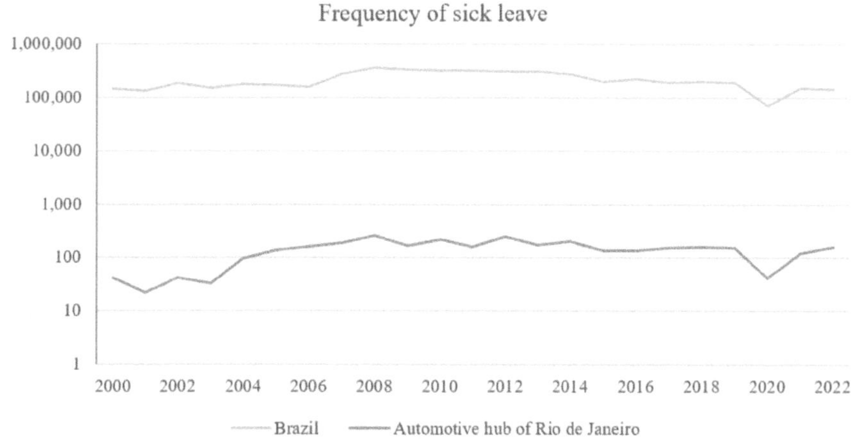

FIGURE 9.1 Historical series between 2000 and 2022 in the granting of social security benefits for all economic activities – Brazil vs. automotive hub of Rio de Janeiro.

Source: Adapted from the Ministry of Labor and Employment/Annual Social Information Report (Brasil, 2023b).

TABLE 9.1
Duration of sick leave in days | Occupational accident sickness benefit

Year	Automotive hub of Rio de Janeiro	Brazil
2012	67,4K	73,2M
2013	55,5K	71,6M
2014	68,4K	65,3M
2015	35,3K	47,9M
2016	43,2K	47,5M
2017	39,9K	35,1M
2018	46,3K	34,4M
2019	33,8K	31,4M
2020	14,0K	14,7M
2021	22,7K	22,5M
2022	20,0K	17,9M
Total Accumulated (days)	446,5K	461,5M
Inhabitants	194,6 K	203,1M
Rates – TA/H dlw*	2,29	2,27

Source: Adapted from the Ministry of Labor and Employment/Annual Social Information Report (Brazil, 2023b) and the National Social Security Institute (INSS, 2023a).

*Rate = Days lost from work/inhabitant.

was used in Table 9.1 to obtain the rate of working days lost per inhabitant using the metric of comparison with different units.

$$(\text{TA/I})\ \text{dlw} = \text{Rate}$$

TA/I dlw = rate days lost from work / inhabitant.

The social security expenditure rate per inhabitant was calculated with the formula below in Table 9.2 in order to compare variant units:

$$(\text{TA/I})\ \text{exp} = \text{Rate}$$

TA/I exp = rate social security expenditure / Inhabitant.

The duration of sick leave (Table 9.1) shows the number of days lost from work in Brazil and the automotive hub per year, between 2012 and 2022. When considering the population of the automotive hub, which is 194,600 inhabitants according to the

TABLE 9.2
Social security expenditure on sickness benefits for accidents at work

Year	Automotive hub of Rio de Janeiro	Brazil
2012	2,5M	2,2B
2013	2,9M	2,5B
2014	3,3M	2,6B
2015	3,7M	2,7B
2016	3,9M	3,0B
2017	4,1M	2,6B
2018	4,0M	2,3B
2019	3,6M	2,0B
2020	3,7M	1,7B
2021	3,6M	1,8B
Total Accumulated R$	35,3M	23,4B
Inhabitants	194,6K	203,1M
Rate – TA/I exp*	181,39	115,21

Source: Adapted from the Ministry of Labor and Employment/Annual Social Information Report (Brazil, 2023b) and the National Social Security Institute (INSS, 2023a).
*Rate = Social Security Expenditure/Inhabitant.

2022 Demographic Census (IBGE, 2022), the result is a rate of 2.29 days lost from work per inhabitant, higher than in Brazil.

The Social Security Expenditure on sick pay for accidents at work (Table 9.2) shows the amount paid out each year, as well as the accumulated amount over the period from 2012 to 2021 according to data availability. Social security spending on this type of benefit amounts to 35.3 million in the automotive hub and 23.4 billion in Brazil. These figures linked to the number of inhabitants show that the automotive hub spends more per inhabitant, R$181.39, while in Brazil it is R$115.21.

The data points to a scenario of negative impacts with losses in various dimensions, such as human, economic, and productivity losses for the local, regional, and national economy, since the work factors that led to the sick leave are related to the subsequent illness processes. Furthermore, when workers return to work, it must be gradual. During this period, the activities assigned to the worker are restricted to less intense tasks until they fully recover.

The profile of sick leave from the economic activities analyzed is shown in Tables 9.3 and 9.4, according to the chapters of the International Classification of Diseases – ICD, the ICD code and the occupation of the workers employed for the period from 2012 to 2022.

The most frequent reasons for sick leave in Brazil and in the automotive hub are the Osteomuscular and Connective Tissue ICD, Fractures, Trauma, and Injuries, which appear in the same order of classification.

TABLE 9.3
Profile of sick leave according to ICD, ICD, code and occupation for the economic activity of manufacturing cars, vans, and utility vehicles from 2012 to 2022

	Automotive hub of Rio de Janeiro		Brazil	
ICD	Osteomuscular and connective tissue	48%	Osteomuscular and connective tissue	70%
	Luxations	24%	Traumas	10%
	Traumas	12%	Nervous	5%
	Fractures	9%	Luxations	5%
	Injuries	3%	Traumas	3%
ICD code	M54- Dorsalgia	29%	M75- Shoulder lesions	41%
	M75- Shoulder lesions	19%	M54- Dorsalgia	9%
	S43- Dislocation, sprain, and strain of joints and ligaments of shoulder girdle	19%	G56- Mononeuropathies of upper limb	5%
	M77 – Other enthesopathies	10%	M51 – Other intervertebral disc disorders	5%
	S62- Fracture at wrist and hand level	5%	M65- Synovitis and tenosynovitis	3%
Occu-pation	Welder	47%	Vehicle assembler (assembly line)	50%
	Vehicle assembler (assembly line)	20%	Vehicle Painter (manufacturing)	7%
	Forklift operator	17%	Welder	4%
	Quality inspector	7%	Forklift operator	4%
	Automotive technician	3%	Quality inspector	3%

Source: Adapted from the Ministry of Labor and Employment/Annual Report of Social Information (Brazil, 2023b); Statistical Yearbook of Accidents at Work of the National Institute of Social Security (INSS, 2023a).

The context identified in Table 9.3 shows that work-related musculoskeletal disorders and fractures have a higher rate of sick leave, close to 70%, in the automotive industry and production lines (Zurada, 2012; Cohen and Shoval, 2020), because many workers are exposed to lifting manual loads, combined with high frequency and body postures with physical overload.

The profile of damage to health in the automotive industry, which accounts for reduced working capacity, is made up of after-effects resulting from accidents at work, especially mutilations and amputations of fingers and hands (Reyes et al., 2015), and lumbar and cervical spine injuries, as well as other musculoskeletal injuries and diseases (Pina, Stotz and Filho, 2018). The upper limbs and spine are the parts of the worker's body most affected by injuries that result in time off work. This condition is associated with work activities that require overloading fingers, hands, and wrists with fine movements (Aguilera et al., 2022).

TABLE 9.4
Profile of sick leave according to ICD, ICD code, and occupation for the economic activity of manufacturing parts and accessories for motor vehicles not previously specified from 2012 to 2022

	Automotive hub of Rio de Janeiro		Brazil	
ICD	Osteomuscular and connective	47%	Fractures	37%
	Fractures	21%	Osteomuscular and connective	32%
	Traumas	16%	Traumas	8%
	Injuries	6%	Injuries	5%
	Mental and behavioral	3%	Luxations	5%
ICD code	M75- Shoulder lesions	18%	S62- Fracture at wrist and hand level	20%
	M54- Dorsalgia	15%	M75- Shoulder lesions	15%
	S62- Fracture at wrist and hand level	13%	M54- Dorsalgia	5%
	M65- Sinovite e tenossinovite	9%	S92- Fracture of the foot, excluding the ankle	5%
	S66- Injury of muscles and tendons at wrist and hand level	7%	S61- Injury at wrist, hand and fingers	4%
Ocupa-tion	Machine operator	28%	Production line feeder	18%
	Machine seamstress in mass production	21%	Machine operator	9%
	Vehicle assembler (assembly line)	7%	Conventional machine tool operator	8%
	Production operator	6%	Welder	6%
	Upholsterer	4%	Press operator	5%

Source: Adapted from the Ministry of Labor and Employment/Annual Report of Social Information (Brazil, 2023b), Statistical Yearbook of Accidents at Work of the National Institute of Social Security (INSS, 2023a).

In Table 9.4, the Osteomuscular and Connective Tissue ICD chapter stands out and accounts for 70% of sick leaves in Brazil and 48% in the automotive hub, followed by the dislocations, trauma, and fractures chapters with values above 10%.

The rates of sick leave in the automotive hub of Rio de Janeiro identified in the economic activities analyzed are similar, with work-related musculoskeletal disorders standing out, with high rates (Zurada, 2012), associated with physical overload, which implies sustained postures with frequencies and loads beyond the natural limits of the human body.

According to the International Classification of Diseases code, the two main causes of sick leave are shoulder injuries (M75) and back pain (M54). Although there are variations in the rates of each ICD code between Brazil and the automotive hub, there is a predominance of sick leaves due to injuries and disorders of the upper limbs and spine, a pattern also identified in the economic activity of manufacturing parts and accessories for motor vehicles.

This scenario portrays the characteristics of the automobile industry, which involves repetitive manual labor, body postures with physical overload and lifting loads (Nordander et al., 2016), requiring the upper limbs to be held above the head and requiring trunk flexion and turns, which can cause injuries and musculoskeletal disorders in the lumbar region and upper limbs (Reyes et al., 2015).

Studies on the automotive industry in Brazil (Silva, Bernardo and Souza, 2016) point out that the body parts most affected by accidents in the operator's occupation are shoulders, wrists, and lower back, characteristics that strengthen the relationship between the most affected body parts and the main occupations. In addition, hybrid production systems in the automotive industry, where the space is shared by the operator and the machine, pose several technological challenges for the industry. However, safety in these systems is still the main issue, as employees are more likely to be exposed to risky situations, resulting in a greater number of accidents (Marvel, Falco and Marstio, 2015; Gopinath et al., 2021).

9.4.1 Accident at Work

Figure 9.2 shows the absolute numbers of occupational accident notifications considering the population with regular employment in Brazil and in the automotive hub of Rio de Janeiro between 2002 and 2022, according to data availability. Variations in occupational accident indicators in the country and the automotive hub are affected by economic cycles, with a close relationship to the Gross Domestic Product and periods of reduced economic activity with a decrease in indicators (Anyfantis, Boustras and Karageorgiou 2018).

The official notifications of accidents at work available from 2012 to 2022 show that the majority of WAC records are related to the occupation of vehicle assembler on the assembly line (Table 9.5) in the economic activity Manufacture of cars, vans and utility vehicles. This may be associated with a greater number of workers in this occupation

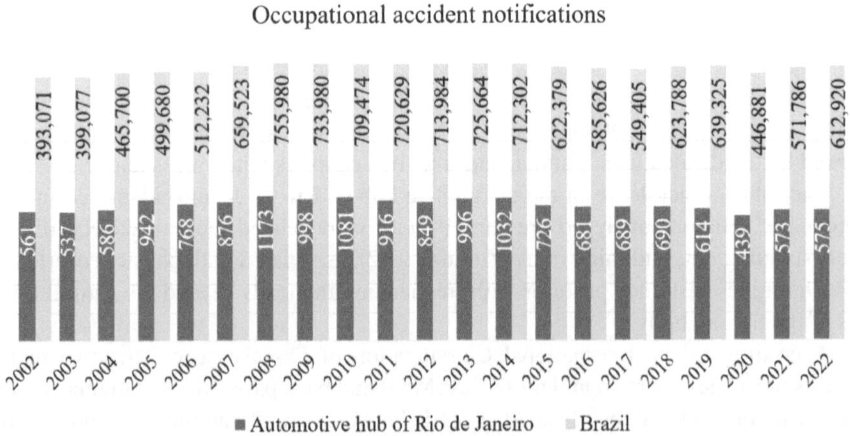

FIGURE 9.2 Historical series of occupational accidents in absolute numbers. Adapted from Ministry of Labor and Employment, and National Institute of Social Security (Brasil, 2023c).

TABLE 9.5
Occupations with the highest accident rates for the economic activity of manufacturing cars, vans, and utility vehicles from 2012 to 2022

Automotive hub of Rio de Janeiro		Brazil	
Vehicle assembler (assembly line)	35,9%	Vehicle assembler (assembly line)	35,3%
Welder	11,2%	Welder	6,0%
Forklift operator	5,3%	Production line feeder	5,5%
Vehicle Painter (manufacturing)	2,3%	Vehicle Painter (manufacturing)	5,3%
Mechanical technician (machines)	2,3%	Forklift operator	2,8%

Note: Adapted from Ministry of Labor and Employment (Brasil, 2023b).

TABLE 9.6
Occupations with the highest accident rates for the economic activity of manufacturing parts and accessories for motor vehicles not previously specified from 2012 to 2022

Automotive hub of Rio de Janeiro		Brazil	
Machine operator	30%	Production line feeder	21,4%
Vehicle assembler (assembly line)	24%	Machine operator	7,5%
Production line feeder	6%	Conventional machine tool operator	6,2%
Toolmaker	3%	Press operator	5,2%
Machine seamstress in mass production	3%	Welder	4,9%

Note: Adapted from Ministry of Labor and Employment (Brasil, 2023b).

compared to other occupations in the companies, as well as the exposure of these workers to a high demand for physical effort, with repetitive, monotonous, manual work, with load handling (Lee and Jung 2020), the imposition of penalizing postures, with overhead work, rotation and flexion of the trunk (Dalle Mura and Dini, 2019; Hermans et al. 1999), complexity of the process of choosing the operator in each assembly task (Zhu et al., 2008) and the industrial physical arrangement of the production line (Comper et al., 2017). These factors are in addition to organizational factors such as the fast pace of production, cycle time, long working hours, and production targets.

In the economic activity of Manufacture of parts and accessories for motor vehicles (Table 9.6), the occupations with the highest accident rates are also those that deal directly with the production process, production line feeders in the case of

TABLE 9.7
Groups of agents causing accidents for the economic activity of manufacturing cars, vans, and utility vehicles from 2012 to 2022

Automotive hub of Rio de Janeiro		Brazil	
Chemical agent	23,9%	Chemical agent	22,6%
Machines and equipment	22,7%	Machines and equipment	22,5%
Transport vehicles	10,2%	Physical effort	11,4%
Queda do mesmo nível	9,6%	Transport vehicles	9,6%
Hand tools	6,4%	Fall from the same level	9,1%

Note: Adapted from Ministry of Labor and Employment (Brasil, 2023b).

TABLE 9.8
Groups of agents causing accidents for the economic activity of manufacturing parts and accessories for motor vehicles not previously specified from 2012 to 2022

Automotive hub of Rio de Janeiro		Brazil	
Chemical agent	33%	Machines and equipment	28,7%
Machines and equipment	28%	Chemical agent	23,3%
Hand tools	8%	Physical effort	9,2%
Impact against person/ object	6%	Transport vehicles	7,2%
Furnishings	5%	Motorcycle	5,8%

Note: Adapted from Ministry of Labor and Employment (Brasil, 2023b).

Brazil. In the automotive hub of Rio de Janeiro, the occupations are machine operators and vehicle assemblers, which can also be associated with the characteristics of the production system, with collaborative work with machines, monotony, handling manual loads, and repetitive work.

The groups of agents causing accidents in the two economic activities studied between 2012 and 2022 were chemical agents and machinery and equipment, as can be seen in Tables 9.7 and 9.8. This is related to workers' exposure to welding dust, gases such as carbon monoxide, paint mists or vapors, which can be absorbed through respiration, ingestion, or contact (Blunt and Balchin, 2002). The second group of accident-causing agents corresponds to Collaborative Operations, where the workspace is shared with industrial robots, and this man-machine proximity increases the likelihood of situations that result in accidents.

The characteristics of the production line are associated with the most frequent injuries in accidents at work in both economic activities and in the same period being of the cut, laceration, blunt wound, or puncture type (Dalle Mura and Dini, 2019; Hermans et al., 1999) (Tables 9.9 and 9.10). The second most reported type was contusion or crushing, which is related to the intense handling of loads and materials, the use of machinery, and storage and stockpiling activities (Lee and Jung, 2020).

The occupations with the highest rates of accidents at work and the causative agents in both economic activities show that the extremities of these workers' upper limbs are the most used in work activities. This aspect is combined with the particularities of the automotive sector, mainly because the work still involves

TABLE 9.9

Most frequent injuries in accidents for the economic activity of manufacturing cars, vans, and utility vehicles from 2012 to 2022

Automotive hub of Rio de Janeiro		Brazil	
Cut, laceration, contused wound, puncture	36,1%	Cut, laceration, contused wound, puncture	28,7%
Bruising, crushing (skin surface I)	25,9%	Bruising, crushing (skin surface I)	22,2%
Distention, twisting	7,9%	Scorching, abrasion (superficial wound)	9,4%
Scorching, abrasion (superficial wound)	6,3%	Fracture	8,9%
Fracture	5,8%	Distention, twisting	6,7%

Note: Adapted from Ministry of Labor and Employment (Brasil, 2023b).

TABLE 9.10

Most frequent injuries in accidents for the economic activity of manufacturing parts and accessories for motor vehicles not previously specified from 2012 to 2022.

Automotive hub of Rio de Janeiro		Brazil	
Cut, laceration, contused wound, puncture	49%	Cut, laceration, contused wound, puncture	29,3%
Bruising, crushing (skin surface I)	22%	Bruising, crushing (skin surface I)	19,0%
Fracture	10%	Fracture	15,5%
Dislocation	5%	Immediate Injury, Nic	4,6%
Scorching, abrasion (superficial wound)	3%	Scorching, abrasion (superficial wound)	7,6%

Note: Adapted from Ministry of Labor and Employment (Brasil, 2023b).

TABLE 9.11
Body parts most affected in work accidents for the economic activity of manufacturing cars, vans, and utility vehicles from 2012 to 2022

Automotive hub of Rio de Janeiro		Brazil	
Finger	25,2%	Finger	24,2%
Hand (except wrist or fingers)	9,1%	Shoulder	8,2%
Head, NIC	7,2%	Hand (except wrist or fingers)	7,5%
Knee	6,3%	Head, NIC	6,0%
Ankle joint	3,0%	Knee	4,5%

Note: Adapted from Ministry of Labor and Employment (Brasil, 2023b).

TABLE 9.12
Body parts most affected in work accidents for the economic activity of manufacturing parts and accessories for motor vehicles not previously specified from 2012 to 2022

Automotive hub of Rio de Janeiro		Brazil	
Finger	42%	Finger	37,0%
Hand (except wrist or fingers)	13%	Hand (except wrist or fingers)	9,3%
Head, NIC	6%	Foot (except toes)	5,9%
Forearm (between wrist and elbow)	4%	Shoulder	3,8%
Ankle joint	3%	Forearm (between wrist and elbow)	3,7%

Note: Adapted from Ministry of Labor and Employment (Brasil, 2023b).

many manual and repetitive activities. Thus, fingers and hands are the parts of the workers' bodies with the highest rates of occupational accidents, as shown in Tables 9.11 and 9.12.

9.5 CONCLUSION

The automotive industry is an industrial and economic force in many countries. It contributes to the global economy and growth and development around the world. The industry is capital-intensive, drives innovation, and generates billions of dollars in investments and millions of jobs and livelihoods (ILO, 2020).

It is an essential sector for achieving the 2030 Agenda for Sustainable Development, in particular for achieving Goal 8 of the Sustainable Development Goal (SDG), which aims to promote sustained, inclusive and sustainable economic growth, full and productive employment and decent work for all (ILO, 2021).

However, the data shows that the working population of the automotive industry in Brazil, including automakers and suppliers, is falling ill, with economic repercussions both for the companies, due to production and quality problems, and for the country, due to social security costs.

There are work-related factors that have a major impact on the process of medium or long-term illness, which are often not recognized or associated with work, and which imply personal and economic losses that are complex to measure and include in risk indicators.

In Brazil, and the automotive hub of Rio de Janeiro, data on sick leave and accidents at work in the automotive industry point to the need for Ergonomics and its principles to be approached broadly and applied to the different workstations and their processes, both in the assembly companies and in the supplier companies. The scenario points to an unsustainable working environment, especially in social terms. On the other hand, Ergonomics can contribute to positive change, with high levels of productivity that take into account the various risks of line balancing, organizational aspects, human factors and working environments combined with technological modernization and innovation to achieve a wide range of full employment and decent work, as well as promoting the well-being of workers in safe and secure working environments and contributing to economic and social development at regional and national levels.

The automotive industry has for many years been a leader in manufacturing research and development, with investments in the development of new production processes and greener, safer vehicles with new technologies and materials. On the other hand, this industry will need to advance in the dimensions of decent and sustainable work to promote a just transition to a future of work that contributes to sustainable development in its economic, social, and environmental dimensions.

The automotive industry is going through a digital revolution unlike any other in its history. The current transformation process is the result of climate change and is marked by the introduction of new powertrains, such as electric and hydrogen, the development of digitalization and the desire of automakers to become mobility service providers. In addition, this industry is affected by technological innovations, which have an impact on the generation of new jobs and positions.

In addition to these challenges and the need for workers to adapt and improve their skills in the human-system interface, we need to consider the ongoing digitalization and automation in the industrial world, which promises unprecedented improvements in productivity, but more insecurity for workers, unless tripartite sustainability solutions can be developed and applied.

REFERENCES

Aguilera, M.V.C., Fonseca, B.B., Santos, M.S.E., Matias, N.T. and Sampaio, N.A.S. (2022). Usability evaluation of exoskeleton systems in automotive industry. In M.M. Soares, F. Rebelo and T.Z. Ahram (Orgs.), *Handbook of Usability and User Experience Research and Case Studies* (pp. 47–57). Boca Raton: CRC Press.

Anfavea. Associação Nacional dos Fabricantes de Veículos Automotores. (2024). Industry yearbook. Accessed 26 de april 2024, available at: https://anfavea.com.br/site/wp-content/uploads/2024/05/ANFAVEA-ANUARIO-DIGITAL-2024-NOVOATUALIZADOalta_compressed.pdf.

Anyfantis, I., Boustras, G. and Karageorgiou, A. (2018). Maintaining occupational safety and health levels during the financial crisis – A conceptual model. *Safety Science*, 106, pp. 246–254. https://doi.org/10.1016/j.ssci.2016.02.014.

Araujo, M.J.F.d., Araújo, M.V.F.d., Araujo Jr, A.H.d., Barros, J.G.M.d., Almeida, M.d.G.d., Fonseca, B.B.d., Reis, J.S.D.M., Barbosa, L.C.F.M., Santos, G. and Sampaio, N.A.D.S. (2021). Pollution credit certificates theory: An analysis on the quality of solid waste management in Brazil. *Quality Innovation Prosperity*, 25(3), pp. 1–17. https://doi.org/10.12776/qip.v25i3.1574.

Barbosa, L.C.F.M., Mathias, M.A.S., Santos, G.M. and De Oliveira, O.J. (2020). How the knowledge of the major researchers is forging the business strategy paths: Trends and forecasts from the state of the art. *Quality Innovation Prosperity*, 24(3), p. 1. https://doi.org/10.12776/qip.v24i3.1404.

Beier, G., Matthess, M., Shuttleworth, L., Guan, T., de Oliveira Pereira Grudzien, D.I., Xue, B., Pinheiro de Lima, E. and Chen, L. (2022). Implications of Industry 4.0 on industrial employment: A comparative survey from Brazilian, *Chinese, and German practitioners. Technology in Society*, 70, p. 102028. https://doi.org/10.1016/j.techsoc.2022.102028.

Blumenfeld, D.E. and Inman, R.R. (2009). Impact of absenteeism on assembly line quality and throughput. *Production and Operations Management*, 18, pp. 333–343. https://doi.org/10.1111/j.1937-5956.2009.01019.x.

Blunt, J. and Balchin, N.C. (2002). *Health and Safety in Welding and Allied Processes*. 5th ed. Cambridge: Woodhead Publishing.

Brasil. (1991, 24 de july). Lei nº 8.213, de 24 july de 1991. Dispõe sobre os Planos de Benefícios da Previdência Social e dá outras providências. Accessed 18 de february 2024, available at: https://www.planalto.gov.br/ccivil_03/leis/l8213compilado.htm.

Brasil. (2020, 2 de abril). Lei nº 13.982, de 2 de abril de 2020. Altera a Lei nº 8.742, de 7 de dezembro de 1993, para dispor sobre parâmetros adicionais de caracterização da situação de vulnerabilidade social para fins de elegibilidade ao benefício de prestação continuada (BPC), e estabelece medidas excepcionais de proteção social a serem adotadas durante o período de enfrentamento da emergência de saúde pública de importância internacional decorrente do coronavírus (Covid-19) responsável pelo surto de 2019, a que se refere a Lei nº 13.979, de 6 de fevereiro de 2020. Recuperado em 01 de março de 2024, de https://www.planalto.gov.br/ccivil_03/_Ato2019-2022/2020/Lei/L13982.htm.

Brasil. Ministério da Previdência Social. (2023a). INSS - Anuário Estatístico de Acidentes do Trabalho. Accessed 21 january 2024, available at: https://www.gov.br/previdencia/pt-br/assuntos/previdencia-social/saude-e-seguranca-do-trabalhador/acidente_trabalho_incapacidade.

Brasil. Ministério do Trabalho. (2023b). Relação Anual de Informações Sociais. Accessed 26 april de 2024, available at: https://www.gov.br/trabalho-e-emprego/pt-br/assuntos/estatisticas-trabalho/rais/rais-2022.

Brasil. Ministério Público do Trabalho. (2023c). Smartlab – Observatório Digital de Saúde e Segurança no Trabalho. Accessed 05 de march de 2024, available at: https://smartlabbr.org/sst.

Brasil. Portaria nº 397. (9 october 2002). Aprova a Classificação Brasileira de Ocupações - CBO/2002, para uso em todo território nacional e autoriza a sua publicação. Accessed 18 february 2024, available at: https://www.mtecbo.gov.br/cbosite/pages/legislacao.jsf.

Cheng, Y.-P., Li, C.-W. and Chen, Y.-C. (2019). Apply computer vision in GUI automation for industrial applications. *Mathematical Biosciences and Engineering*, 16(6), pp. 7526–7545. https://doi.org/10.3934/mbe.2019378.

Civil House. (1999). Decree nº. 3.048, *From* May 6, 1999. [online] BRASIL. Available at: https://www.planalto.gov.br/ccivil_03/decreto/d3048.htm.

Cohen, Y. and Shoval, S. (2020). A new cobot deployment strategy in manual assembly stations: Countering the impact of absenteeism. *IFAC-PapersOnLine*, 53(2), pp. 10275–10278. https://doi.org/10.1016/j.ifacol.2020.12.2760.

Comper, M.L.C., Dennerlein, J.T., Evangelista, G. dos S., Rodrigues da Silva, P. and Padula, R.S. (2017). Effectiveness of job rotation for preventing work-related musculoskeletal diseases: A cluster randomised controlled trial. *Occupational and Environmental Medicine*, 74(8), pp. 543–544. https://doi.org/10.1136/oemed-2016-104077.

Conti, R.F. (1996). Variable manning JIT: An innovative answer to team absenteeism. *Production & Inventory Management Journal*, 37(1), pp. 24–27. https://www.proquest.com/openview/c09191c8c0a3bb80f9c9270dcf03cccc/1?cbl=36911&pq-origsite=gscholar.

Dalle Mura, M. and Dini, G. (2019). Optimizing ergonomics in assembly lines: A multi objective genetic algorithm. *CIRP Journal of Manufacturing Science and Technology*, 27, pp. 31–45. https://doi.org/10.1016/j.cirpj.2019.08.004.

De Looze, M.P., Bosch, T., Krause, F., Stadler, K.S. and O'Sullivan, L.W. (2015). Exoskeletons for industrial application and their potential effects on physical work load. *Ergonomics*, 59(5), pp. 671–681. https://doi.org/10.1080/00140139.2015.1081988.

Doner, R.F., Noble, G.W. and Ravenhill, J. (2021). *The Political Economy of Automotive Industrialization in East Asia*. Oxford University Press. https://doi.org/10.1093/oso/9780197520253.001.0001.

Falck, A.C., Örtengren, R. and Rosenqvist, M. (2014). Assembly failures and action cost in relation to complexity level and assembly ergonomics in manual assembly (part 2). *International Journal of Industrial Ergonomics*, 44(3), pp. 455–459. https://doi.org/10.1016/j.ergon.2014.02.001.

Gopinath, V., Johansen, K., Derelöv, M., Gustafsson, Å. and Axelsson, S. (2021). Safe collaborative assembly on a continuously moving line with large industrial robots. *Robotics and Computer-Integrated Manufacturing*, 67. https://doi.org/10.1016/j.rcim.2020.102048.

Hermans, V., Hautekiet, M., Spaepen, A., Cobbaut, L. and De Clerq, J. (1999). Influence of material handling devices on the physical load during the end assembly of cars. *International Journal of Industrial Ergonomics*, 24(6), pp. 657–664. https://doi.org/10.1016/S0169-8141(98)00070-5.

IBGE (Instituto Brasileiro de Geografia e Estatística). (2022). Painel Pesquisa Nacional por Amostra de Domicílio. Accessed 28 January 2024, available at: https://painel.ibge.gov.br/pnadc/.

ILO. (2020). The future of work in the automotive industry: The need to invest in people's capabilities and decent and sustainable work, Issues paper for the Technical Meeting on the Future of Work in the Automotive Industry (Geneva, 15–19 February 2021), International Labour Office, Sectoral Policies Department, Geneva.

ILO, 2021. The future of work in the automotive industry, technical meeting on the future of work in the automotive industry (Geneva, 15–19 February 2021), Note on the proceedings, International Labour Office, Sectoral Policies Department, Geneva.

Lee, K. and Jung, M. (2020). Investigation of common insertion hand postures and directions and determination of the voluntarily maximal and preferred insertion forces in automotive assembly tasks. *Human Factors and Ergonomics in Manufacturing & Service Industries*, 30(2), pp. 93–102. https://doi.org/10.1002/hfm.20825.

Marvel, J.A., Falco, J. and Marstio, I. (2015). Characterizing task-based human-robot collaboration safety in manufacturing. *IEEE Transactions on Systems, Man, and Cybernetics: Systems*, 45(2), pp. 260–275. https://doi.org/10.1109/TSMC.2014.2337275.

Mateo, R., Tanco, M. and Santos, J. (2014). Less expert workers and customer complaints: Automotive case study. *Human Factors and Ergonomics in Manufacturing & Service Industries*, 24, pp. 444–453. https://doi.org/10.1002/hfm.20396.

Nordander, C., Hansson, G., Ohlsson, K., Arvidsson, I., Balogh, I., Strömberg, U., Rittner, R. and Skerfving, S. (2016). Exposure-response relationships for work-related neck and shoulder musculoskeletal disorders - Analyses of pooled uniform data sets. *Applied Ergonomics*, 55, pp. 70–84. https://doi.org/10.1016/j.apergo.2016.01.010.

Pina, J.A., Stotz, E.N. and Filho, J.M.J. (2018). Trabalhador "compatível" fratura exposta no processo de produção da indústria automobilística: Intensificação do trabalho e saúde em questão. *Cadernos de Saúde Pública*, 34(7). https://doi.org/10.1590/0102-311X00114017.

Ramaswamy, S.K. and Mosher, G.A. (2017). Using workers' compensation claims data to characterize occupational injuries in the commercial grain elevator industry. *Journal of Agricultural Safety and Health*, 23(3), pp. 203–217. https://doi.org/10.13031/jash.12196.

Reyes, R.M., Riva, J., Maldonado, A., Woocay, A. and Rodolfo, O. (2015). Association between human error and occupational accidents' contributing factors for hand injuries in the automotive manufacturing industry. *Procedia Manufacturing*, 3, pp. 6498–6504. https://doi.org/10.1016/j.promfg.2015.07.936.

Sales, J.P.d., Reis, J.S.d.M., Barros, J.G.M.d., Fonseca, B.B.d., Junior, A.H.d.A., Almeida, M.d.G.D.d., Barbosa, L.C.F.M., Santos, G. and Sampaio, N.A.d.S. (2022). Quality management in the contours of continuous product improvement. *International Journal for Quality Research*, 16(3), pp. 689–702. https://doi.org/10.24874/IJQR16.03-02.

Santos, E.F. (2024). O que é Ergonomia 5.0? Blog do Professor Ferro [online].

Silva, A.C.P., Pasini, V.D., Aguilera, M.V.C., Fonseca, B.B., Sampaio, N.A., Reis, J.S.M., Santos, G. and Barbosa, L.C.F.M. (2023). Mapping the accidents and unsafe conditions of workers in the automotive sector. *Quality Innovation Prosperity*, 27(2), pp. 139–157. https://doi.org/10.12776/qip.v27i2.1849.

Silva, H.d.O.G.d., Costa, M.C.M., Aguilera, M.V.C., Almeida, M.d.G.D.d., Fonseca, B.B.d., Reis, J.S.d.M., Barbosa, L.C.F.M., Santos, G. and Sampaio, N.A.d.S. (2021). Improved vehicle painting process using statistical process control tools in an automobile industry. *International Journal for Quality Research*, 15(4), pp. 1251–1268. https://doi.org/10.24874/IJQR15.04-14.

Silva, M.P., Bernardo, M.H. and Souza, H.A. (2016). Relação entre saúde mental e trabalho: A concepção de sindicalistas e possíveis formas de enfrentamento. *Revista Brasileira de Saúde Ocupacional*. https://doi.org/10.1590/2317-6369000003416.

Sterud, T. and Johannessen, H.A. (2018). Influence of occupational factors on regional differences in sick leave: A prospective population study. *Scandinavian Journal of Public Health*, 46(3), pp. 314–320. https://doi.org/10.1177/1403494817748263.

Stoyneva, Z.B., Dermendjiev, S., Dermendjiev, T. and Dobrev, H. (2015). Work-related upper limb disorders: A case report. *Open Access Macedonian Journal of Medical Sciences*, 3(1), pp. 146–150. https://doi.org/10.3889/oamjms.2015.033.

Trstenjak, M., Benešova, A., Opetuk, T. and Cajner, H. (2025). Human factors and ergonomics in Industry 5.0—A systematic literature review. *Applied Sciences*, 15(4), p. 2123. https://doi.org/10.3390/app15042123.

Zhu, X., Hu, S.J., Koren, Y. and Marin, S.P. (2008). Modeling of manufacturing complexity in mixed-model assembly lines. *Journal of Manufacturing Science and Engineering*, 130(5). https://doi.org/10.1115/1.2953076.

Zurada, J. (2012). Classifying the risk of work related low back disorders due to manual material handling tasks. *Expert Systems with Applications*, 39(12), pp. 11125–11134. https://doi.org/10.1016/j.eswa.2012.03.043.

10 Ergonomics 5.0 and Artificial Intelligence in Industry

Impacts of Occupational Stress and Strategies for Workers' Health and Organizational Efficiency

Suzi Maria Carvalho Mariño,
Marcelo M. Soares, Carina Santos Silveira,
and Vera Lopes de Abreu Lima

10.1 INTRODUCTION

Industry 5.0 represents a new paradigm of the industrial revolution, characterized by the integration of artificial intelligence and human creativity, promoting a more customizable, sustainable, and human-centered industrial environment. Unlike Industry 4.0, which emphasizes automation and production efficiency, Industry 5.0 aims to strike a balance between technological advancements and the well-being of workers (Silva, 2015). However, Sarmento (2024) points out that, despite seeking industrial environments that are more customizable, sustainable, and human-centered, the implementation of intelligent systems in industries has had significant impacts on the labor market, including transformations in occupational dynamics and an increase in stress-related disorders.

According to Lanna (2018), artificial intelligence, by automating tasks and improving production, also triggers challenges such as structural unemployment and increased pressure on workers to adapt to new demands. This scenario has been identified as one of the risk factors for occupational stress since professional insecurity and the need for constant retraining increase workers' levels of anxiety and mental exhaustion (Khogali and Mekid, 2023, apud Sarmento, 2024). Araújo and Rayol (2024) note that to mitigate these impacts, it is essential to implement training programs and ergonomic measures that promote workers' health and facilitate their adaptation to the new production processes.

DOI: 10.1201/9781003559399-12

Ergonomics 5.0 has emerged as a solution to mitigate the damage caused by stress resulting from the implementation of AI in the industry by integrating strategies focused on safety, sustainability, and organizational resilience. It focuses on optimizing working conditions by analyzing the interaction between humans, machines, and the environment to ensure operational efficiency and the quality of life for workers (Pinto, Tereso and Abrahão, 2018). From this perspective, Veiga and Pires (2018) state that this approach is fundamental to preventing physical and mental exhaustion among professionals, considering that prolonged exposure to stressful situations can trigger occupational diseases, such as musculoskeletal disorders and psychological disorders.

Given this context, this chapter aims to present the impacts of occupational stress resulting from the implementation of Industry 5.0 artificial intelligence and outline the ergonomic strategies needed to promote workers' health and organizational efficiency. The approach adopted emphasizes Ergonomics 5.0 as a crucial mechanism for mitigating the challenges posed by digital transformation, promoting a more humanized and sustainable work environment. The results obtained are expected to contribute to the implementation of innovative ergonomic solutions in line with the demands of the new industrial revolution, thereby guaranteeing the well-being of workers and the competitiveness of companies in the digital age.

10.2 INDUSTRY 5.0: A NEW PARADIGM IN THE INDUSTRIAL REVOLUTION

According to Silva (2015), Industry 5.0 is an innovative concept that aims to optimize the use of technology, boost the creation of innovative products and solutions, and improve operational efficiency in companies. Based on Industry 4.0, this new industrial phase emphasizes collaboration between humans and intelligent machines, highlighting the value of personalization and sustainability, and transforming organizational interactions with the environment. The author points out that its pillars are customer experience and value generation for partners using advanced technologies such as artificial intelligence.

Industry 5.0 is a significant evolution of industrial paradigms. While Industry 4.0 was characterized by the total automation of production processes and the integration of cyber-physical systems, Industry 5.0 aims to strike a balance between technology and human creativity. According to the European Commission (2021), this new industrial model is based on social objectives, including the prosperity and well-being of workers, as well as economic growth, to promote a sustainable and human-centered industry.

As such, this model represents a departure from the previous paradigm, as it encompasses not only productive efficiency but also the inclusion of workers in the production process (Silva, 2015). Collaboration between humans and intelligent systems is one of the pillars of the new industrial revolution, enabling the full utilization of human creativity and critical thinking to optimize production across various sectors.

One of Industry 5.0's primary differentiators is its ability to customize products on a large scale. The global trend for customizing products and services has driven companies to develop solutions that meet consumers' needs. Østergaard (2018) emphasizes that Industry 5.0 seeks to offer consumers products that meet their needs and provide workers with more fulfilling jobs.

In addition, the author states that it aims to meet the growing demand for mass customization, which is driven by both psychological and cultural factors. Technology plays a crucial role in this context, allowing personalized production to be carried out efficiently without compromising companies' productivity and profitability.

Another fundamental aspect of Industry 5.0 is sustainability. The need to reduce the environmental impact of industrial activities has driven the development of greener and more efficient processes. Resilience and sustainability are essential pillars of this new industrial era. According to Berg (2022), Industry 5.0 promotes flexible and agile technologies that can adapt to global changes, striking a balance between business competitiveness and environmental responsibility. The emphasis on sustainability is reflected in the ability to maintain operational efficiency in the face of unforeseen challenges, as well as a commitment to minimizing environmental impact.

Similarly, Tipan and Garzon (2023) highlight three fundamental elements: human-centrism, which considers the human being as the central axis of development, aimed at promoting well-being; and economic and social sustainability, also focused on well-being and resilience. Berg (2022) adds that its principles include a human-centered approach, resilience, sustainability, and collaboration between humans and machines.

Active collaboration between humans and machines is one of the most innovative principles of Industry 5.0. In contrast to the total automation promoted by Industry 4.0, the new approach recognizes both the advantages and limitations of automation. The aim is to create an ecosystem where humans and technology can complement each other, maximizing their potential. In the words of Berg (2022), this synergy enables the capitalization on the advantages of precision and robotic efficiency while addressing the limitations of inflexibility and lack of adaptability.

From this perspective, Østergaard (2018) notes that this new industrial paradigm also has implications for the labor market. The revaluation of human beings in industrial production provides workers with new job opportunities and more rewarding roles. Rather than replacing humans, Industry 5.0 seeks to reallocate them to strategic positions where their cognitive and emotional skills are essential for innovation and the development of production processes.

Industry 5.0 revolutionizes production and redefines the relationship between companies and consumers. The hyper-personalization of products and services is one of the distinguishing features of this new industrial model, as it allows organizations to meet customers' specific needs more effectively. Berg (2022) affirms that companies adopting this approach can exceed consumer expectations, setting a new standard of excellence.

The author continues, "Industry 5.0 promotes a safer, more inclusive, and stimulating working environment. By placing human beings at the center of industrial operations, this approach seeks to improve workers' quality of life and create more favorable conditions for their professional development."

Tipan and Garzon (2023) note that Industry 5.0 seeks to integrate machines and human beings to enhance production value by creating customized and innovative products and solutions that cater to the specific needs of customers and society.

In this sense, Latam (2021) emphasizes that Industry 5.0 entails the rehumanization of industry, as humans assume the role of managers, not just workers. This model acknowledges that advances in robotics, digital technology, and automation must be leveraged to complement human skills. In addition, the author notes that Industry 5.0 is characterized by the integration of machines and human beings to add value to production, enabling the creation of innovative and personalized solutions.

Vural and Neziyh (2018 as cited in Tipan, 2023), note that Industry 5.0 aims to integrate robots' artificial capabilities with human skills, thereby enhancing production efficiency and precision. This integration enables faster and more tangible results, thereby optimizing industrial processes. Londoño and Ussa (2020) also emphasize that this new paradigm allows the execution of hazardous and demanding tasks through collaboration between humans and machines, thereby ensuring greater safety and efficiency.

Industry 5.0 marks a new chapter in the industrial revolution, fostering collaboration between humans and machines, promoting sustainability, and enabling the hyper-personalization of products and services. This new paradigm acknowledges the importance of technology while also valuing creativity and human intelligence as essential elements for innovation.

With human beings relocated to the epicenter of industrial production, Industry 5.0 seeks to optimize production processes, improve workers' quality of life, and more effectively meet consumer demands (Østergaard, 2018).

Given this scenario, adopting Industry 5.0 presents an opportunity for companies seeking to remain competitive in the digital age while promoting a more sustainable and human-centered production model. Industry 5.0 revolutionizes production processes and redefines the way society interacts with technology and the industrial environment.

10.3 DEPLOYMENT OF ARTIFICIAL INTELLIGENCE IN INDUSTRY 5.0

Industry 5.0, which has evolved from Industry 4.0, emphasizes collaboration between humans and intelligent machines, garnering global attention and transforming the job market while sparking debate about replacing human activities with machines (Sarmento, 2024). The author continues that while Industry 4.0 was based on automation and the interconnectivity of cyber-physical systems, Industry 5.0 seeks to integrate artificial intelligence to increase the synergy between humans and intelligent technologies, generating debates about replacing human activities with machines. According to Araújo and Rayol (2024), artificial intelligence aims to enhance production efficiency and humanize industrial processes, thereby optimizing the utilization of the human workforce in conjunction with artificial intelligence. Sarmento (2024) corroborates this view, adding that artificial intelligence redefines various sectors, including manufacturing, health, and logistics, by optimizing routine and repetitive tasks.

Lanna (2018) notes that incorporating artificial intelligence into Industry 5.0 has both positive and negative impacts. The positive effects include improved automation, reduced operational errors, increased productivity, and the creation of new roles in the labor market. The author emphasizes that artificial intelligence enables more personalized and flexible production, allowing for mass customization and thereby improving the consumer experience. Additionally, he notes that it can significantly contribute to industrial sustainability by optimizing resources and reducing waste.

Lanna (2018) continues to affirm that the challenges include the replacement of conventional jobs by automated systems and the increase in structural unemployment. Khogali and Mekid (2023, as cited in Sarmento, 2024) note that increasing automation leads to a decrease in jobs and salaries, particularly for professionals with intermediate skills. In addition, Lanna notes, excessive reliance on AI can create cyber vulnerabilities and ethical challenges associated with autonomous decision-making by machines that Araújo and Rayol (2024) state that, to mitigate these effects, it is necessary to implement training and retraining programs for the workforce, promoting the adaptation of workers to the new realities of the market, as this is a reality of no return. Adaptation is necessary to eliminate the negative impacts. Sarmento (2024) continues to affirm that it is essential for professionals to adapt and acquire new skills and abilities to meet new needs. He points out that interpersonal skills, such as problem-solving, analytical ability, negotiation, communication, and human resource management, are essential.

Artificial intelligence in Industry 5.0 is a significant advance, bringing opportunities and challenges. Thus, in this transition, it will be essential to ensure a sustainable and inclusive industrial evolution, enhancing its benefits while minimizing its negative impacts and contemplating strategic management (Lanna, 2018). Ashrafi et al. (2023, as cited in Sarmento, 2024) suggest using tools to re-educate workers and employing automated analysis of CVs to recommend more suitable jobs with better salaries. Sakib, Anik and Li (2023, as cited in Sarmento, 2024) highlight the benefits of education, including increased interactivity and stimulation of creativity.

Despite the challenges, Sarmento (2024) argues that artificial intelligence can optimize tasks, increase efficiency, and enhance productivity in various areas. Therefore, it is crucial to balance the benefits of artificial intelligence with reducing potential harm to people and society by implementing ethical standards and strict regulations.

10.4 IMPACTS OF IMPLEMENTING ARTIFICIAL INTELLIGENCE ON THE JOB MARKET

In the opinion of Sarmento (2024), artificial intelligence has made its mark globally, transforming the job market and generating debate about the replacement of human activities by machines. Research shows that artificial intelligence is redefining various sectors, including manufacturing, healthcare, and logistics, by optimizing routine and repetitive tasks.

10.4.1 Transformations and Task Optimization

- Gomes (2010, as cited in Sarmento, 2024) states that artificial intelligence is a subfield of computer science focused on developing techniques that enable computers to think and behave intelligently, adapting to human behaviors such as reasoning and perception. This opinion is corroborated by Gongora (2012, as cited in Sarmento, 2024).

- As reinforced by Sarmento (2024), artificial intelligence's ability to automate procedures accelerates activities, enabling it to perform cognitive, repetitive, and even hazardous tasks.
- Automating repetitive tasks is a positive aspect of AI, as it allows activities to be carried out quickly and efficiently, optimizing resources and increasing productivity (Sarmento, 2004).

10.4.2 Impacts Sectoral (Sarmento, 2024)

- **Manufacturing**—Automation, driven by artificial intelligence, improves processes; however, it could lead to workers being replaced by machines, requiring new skills and adjustments to social security laws.
- **Logistics and health**—In logistics, automation is transforming the job market, increasing the demand for information technology and intelligent stock control. In healthcare, artificial intelligence is boosting the interpretation of medical diagnostic images.
- **Transportation**—Vehicle automation and intelligent control systems are transforming the industry and job market, enhancing safety and lowering operating costs.
- **Education**—Artificial intelligence personalizes teaching, changing the interaction between teachers and students, and demanding changes in traditional education models.

10.4.2.1 Positive Impacts

Implementing artificial intelligence in Industry 5.0 is a significant milestone in the evolution of production processes. It provides notable advances in productivity, efficiency, safety, and the creation of new job opportunities. Artificial intelligence enables the automation of repetitive and hazardous tasks, allowing workers to focus on activities that require human intellect. This technology is transforming production and labor relations dynamics, boosting economic and social development.

10.4.3 Increased Productivity and Efficiency

Artificial intelligence has been widely used in Industry 5.0 to increase productivity and operational efficiency. According to Araújo and Rayol (2024), artificial intelligence-driven automation reduces errors, accelerates task execution, and enables greater precision in production processes. This technology allows professionals to focus on more strategic tasks that require critical thinking and creativity while delegating cognitive, repetitive, and exhausting tasks. In addition, introducing new technologies in the industry has historically been associated with increased productivity and economic growth, thereby raising workers' standards of living and stimulating innovation (Lanna, 2018).

10.4.4 Reduced Operating Costs

Automating repetitive and dangerous tasks significantly reduces companies' operating costs. Araújo and Rayol (2024) note that intelligent robots can perform manual tasks, thereby reducing the need for human labor in mechanical functions, which contributes to the efficiency of production processes. Additionally, the reduction in operational errors and the optimization of resources lead to time savings and an improvement in product quality. In this way, artificial intelligence reduces production costs and improves companies' competitiveness in the global market.

10.4.5 Improving Safety and Health at Work

Utilizing automated systems to replace workers in hazardous activities is one of the primary advancements enabled by artificial intelligence. Studies show that robotization significantly reduces the risk of accidents and occupational illnesses, thereby promoting a safer working environment (Veiga and Pires, 2018). Additionally, replacing repetitive and exhausting activities with artificial intelligence contributes to workers' well-being, thereby reducing excessive fatigue and musculoskeletal disorders. In this way, implementing artificial intelligence is a crucial strategy for preventing accidents and enhancing working conditions.

10.4.6 Creation of New Job Opportunities

While automation can replace some functions, it also creates job opportunities in sectors that require specialized knowledge. According to Wisskirchen et al. (2017, as cited in Araújo and Rayol, 2024, apud Veiga and Pires, 2018), the transition to an economy based on artificial intelligence must be accompanied by professional training and new management models. Professions related to software engineering, data analysis, computer science, and intelligent systems management are on the rise, offering new job prospects for workers. The coexistence of humans and machines can benefit technological development and improve professionals' quality of life (Araújo and Rayol, 2024).

10.4.7 Time and Optimization Task

Implementing artificial intelligence enables a more efficient distribution of work activities, thereby increasing workers' productive capacity. Intelligent machines can perform repetitive tasks more efficiently, freeing professionals to dedicate themselves to more complex and creative activities (Araújo and Rayol, 2024). Studies have shown that systems based on artificial intelligence can analyze large volumes of data quickly, providing valuable insights for informed decision-making *and* enhancing business strategies. Thus, artificial intelligence enhances productivity and leverages human skills more effectively.

10.4.7.1 Negative Impacts

Industry 5.0 is characterized by the interaction between humans and intelligent machines, promoting greater efficiency and personalization in production. However, implementing artificial intelligence in this context can harm the labor market, workers' health, and exacerbate socio-economic inequality.

10.4.8 Unemployment and Fear of Replacement

Araújo and Rayol (2024) state that the automation of tasks previously carried out by humans can lead to the extinction of many jobs, particularly those involving manual and repetitive activities, as well as those requiring low levels of qualification. This phenomenon causes insecurity among workers, who fear for the stability of their careers. According to Veiga and Pires (2018), it is estimated that a third of current occupations could be performed by machines, rendering them obsolete. This scenario highlights the importance of ongoing training, enabling professionals to adapt to the evolving market demands. Areias and Comandule (2020) report that when faced with this scenario, workers may dedicate excessive hours to work, feeling insecure and afraid of losing their jobs or benefits.

10.4.9 Increased Stress and Mental Health Problems

Bossmann (2016, as cited in Araújo and Rayol, 2024, as cited in Veiga and Pires, 2018) affirms that the pressure to adapt to new technologies and uncertainty about the future of employment can increase stress and anxiety levels among workers. These factors are very damaging to them in the long term, with consequences of premature aging for specific areas of the brain responsible for neurogenesis and anticipating diseases such as dementia, Alzheimer's, and others (Girardello, 2004; Aesst 2017, as cited Veiga and Pires, 2018) and Areias and Comandule (2020) add that the growing dependence on automated systems can lead to alienation and social isolation, compromising mental health. From the same perspective, Pena and Araujo Faria (2024) note that precarious work intensified by artificial intelligence can harm working conditions, contributing to psychological disorders and reduced job satisfaction. According to Areias and Comandule (2020), researchers have shown that the growing use of information and communication technologies is one of the changes in the world of work that has led workers to release stress hormones, such as catecholamines, adrenaline, and noradrenaline, in their homes.

10.4.10 Social Inequality

Adopting artificial intelligence primarily benefits highly qualified workers, while those with lower and medium qualifications face stagnation or wage cuts (Lanna, 2018). Cowen (2018, *cited in* Lanna, 2018) notes that this phenomenon, known as the polarization of work, exacerbates income inequality, creating a chasm between a highly qualified elite and most workers, whose routine tasks are increasingly being replaced by machines.

10.4.11 Job Insecurity

Pena and Araujo Faria (2024) reiterate that introducing artificial intelligence into the workplace can lead to job insecurity, with formal contracts being replaced by temporary or freelance models, which compromise job security and labor rights, increasing workers' economic vulnerability. In addition, they conclude that low wages and a lack of benefits exacerbate the instability of working conditions.

10.4.12 Physical Problems

In addition to the issues mentioned above, Araújo and Rayol (2024) note that the continuous use of artificial intelligence can also lead to physical problems for workers. Operating and maintaining machines require highly qualified professionals, which can result in cognitive overload and physical exhaustion. Prolonged use of electronic devices can also lead to occupational illnesses, such as repetitive strain injuries and visual problems.

Technological changes in the world of work necessitate constant adaptation, and how these changes are implemented and managed can have a significant impact on workers' health.

10.5 ERGONOMICS 5.0: FOCUS ON THE HUMAN BEING, SUSTAINABILITY, AND RESILIENCE

10.5.1 Ergonomics: A Quick Review

Ergonomics, as an interdisciplinary field, has its origins in the human attempt to improve tools and environments by addressing the limitations and capabilities of the human body. From the point of view of Moraes and Mont'Alvão (2012), this concern has been observed since ancient civilizations, as evidenced by the handles of scythes, which were adapted to fit the human hand. According to Pinto, Tereso and Abrahão (2018), Leonardo da Vinci already considered physiological characteristics in his inventions. Chapanis (1965) reinforces that ergonomics has existed since the creation of the first tools, highlighting the studies of Taylor and Gilbreth as milestones in the analysis of work efficiency. As Witzel (2005) points out, together with Lillian Gilbreth, several engineers founded classical ergonomics, which aimed to reduce physical effort and prevent accidents.

According to Chapanis (1965), the Second World War redefined the focus of ergonomics, directing it toward cognitive skills and demanding greater integration between humans and machines. As Moraes and Mont'Alvão (2012) state, technological acceleration has exposed human limitations, making it essential to consider human factors in the design of systems. Ergonomics was formalized in 1949 with the founding of the Ergonomic Research Society by Hywell Murrel (Pheasant, 2005).

According to Karwowski (1996, as cited in Moraes and Mont'Alvão, 2012), ergonomics has evolved to encompass the adaptation of products and environments to human capabilities, thereby promoting safety and well-being. Methodologies vary depending on the context, ranging from physical ergonomics to approaches such as macroergonomics (Hendrick and Kleiner, 2006) and classifications, including Iida's (2016), which encompass conception, correction, awareness, and participation.

In this way, ergonomics reaffirms its relevance to the contemporary challenges of Industry 5.0, which focuses on the human being, sustainability, and resilience. In this scenario, the discipline is evolving to meet today's emerging pillars and align its purposes with the social, environmental, and technological demands of the present. Ergonomics 5.0 is a conceptual expansion of traditional ergonomics, incorporating the core principles of Industry 5.0.

As Karwowski (2006) points out, this new approach distinguishes itself from previous generations by prioritizing the centrality of the human being in increasingly complex and intelligent socio-technical systems, encompassing not only the physical and cognitive aspects but also the emotional, social, and environmental dimensions of the human experience. Influenced by guidelines proposed by the European Union, Ergonomics 5.0 is designed to create more ethical, resilient, and humanized production environments (European Commission, 2021).

In addition, Ergonomics 5.0 articulates dimensions such as diversity, accessibility, mental health, and co-creation with users, consolidating a truly human-centered design paradigm anchored in a transdisciplinary perspective. It is therefore an innovative and transformative approach that aims to reconfigure organizational and technological systems in order to fully value the human condition, promoting socially committed innovation guided by principles of collective responsibility.

With regard to the centrality of the human being, the main objective of ergonomics, Brunoro, Bolis and Sznelwar (2017) point out that workers' health should be an end, not a means to progress, and add that ergonomics seeks positive results for the organization and the workers, considering the impact of the content and design of the work, as well as the need for regulation of the production system by the worker. Hernández et al. (2019) follow this line of thought and point out that ergonomics prioritizes the well-being of human beings. In this way, ergonomics and sustainability share a joint interest in satisfying human needs, with a systematic perspective that encompasses the interdependence and interrelationships between human activities and their systems. One of the ideas on which sustainability is based is that human beings have the right to a healthy life in harmony with nature.

Ergonomics contributes significantly to working in a context of sustainability (Brunoro, Bolis and Sznelwar, 2017). The author reinforces that corporate sustainability, although still being defined, considers the issue of work to be a fundamental element for organizations that are aligned with sustainable development and that consider environmental, economic, and social aspects, seeking production systems sustainable and healthy work environments. With this in mind, Hernández et al. (2019) state that ergonomics comes up with innovative methods that aim to solve problems that hinder the effective development of sustainability, helping to ensure appropriate sustainable development.

Falzon and Mollo (2009) state that constructive ergonomics aims at work efficiency and the continuous development of workers and organizations. Gonçalves (2014) highlights the importance of well-implemented ergonomic models for improving occupational health.

In this direction, Brunoro, Bolis and Sznelwar (2017) advocate that the analysis of the production system must involve the social aspect, valuing work and its contribution to the quality, productivity, and social and cultural development, and that corporate sustainability must seek to design production or work systems in line with the concepts of sustainability. It considers that a sustainable work system aims to regenerate the resources it uses, returning them to society preserved or improved, with responsibility for external costs. The authors add that the shift toward sustainability involves understanding interdependence and addressing externalities with actions focused on the main sources.

Occupational ergonomics and work psychodynamics contribute to sustainable work, including understanding work to transform it, identifying sustainable work, and aligning the organization on work-related issues. From a sustainability perspective, work inducers promote efficiency by recognizing the engagement of intelligence at all levels of the organization and collaborating to build health and professional development (Brunoro, Bolis and Sznelwar, 2017).

Organizational resilience is fundamental to accident prevention. This approach aims to analyze weak signals and minor incidents to avoid adverse events. Integrating ergonomics into safety strategies makes it possible to anticipate risks and improve processes (Pinto, Tereso and Abrahão, 2018).

According to Fernandes et al. (2014), resilience studies human behavior in stressful situations and how to develop the ability to obtain the best results in the face of adversity, improve organizational performance in various conditions, and avoid undesirable results. Along these lines, Fernandes, Anacleto et al. (2014) point out that resilience enables organizations to overcome adverse situations and achieve better results. The authors reiterate that applying resilience principles to ergonomics management enables flexible actions when facing day-to-day pressures. Resilience is the ability to adjust the functioning of organizations before, during, and after changes or disturbances, as they need to develop skills to deal with unexpected situations.

Knowledge about resilience can help workers understand the factors that strengthen their resilience (Cruz et al., 2018). It can be learned, promoted, and developed, allowing workers and institutions to build alternatives to reduce risks to health. Knowledge of the factors related to resilience can help workers understand the determinants that strengthen their ability to cope with adversity, adding to their efforts to improve their health. With this in mind, Fernandes, Bini et al. (2014) state that knowledge and awareness of ergonomic risks favor applying resilience principles, creating benchmarks for judging risks, making decisions, responding to occurrences, monitoring, and anticipating.

Ergonomics plays a crucial role in promoting workers' safety, health, and well-being. Throughout its evolution, this science has contributed to creating sustainable and resilient work environments in line with the principles of Industry 5.0. Integrating ergonomics, sustainability, and resilience reinforces the importance of organizational models that value the worker experience, anticipate risks, and promote innovation in working practices.

The analysis of work activity helps us understand the interactions between possible technology and organization, making it possible to prevent accidents and continuously improve processes. In this way, ergonomics not only improves working conditions but also contributes to the sustainable development of companies and society as a whole human beings.

10.5.2 Cognitive Ergonomics

In the context of industrial production, physical ergonomics initially focused predominantly on the physical strength of workers (Abrantes, 2011). However, intellectual work gained prominence with technological evolution and increasing automation, driving cognitive ergonomics. In the past, the author continues, work was essentially physical, with little or no intellectual participation from those who worked on the factory floor. The term "manual labor" comes from this activity, in which the hands were used and not the "head." With the advancement of technology, physical work was reduced while intellectual work intensified, making it possible to coin the term "brainpower."

From this perspective, Kroemer and Grandjean (2005, as cited in Abrantes, 2011) report that after the end of the Second World War in 1945, technological developments intensified the need for more significant cognitive interaction between human beings and machines, especially in the use of displays and panels. Workers began to visualize data, such as speed, and make decisions based on processing this information, requiring attention, decision-making, memory, motor responses, and information retention for this interaction. This is a fruitful field, as it addresses human factors involved in the cognitive process of human-system interactions (Winkelmann and Mager, 2019).

Thus, cognitive ergonomics was consolidated with the contribution of renowned researchers such as Swiss physician Etienne Grandjean and German engineer Karl H. E. Kroemer. It has emerged with the role of making technological solutions compatible with the characteristics and needs of users, analyzing how work affects human cognition and how it is affected by it, and applying theories, principles, data, and methods to projects aimed at optimizing human well-being and overall system performance. It seeks to explain how cognitive processes are articulated in problem-solving situations at different levels of complexity, which results in the mental load that can manifest as underload when workers' capacities are underused or overload when demands exceed their capacities, which are detrimental to performance and health. This is a field of application of ergonomics aimed at understanding the interaction between cognitive processes and problem-solving situations at their different levels of complexity (Régis et al., 2008; Malvestiti, Maas and Gontijo, 2017).

Kontogiannis (1989, as cited in Abrahão et al., 2005) and Cunha et al. (2017) explain that the role of cognitive ergonomics is to make technological solutions compatible with the characteristics and needs of users. In this sense, it contributes to a theoretical and methodological framework that makes it possible to analyze how work affects human cognition and, at the same time, is affected by it, creating service strategies and helping to combat the mental exhaustion caused by the complexity of some tasks. Hollnagel (1993, as cited in Abrahão et al., 2005) refers to mental processes such as perception, attention, cognition, motor control, memory storage, and retrieval and highlights how these elements affect the interaction between human beings and other components of a system.

In this view, Abrantes (2011), Cunha et al. (2017) and Araujo and Tacconi (2023) points out that cognitive ergonomics considers relevant aspects, which include mental workload, vigilance, decision-making, precision work, skill performance, human error, the interaction between humans, machines and computers, stress and fatigue, which can cause difficulties in perception, absorption, and retention of information, aiming to maintain mental and cognitive balance to avoid stress, wear and tear. Routine and repetitive activities reduce productivity and wear down cognitive abilities. Abrantes (2011) emphasizes that, when performing their duties, workers need to process information and make decisions that require attention, situational analysis, and coordination in complex tasks.

Considering Bossmann's statement (2016, apud Veiga and Pires, 2018) that the pressure to adapt to new technologies and uncertainty about the future of employment can significantly raise stress levels among workers. This is a factor that causes damage long-term and can lead to premature aging of specific areas of the brain responsible for neurogenesis, anticipating diseases such as dementia and Alzheimer's (Girardello, 2004). The author focused on describing the concepts, causes, and impacts of stress in the workplace and sought to deepen the understanding of stress and identify strategies to minimize the damage caused by this condition.

10.5.3 Concepts, Causes, and Effects of Stress

Medical interest in stress dates to Hippocrates (470-377 BC), but it was only in 1929, after the work of physiologist Cannon, who formulated the concept of homeostasis, that the stress reaction came to be understood as a phenomenon related to the interaction between body and mind (Myers, 1999, as cited in Oliveira, 2020). Homeostasis corresponds to the ability of the various systems in the body to create a self-regulating situation in which, when faced with changes in their internal balance, they react in such a way as to seek to re-establish a position like that before the occurrence of the imbalance (Oliveira, 2020).

The author adds that the relationship between homeostasis and stress lies in the body's search for balance. Homeostasis seeks to re-establish lost equilibrium by activating physiological mechanisms. On the other hand, stress is a state of activation of the organism that prepares it to react to changes identified in the environment.

In addition to formulating the concept of homeostasis, Walter B. Cannon, according to Lipp (1984), defined the "fight or flight" response around 1920 as immediate internal physical reactions to threatening events, which are essential for the survival of the organism.

Silva (2020) points out that Hans Selye introduced the term "stress" in 1936 to describe the body's general and non-specific response to a stressor. This response is triggered by the perception of stimuli that disturb homeostasis, leading to systemic manifestations, including physiological and psychological disorders. Stressors can be physical, chemical, biological, psychological, or social. The predictability of stress is also important since unpredictable stressors can have a more negative impact than predictable ones.

With this in mind, Oliveira (2020) points out that the concept of stress is defined as a state of activation of the organism in the face of changes that generate instability. It manifests itself in the body and is independent of the meaning attributed to the triggering event. This stress reaction is followed by emotional reactions that overlap and coexist, depending on the meaning attributed to the event and constituting potential sources of stress, the author points out.

Lipp (1984) stresses that stress is a psychological reaction with emotional, physical, mental, and chemical components to stimuli that irritate, frighten, excite, or confuse. Any change in life generates stress, and when it is excessive and/or negative, it becomes "distressing," and when it is positive, it becomes "*eustress*". The author adds that there is a "comfort zone" where efficiency is more significant, allowing quick decisions and greater energy, which varies from person to person. Depending on the individual's interpretation, an event can cause *distress* or *eustress*. For example, aversive stimuli can generate *distress*. Most excessive stress responses are self-produced, depending on the individual's interpretation of the harmless stimuli.

As mentioned above, the "fight or flight" reactions defined by Cannon are part of the alarm stage, proposed by Selye as the first of the three ways the body reacts to stress, described below.

According to Oliveira (2020), Hans Selye, a Canadian doctor considered the "father of stressology," expanded on Cannon's work over four decades of research (1936–1976). Selye described a unique and general adaptive reaction of the body when subjected to stressors, which he called the general adaptation syndrome, characterized by three phases.

Lipp (1984) and Oliveira (2020) state that the first alarm phase activates the sympathetic nervous system, preparing the body to face the challenge. The second, the resistance phase, is marked by the attempt to make the body remain activated, albeit at a lower intensity, to ensure adaptation to the stressor. At this stage, the body tries to repair the damage caused by the stress. If stress persists, the body remains alert, which can lead to the third phase, exhaustion, with harmful reactions and possibly serious illnesses. The exhaustion phase occurs when the body collapses after being stressed for a prolonged time, becomes vulnerable to disease, shows a drop in the ability to think, remember, and act, and damages the immune system. The stress reaction should not be considered pathological but part of the individual's normal functioning. Only long-lasting imbalances can be regarded as pathological states, says Oliveira (2020).

In this vein, Cortez (2007) adds that stress causes physiological responses that can lead to psychological imbalance, affecting mental health. It can be understood as both a stimulus and a response. Biologically, a stressor is any agent that triggers the stress reaction, and the physiological response seeks to adapt the organism to the new condition. Increased production of glucocorticoids, such as cortisol, is one of the main mechanisms involved in this process; the author warns that excess cortisol may be related to the development of Alzheimer's disease. According to Khalsa and Stauth (1997, as cited in Girardello, 2004), cortisol is considered toxic to the brain and can damage or even kill billions of brain cells.

Hespanhol (2005) and Lipp (1984) detail that the manifestations of stress can be physical, such as headaches, teeth grinding, muscle pain, weight changes, digestive problems, heart changes, increased blood pressure, increased sweating and nausea, or psychological, which can cause anxiety, depression, panic, tension, anguish and insomnia, interpersonal difficulties, excessive worry, etc. In behavior, it can lead to nervous tics, rapid speech, inability to relax, chronic fatigue, and even substance abuse etc. These reactions prepare the body for the "fight or flight" response, described by Cannon as an essential defense mechanism for survival, adds Lipp (1984). In this context, Cortez (2007) states that stress can be acute when it ceases after the stressor is removed, or chronic, when the body's adaptation is prolonged and exhausting.

In this scenario, Hespanhol (2005) adds that, in the occupational context, Cooper classifies the causes of stress into five categories. The first is intrinsic to the job, including physical conditions (noise, inadequate lighting, occupational hazards) and excessive or insufficient workload. The second category is the role in the organization, covering role conflicts and ambiguity of functions. The third refers to career progression, including job insecurity and professional frustrations. The fourth involves interpersonal relationships at work and incredibly hierarchical conflicts. The fifth and final category includes the organizational structure and climate of the company, which can directly influence worker well-being. In addition, extra-organizational factors, such as family problems and financial difficulties, can also intensify occupational stress, the author points out.

10.5.4 Impacts of Excess Cortisol Resulting from Chronic Stress

Considering that excess cortisol is directly related to stress and that, when it becomes chronic, it can cause various physiological and neurological alterations, it is essential to discuss the impact of cortisol on the brain and its consequences. In this sense, Girardello (2004) corroborates this perspective by highlighting the importance of studying cortisol as a predictor of stress since understanding its action has proved essential to mitigating its harmful effects on quality of life.

In the same vein, Louro, Louro and Duarte (2020) state that prolonged exposure to stress can cause chemical changes in both the brain and the body as a whole, affecting fundamental brain structures such as the hypothalamic-pituitary-adrenal (HPA) axis, the amygdala, and the hippocampus. As established in the literature, cortisol is the main neurochemical agent involved in the stress response, released by the adrenal gland in response to activation of the HPA axis. In stressful situations, the hypothalamus sends signals to the adrenal gland, stimulating the release of hormones such as cortisol and adrenaline into the bloodstream and preparing the body to react to such stimuli, reinforcing the authors.

In addition, as mentioned by Girardello (2004), there is evidence associating excess cortisol with the development of neurodegenerative such as Alzheimer's, which reinforces the need for further studies into the impact of this hormone on brain health.

In the same vein, Lima et al. (2021) state that hypersecretion of cortisol, in response to chronic stress, is the leading cause of physiological disorders. Excess cortisol has been associated with metabolic syndrome, which includes disorders such as glucose intolerance, insulin resistance, central obesity, dyslipidemia, and hypertension, in addition to the alterations above. Cortisol, the authors continue, is important and plays several roles in the body, including adapting to stress and maintaining adequate glucose levels. However, prolonged exposure to high levels of this hormone can result in negative changes in glucocorticoids, activating the degradation and inhibition of protein synthesis in muscles, adipose, lymphoid, and connective tissues, as well as in the skin, in addition to suppressing DNA duplication. In higher doses, it can promote its degradation.

As mentioned, the persistence and exaggerated intensity of cortisol can make the HPA axis hyper-reactive, damaging the body. Louro, Louro and Duarte (2020) also add that excess cortisol in the bloodstream can also affect the hippocampus, a structure in the limbic system responsible for conscious memory, reducing its function. This alteration in memory, the authors continue, caused by deregulation of the HPA axis, has led to adverse effects of cortisol on the hippocampus. Long-term excess can cause memory problems, impacting cognition and learning. Excessive stress can also decrease attention and concentration, increase task errors, and delay reacting to stimuli, thus impairing work performance.

Belo-Silva (2015) details this scenario by highlighting the negative impacts of chronic stress on various brain regions, including those responsible for neuromodulation. The author explains that the prefrontal cortex and hippocampus play a fundamental role in regulating the stress response, influencing the HPA axis, the main system responsible for the neuroendocrine control of stress. These brain regions exert an inhibitory effect on releasing glucocorticoids such as cortisol, regulating the physiological response to stress and promoting the body's homeostatic balance.

In particular, the medial prefrontal cortex is directly involved in executive control and the inhibitory modulation axis activation of HPA. At the same time, the hippocampus participates in the negative feedback mechanism, limiting hormone secretion in response to stressful stimuli. On the other hand, the author points out that the amygdala acts as a stimulus for releasing cortisol, establishing a dynamic balance between these brain structures. Alterations in this system have been strongly associated with the development of anxiety disorders since prolonged activation of the HPA axis can result in deregulation of the stress response.

In experimental models developed by the author, it was found that plasma corticosterone levels increased anxious behaviors, suggesting that chronic exposure to stress can negatively impact neural function. Furthermore, according to Belo-Silva, the degeneration of hippocampal neurons induced by prolonged stress can further compromise the inhibition capacity of the HPA axis, generating a vicious cycle of neuroendocrine hyperactivation. Similarly, dysfunction of the prefrontal cortex can lead to deficits in emotional and cognitive regulation, exacerbating vulnerability to the development of psychiatric disorders.

Thus, the author concludes that the functional integrity of the prefrontal cortex and hippocampus is essential for an adaptive stress response, and its deregulation can contribute significantly to the onset of anxiety.

Stress is also related to anxiety, which is a variation of fear. Louro, Louro and Duarte (2020) explain that fear is a primitive emotion and protects us from danger. Neurologically, fear is linked to the limbic system, specifically the amygdala, responsible for detecting, generating, and maintaining related emotions. The authors add that when faced with danger, the amygdala triggers signals. To prepare the individual for the situation. Anxiety leads to anticipatory thoughts and physiological and emotional symptoms that can also impair the individual's social life.

10.5.5 Stress and Mental Load at Work

Occupational stress has become a growing concern due to its significant impact on workers and organizations. The high incidence and prevalence of mental suffering among workers have generated high costs for companies and governments, which is reflected in low productivity, sick leave, and absenteeism (Areias and Comandule, 2020). The study of the interaction between mental and behavioral disorders related to work has gained visibility both nationally and internationally, with political, economic, and social changes having an impact on labor relations, quality of life, and workers' mental health. The authors add that these changes highlight occupational stress as a problem of great relevance to society.

Stress at work can have a wide-ranging negative effect on workers' mental health and can lead to violence and risky behavior. Factors contributing to this scenario are the increasing use of information technologies, staff reductions, and the demand for flexibility in the corporate environment. Data from the European Union indicates that 28% of workers report stress-related health problems, resulting in the loss of millions of working days and high costs for the economy. Concern about these consequences has led the European Agency to conduct studies to assess the nature of stress, its causes, and its impact on workers and companies (Areias; Comandule, 2020).

From this perspective, Hirschle and Gondim (2020) point out that the weight of responsibility, adaptation to stressful contexts, and pressure for performance, impacting workers' family and social lives, results in stress in the workplace. The consumer economy demands more and more productivity, generating overload and amplifying mental and physical health risks, according to Silva, Dias and Teixeira (2012). Hirschle and Gondim (2020) point out that these factors reduce workers' well-being and include overload, high demands combined with low control, monotony, low decision-making power, negative social interactions, and lack of support as aggravating factors. In addition, negative affective events in the workplace and conflicts between work and family also have a negative influence on workers' mental health, the authors conclude.

This situation, according to Paschoal and Tamayo (2005), results in chronic stress, which can cause physical and psychological symptoms, contributing to the development of disorders such as anxiety, depression, and insomnia, as well as cardiovascular diseases and type two diabetes. Prolonged exposure to psychosocial risk factors at work can also cause illnesses and work-related disabilities, a situation made worse by performing multiple tasks simultaneously, under high pressure, short deadlines, and the difficulty of disconnecting from work, which is among the primary sources of tension, according to Maffia and Pereira (2014). Occupational stress, according to cybernetic theory, is understood as a discrepancy between the situation perceived and that desired by the worker, which is influenced by individual and environmental factors

In this sense, Silva, Dias and Teixeira (2012) point out that the permanence of stressors in the individual's life history, the accumulation of stressors, and recurring stress can lead to exhaustion, compromising organic functioning and generating problems such as those mentioned above and more: hypertension, immunological alterations, digestive disorders, musculoskeletal pain, and sleep disorders. In addition, they add that there can be an impact on memory and concentration, increased irritability, the consumption of harmful substances, and deteriorating working relationships. Blanco and Souza (2018) point out that chronic stress can also result in neuroinflammation, excitotoxicity, and oxidative damage, affecting cognitive and behavioral functions and leading to the development of psychiatric disorders, including anxiety and depression.

Among the occupational impacts cited are psychological distress, symptoms of depression, and post-traumatic stress disorder, as reported by Areias and Comandule (2020). The authors state that exposure to traumatic events at work can lead to negative adaptations and alcohol abuse, as well as increase the risk of mental disorders. Conditions such as discrimination, harassment, and violence in the workplace contribute to greater psychological suffering and compromise workers' mental health, according to Coimbra, Ferreira and Araðjo (2020). Occupational stress has become a problem with a significant economic and social impact, influenced by macroeconomic, political, and structural factors, such as unemployment, insecurity, and new forms of work organization, such as lean production, among others, say Areias and Comandule (2020).

Silva, Dias and Teixeira (2012) point out that adaptation to stress requires neuroendocrine and immunological mechanisms. When overloaded, these can result in physiological imbalances and chronic diseases. Sleep irregularity, cognitive alterations, and greater vulnerability to psychological disorders are also evident, demonstrating the importance of strategies to mitigate the impacts of occupational stress, according to Cortez and Silva (2007).

Therefore, in this scenario, occupational exposure to continuous stress directly interferes with workers' well-being and compromises organizational dynamics, leading to increased absenteeism, decreased productivity, and staff turnover. The presence of toxic work environments and precarious working conditions amplifies the adverse effects of stress, impacting not only professional life but also the overall health of the worker. Recognizing occupational stress as a public health issue is essential for implementing organizational policies and practices that promote healthier and more sustainable work environments.

10.6 PREVENTIVE STRATEGIES FOR WORKERS AND ORGANIZATIONS

Prevention in the workplace is an essential factor in promoting occupational health and organizational sustainability. With the increase in occupational stress and sick leave due to common mental disorders, effective strategies aimed at mitigating these problems are necessary. This will provide a better quality of life for workers and greater productivity for organizations.

Almeida et al. (2024) highlight the importance of preventive interventions in reducing absenteeism and sick leave due to occupational stress. Workers' well-being is influenced by multiple factors, including the interface between work and personal life, personal resources, social support, and autonomy in the workplace, according to Hirschle and Gondim (2020). In this sense, Chagas (2010) points out that resilience, self-efficacy, and emotional competencies are key elements in maintaining workers' mental health, enabling them to better adapt to organizational challenges

Concerning preventive strategies, Almeida et al. (2024) highlight approaches that promote both mental and physical health in the workplace, including physical activity, mental health promotion programs, enhanced socio-professional relations, effective conflict management, and organizational support. In addition, professional appreciation and creating a healthy working environment are fundamental aspects of a positive and motivating organizational climate.

In organizational sustainability, Andrade (2020) emphasizes that sustainable development must meet the needs of present generations without compromising future ones. To achieve this, new consumption practices, technological innovation, and sustainable project management must be implemented. According to Pimenta, Rosa and Pimenta (2023), the UN's 2030 Agenda reinforces this global commitment, proposing goals for a more equitable and environmentally responsible future.

Organizations must integrate sustainable strategies into their processes, becoming more socially and environmentally responsible.

Resilience 2.0, also known as antifragility, is an evolution of the traditional concepts of adaptation. Blečić and Cecchini (2019) argue that while resilience seeks to absorb shocks and return to equilibrium, adversity strengthens antifragility. This approach allows organizations to withstand unexpected events and use them as opportunities for growth, as James and George (2014) state.

With this in mind, Jones (2014) points out that implementing antifragility in organizations requires structural and cultural changes, including promoting diversity, creating continuous learning environments, and flexibility management practices, such as self-organization, systems health management, and the use of materials that are self-repairing, which have been applied to increase organizational resilience. Tseitlin (2013) adds that a culture of continuous learning and accepting mistakes as opportunities for improvement is essential element in building antifragile organizations.

Therefore, preventive strategies for workers and organizations should range from measures to promote mental and physical health to structural changes aimed at sustainability and antifragility. Such measures not only guarantee the well-being of workers but also strengthen organizations in the face of future challenges, allowing them to thrive amid adversity and become more sustainable and innovative.

10.6.1 Workers' Health: Overcoming Stress and Improving Resilience

According to Almeida et al. (2024), preventive interventions significantly reduce absenteeism and sick leave due to occupational stress, a growing phenomenon that compromises workers' health, quality of life, and productivity. This highlights the importance of adding knowledge to the sector of occupational health and safety, with a view to the prevention and early identification of common work-related mental disorders, thus reducing occupational sick leave rates.

Hirschle and Gondim (2020) point out that several factors influence workers' well-being, including work-related aspects, personal resources, the interface between work and personal life, and group-level factors. Personal resources include resilience, self-efficacy, emotional competencies, and psychological detachment, essential for individual well-being. In addition, the perception of social support from colleagues and management and autonomy at work can mitigate the negative impacts of stress on workers' mental health.

According to Chagas (2010), adaptability and resilience are fundamental for individuals to be able to cope with the changes and challenges arising from disability, which can result in stress and other damage. The rehabilitation process aims to help recover or compensate for altered functions by acquiring new skills, thus promoting social inclusion. Emotional adaptation to the "scars of the past" and future prospects is essential in this process. Coping with disability can lead to the development of characteristics such as tolerance, patience, perseverance, and acceptance, which contribute to adaptation and well-being. The search for meaning and purpose in life can also act as a motivating factor during this process.

In this context, resilience is an essential psychological resource and can be strengthened through appropriate support, says Chagas (2010). Through resilience, the individual can "relearn how to live," reconstructing their life narrative and finding a new balance.

In addition to developing resilience to cope with adversity in the workplace, preventive strategies can be implemented to reduce stress-related absenteeism. Almeida et al. (2024) highlight the main techniques that involve a multifaceted approach aimed at both mental and physical health in the workplace:

- **Physical activity**—Adoption of interventions in the workplace, through micro-exercises, aimed at improving somatic health symptoms, such as musculoskeletal pain, especially in workers in sedentary or physically demanding jobs. Occupational and breathing exercise programs are also recommended.
- **Mental health promotion**—Enable the installation of mental health promotion programs and use cognitive-behavioral therapies in person or online. Psychoeducation is also an effective tool for integrating other interventions. Psychotherapeutic support and emotion management are other strategies identified.
- **Creating a healthy working environment**—Making the business environment pleasant and stimulating, with a more worker-focused organizational climate and favorable environments.
- **Awareness programs for healthy habits.**
- **Permanent education.**
- **Effective communication.**
- **Conflict management.**
- **Organizational and occupational**—Provide satisfaction with organizational support is a consistent antecedent of well-being. Strengthening supervisor support for employees with common mental health problems is crucial.
- **Continuous training and professional development**—Companies that value their workers and invest in their health tend to create a healthier working environment.
- **Application of *feedback* and professional recognition.**
- **Adjustments to the work environment, making it more ergonomic, and proposing work exercises**—Participatory ergonomics programs are relevant. Mainardi et al. (2014) also add that environmental enrichment increases brain plasticity through an upward shift in the balance between cortical inhibition and excitation.
- **Improved socio-professional relations**—The importance given to the climate between colleagues, trust, and mutual respect are important for work.
- **Encouraging workers' autonomy and creativity.**
- **Provision of dental care.**
- **Consideration of reducing working hours.**

The strategies mentioned above have proven to be effective in the organizational context, promoting the health and well-being of workers. In addition to the micro-exercises suggested by Almeida et al. (2024) and Mainardi et al. (2014), it should be noted that both motor exercise and cognitive activity modulate brain physiology and influence its susceptibility to pathologies associated with aging.

From this perspective, Gąsiewski (2017) adds that different types of exercise contribute to improving the functional efficiency and health of the brain, showing a strong correlation between the regularity of physical activity throughout life and the resulting benefits for brain function and volume. According to the author, aerobic physical activity, widely investigated in literature, is the most effective for increasing brain volume and improving cognitive functions.

In the same vein, Leandro et al. (2007) state that physical training induces greater tolerance to adverse situations, such as stress and aggression. In addition to the benefits mentioned above, Santiago et al. (2022), through functional analysis, show that physical activity can provide neuroprotection in neurodegenerative diseases such as Alzheimer's, Parkinson's, and Huntington's, through the positive regulation of synaptic signaling pathways.

Freitas (2017) also notes that regular exercise plays a crucial role in neuroplasticity and neuronal survival, serving as an essential factor for the proper functioning of the nervous system. He points out that strength exercise contributes to a chronic increase in the levels of Brain-derived neurotrophic factor recognized for its functions in neurogenesis, protection against neuronal death, development of the Central Nervous System, neuroplasticity, and Neurotrophic synaptogenesis.

Also, as a resource for promoting well-being, studies by Fava et al. (2019) show that practicing yoga can reduce perceived stress levels. It provides physical and mental benefits, such as reducing pain and increasing flexibility, in addition to feelings of calm and tranquility.

Techniques for coping with and reducing occupational stress have been discussed for decades. Lipp (1984) points out that an effective therapeutic plan can include the identification of symptoms, the use of stress and anxiety inventories, the evaluation of stressors, and the analysis of the individual's perception of their working conditions. Treatment can include relaxation techniques, restructuring irrational beliefs, developing problem-solving strategies, physical practices, and nutritional guidance.

It can therefore be concluded that various techniques can be implemented to promote the well-being of workers in their working environments. A well-structured intervention plan is essential to help individuals cope with excessive stress, teaching effective strategies for managing stressors, without the need for profound changes in personality structure.

10.6.2 Sustainability: Challenges and Opportunities in Sustainable Projects

The concept of sustainable development, established in 1987 in the "Our Common Future" development report, cited by Andrade (2020), refers to meeting the needs of the present without compromising the ability of future generations to meet their own needs. This concept opposes the unsustainability of growth and demands new approaches to consumption and production habits. It is therefore imperative to implement research and technological innovations that promote paradigmatic changes in the quality of life and the legacy for future generations.

According to Andrade (2020), sustainable development should be understood not as an isolated action, but as a set of strategies and resources aimed at meeting human needs. It is a shared mission between organizations and nations, seeking to engage all segments of society in concrete initiatives. In this context, the UN's 2030 Agenda represents a global commitment with clear goals for 2030, aimed at promoting sustainable development.

The future prospects of project management and its relationship with organizational sustainability are widely discussed by Pimenta, Rosa and Pimenta (2023). According to the authors, building a sustainable future is a unique task, with great potential for scientific and technical advances, as well as driving the evolution of organizations and society. Sustainability, due to its multidimensional nature, functions as a comprehensive concept. In this sense, sustainable project management aims to see the project life cycle as a sequence of decisions that enable the expected benefits through responsible social and environmental practices.

Blak et al. (2023, apud Pimenta, Rosa and Pimenta, 2023) emphasize that this approach challenges traditional project management, which often neglects fundamental aspects of sustainability by not considering the needs of future generations. Andrade (2020) corroborates this perspective by stating that the promotion of human well-being, in a fair, democratic, and equitable manner, must respect the ecological limits of the planet.

The transition to a sustainable future, through project management, depends critically on changing the mindset of managers. As Pimenta, Rosa and Pimenta (2023) point out, project managers play a fundamental role, as they have the authority to lead initiatives and manage resources that are essential for generating sustainable value. Toljaga-Nikolić et al. (2020, apud Pimenta, Rosa and Pimenta, 2023) complement this view by pointing out that sustainable project management changes the relationship with stakeholders, making projects more comprehensive and demanding new responsibilities from managers, such as greater attention to ethical issues, the acquisition of specialized knowledge, and the adoption of innovative and environmentally responsible technological solutions.

In this context, Ferrarez et al. (apud Pimenta, Rosa and Pimenta, 2023) clarify that, in the face of growing social demand for better socio-environmental conditions it is essential to transcend purely economic objectives to incorporate ecological and social aspects in this implies significant transformations in the strategic, economic, environmental, and social objectives of organizations. The authors identify fundamental practices for incorporating sustainability into project management, classified as environmental efficiency, regulatory compliance, social responsibility, continuous improvement, and project success. In this way, organizational maturity and the ability of balanced way managers to transform socio-environmental practices into driving factors for sustainable business are essential for advancing this paradigm.

In line with the aforementioned authors, Andrade (2020) points out that technological tools and sustainable development offer ethical and practical guidance to guarantee the continuity and quality of life on the planet. The convergence of these principles reinforces the author, combined with a vision focused on human well-being in its various dimensions, is essential to optimize the quality of life globally. This requires an active commitment from all sectors of society to concrete actions to substantially change the ways in which we produce and consume.

Finally, although there are significant challenges, such as the lack of priority given to the issue and prejudices related to generational diversity, the opportunities for integrating experienced professionals into sustainable projects and organizations are significant. Recognizing and valuing the skills and experience of this part of the workforce can have a positive impact on organizational sustainability and on society as a whole.

10.6.3 Resilience 2.0: Antifragile Organizations

According to Blečić and Cecchini (2019), antifragile planning can be defined as that which actively seeks to make gains from disorder, in contrast to resilience, which aims to return to a state of equilibrium. The essential principles of antifragile planning involve understanding the system's possible responses to disturbances, as well as detecting its fragility, robustness, resilience, or antifragility.

According to James and George (2014), antifragile approaches complement or replace traditional, deterministic scenario planning, since the latter has limitations in dealing with uncertainty and unexpected events. In this way, an "antifragile" approach was developed and illustrated to prepare for the future, characterized as a non-deterministic methodology that can complement or replace scenario planning based on causal relationships. This methodology recognizes uncertainty as deriving from indeterminacy, and not just as a difficulty in identifying causes and effects.

Blečić and Cecchini (2019) point out that the fundamental difference between resilience and antifragility in planning lies in the way each concept deals with disorder and uncertainty, as well as other essential distinctions (Table 10.1).

According to Table 10.1, the properties of antifragility include optionality, i.e., the availability of options without obligations, allowing unexpected opportunities to be seized and damage to be limited. Blečić and Cecchini (2019) state that antifragile planning should avoid practices that weaken systems, such as plans based on fragile forecasts, excessive centralization and micromanagement, excessive focus on efficiency and optimization, exaggerated specialization, and the maintenance of extractive political and economic institutions.

James and George (2014), as well as Blečić and Cecchini (2019), point out that the antifragile approach shifts the focus to identifying the factors that make an organization fragile or vulnerable to the future, focusing not on eliminating the causes, but on the ability to withstand high-impact events and benefit from their occurrence. The main components of this approach include the creation of optionality, aimed at building flexibility into the organization; the control of the dispersion of results, so that negative impacts of future events are limited; and the adoption of strategies that reject the concept of moderate risk, prioritizing the avoidance of risks in areas subject to large-scale losses, while taking smaller risks in areas with high earnings potential.

From this perspective, Jones (2014) argues that antifragile systems aim to solve problems that conventional design methods cannot effectively address. Traditional systems are often designed to meet specific requirements and operate in predicted environments, becoming vulnerable when subjected to unexpected conditions or stresses that exceed their design parameters.

The author compares the challenges antifragile systems face by contrasting them with traditional systems (Table 10.2).

TABLE 10.1
Fundamental differences between resilience and antifragility

	Resilience	Antifragility
Concept	It seeks to absorb shocks and return to the previous state of equilibrium or to a functionally equivalent state [...] it can recover from disturbances, but necessarily it does not or improve as a result [...] it does not contemplate the possibility that institutions, services, or infrastructures will improve or gain from unexpected opportunities over time evolve.	It gains from disorder, becoming stronger and evolving in response to events, disturbances, and volatility [...] it benefits from errors and randomness, allowing it to deal with the unknown and improve continuously.
Optional	It doesn't offer this option.	It is based on optionality, [...] the availability of options, and flexibility to adapt to different scenarios.
Forecasts	Look for accurate forecasts.	It focuses on "soft predictions," examining the possible responses of the system to disturbances and detecting its fragility, robustness, resilience, or antifragility.
Adaptation	It involves reactive adaptation.	It seeks transformative, evolutionary, and proactive adaptation, promoting innovation.
Objectives	It aims to protect a system against shocks.	It seeks to create favorable conditions for the system to evolve and benefit from disorder.

Source: the authors (2025), based on Blečić and Cecchini (2019) and James and George (2014).

According to Jones (2014), antifragile systems represent a significant philosophical shift away from the rigidity of systems designed to meet fixed requirements and toward structures that can learn, adapt, and improve in the face of uncertainty and change.

James and George (2014) describe how antifragility relates to scenario planning (Table 10.3).

James and George (2014) point out that while traditional planning seeks to predict and prepare for specific futures based on causal relationships, antifragility adopts a more comprehensive approach to building systems and strategies capable of thriving in uncertain and unpredictable environments.

According to Passos (2018), several factors influence the degree of antifragility in organizations, including:

- The complexity of the organizational environment.
- Capacity for self-organization.

TABLE 10.2
Differences between problem-solving and adaptation in traditional and antifragility systems

	Traditional system	Antifragility
Unsustainable costs and schedules	The costs and time required to design, test, and maintain complex systems are increasing at an unsustainable rate.	They can potentially reduce these costs by adapting to new situations and improving performance over time.
Underperformance	They don't work as expected.	They can learn from experience, adapt to unforeseen events, and improve their performance in the face of adversity.
Maintenance and operations problems	They can present intermittent problems that are difficult to correct.	They can be designed to self-repair and adapt to changing conditions, reducing the need for maintenance and improving reliability.
Need for new technologies for autonomous systems	The need to develop technologies for truly autonomous [...] systems.	They can observe their environment, make decisions, and take action, learning from their experiences and adapting to new circumstances.
Inability to deal with uncertainty	Based on known requirements and forecasts of the operating environment.	They are designed to adapt and thrive in the face of uncertainty, adjusting their functionality and performance to meet the unknown.

Source: the authors (2025), based on Jones (2014).

- The ability to learn from mistakes and adapt to new situations.
- Interaction and the exchange of information between agents.
- Position of each agent in the network of relationships.
- Flexibility.
- Redundancy.
- Diversity.

For Passos (2018), antifragility is an advanced stage of resilience since it incorporates systems' ability to adapt and learn.

Jones (2014) lists some ways antifragile systems deal with the unexpected:

- **Learning from experience**—They are designed to learn from their experiences, allowing them to improve their performance over time, identify patterns, and adjust their behavior to better deal with future situations.
- **Adaptation to new circumstances**—They can adapt to new circumstances and unforeseen events and adjust their functionality and performance to meet the unknown rather than relying on predefined plans or models.
- **Real-time environment awareness**—Rather than being designed for what is expected, they are designed to access the operating environment in real

TABLE 10.3
Differences between traditional and antifragility systems

	Traditional system	Antifragility
Vision of uncertainty	Traditional scenario [...] is often based on a deterministic view of uncertainty, planning in which events are seen as determined by causes, even if they are difficult to identify.	It adopts a non-deterministic view, recognizing that events can be influenced by chance and indeterminacy.
Focus on cause and effect	It depends on narratives that describe chains of cause and effect.	It does not require causal constructions and therefore mitigates the potential fallibility in identifying the cause.
Dealing with the "surprising"	It has a limited capacity to deal with "surprise events."	It aims to thrive on chaos and be strengthened by unexpected events, rather than just protecting itself against them.
Strategies	Scenario planning approach, which often focuses on identifying weak signals and preparing contingencies.	It employs strategies such as optionality, control of the dispersion of results, redundancy, and small-scale experimentation to benefit from uncertainty.
Complementarity	It seeks to predict and prepare for specific futures based on causal relationships.	It can complement scenario planning by allowing organizations to position themselves to benefit from the unexpected, taking a broader approach and aiming to build systems and strategies that can thrive in an uncertain and unpredictable environment.

Source: the authors (2025), based on James and George (2014).

time and adapt in response to current events that do not need to be fully known in the design.

- **Capacity for self-organization and cooperation**—They can self-organize and cooperate to complete a mission, even if they are not centrally controlled. They learn from experience and adapt their behavior to better complete the mission.
- **System health management**—Collecting data relevant to condition and performance and transforming it into information that can be used to support operational decisions. They can use artificial intelligence to make autonomous decisions about preventive maintenance.
- **Self-repairing materials**—Inspired by biological systems, they can use self-repairing materials that can autonomously repair damage without human intervention; such materials can even adapt to increase strength or growth in response to stress.

In short, antifragile systems deal with the unexpected and are designed to learn, adapt, and improve in the face of uncertainty and change. They access the environment in real-time, self-organize and cooperate, use system health management and materials self-healing to thrive in chaos and disorder, concludes Jones (2014). In this view,

Tseitlin (2013) adds that failure induction plays a crucial role in building resilient and antifragile organizations, allowing systems to withstand failures and be strengthened by them. He describes practices to create a more antifragile system and organization:

- **Each employee acts as a service operator**, combining development and operations to respond to failures, making the service more resilient and tolerant of future failures.
- **Each failure is seen as a learning opportunity**. The aim is to detect failures more quickly, increase the system's resilience, and induce failure on a regular basis.
- **A blame-free culture is promoted**, where mistakes are accepted as part of the innovation process and used as learning opportunities.

In agreement with Jones (2014) and Tseitlin (2013), Blečić and Cecchini (2019) add that incorporating antifragility into the practice and content of planning can be achieved through three operational plans:

- **Negative way**: establishing general rules that restrict actions, avoiding weakeners, and removing the superfluous.
- **Shared vision**: deliberation on desirable futures, using the capability approach to expand agents' freedoms and capacities.
- **Project space**: creating a flexible space for actions and projects, combining top-down and bottom-up approaches with reversible and modular actions.

They go on to say that the legitimacy of antifragility planning can be built by ensuring that the antifragility of the system does not occur at the expense of the fragility of individuals; it must define and measure the gains in terms of capabilities, stipulating acceptable trade-offs about potential individual fragility:

- Stress tolerance.
- Non-monotonicity—not following a single fixed pattern.
- Absorption capacity.

Thus, the author concludes that instead of focusing on predicting and avoiding crises, organizations should seek to develop their antifragility, becoming more malleable and adaptable to any scenario.

Blečić and Cecchini (2019) conclude that systems should be designed to be antifragile whenever possible, ensuring that they can learn from experience, adapt to unforeseen events, and become stronger in the face of adversity. Rather than designing systems to meet fixed requirements, adopting an approach that allows these systems to thrive in chaos and uncertainty is essential.

10.7 CONCLUSION

Industry 5.0 is an evolution of industrial paradigms, balancing automation and artificial intelligence with human creativity and critical thinking, promoting personalization, sustainability, and workers' well-being (Silva, 2015). However, implementing artificial intelligence in industries has significant positive and negative impacts, which need to be assessed and mitigated to ensure a healthy and productive working environment.

According to Lanna (2018), the introduction of artificial intelligence allows for advances in productivity and operational efficiency, optimizing processes, reducing errors, and improving occupational safety. However, Sarmento (2024) points out that it also brings challenges, such as the replacement of jobs, job insecurity, and the growing pressure for technological adaptation, which can have a significant impact on workers' mental health, Fear of professional replacement, and uncertainty about the future increase levels of stress, anxiety, and depression, compromising well-being and quality of life (Veiga and Pires, 2018). In addition, the chronic stress caused by these changes is related to the premature aging of brain structures and predisposition to neurodegenerative diseases such as Alzheimer's (Girardello, 2004).

Faced with these challenges, adopting workers' health and safety strategies has become essential. As pointed out, Ergonomics 5.0 has emerged as a fundamental concept for balancing production demands with human needs, promoting a more resilient and sustainable work environment. Pinto, Tereso and Abrahão (2018) and Araújo and Rayol (2024) state that practices such as environmental enrichment, psychosocial support, mental health promotion, and continuous training should be implemented to mitigate the negative impacts of artificial intelligence. Training and retraining programs are also necessary to ensure the inclusion of workers in the new industrial model.

From this perspective, Blečić and Cecchini (2019) add that organizational resilience and antifragility must also be promoted so that companies and their workers can adapt and thrive in technological transformations. The authors advocate that companies that adopt antifragile strategies manage to minimize risks, strengthen their employees, and improve their performance in the long term. In this way, Industry 5.0, when well planned, has the potential to revolutionize industrial production and create a more humane and innovative working environment.

The conclusion is that introducing artificial intelligence in Industry 5.0 must be handled responsibly, considering the psychological and social impacts on workers. Implementing support programs, promoting an ergonomic and resilient work environment, and providing professional training are indispensable strategies to ensure that the benefits of this technological revolution are maximized without compromising workers' health and well-being. It is, therefore, essential that Industry 5.0 is structured inclusively and sustainably, guaranteeing prosperity for both organizations and their professionals.

REFERENCES

Abrahão, Júlia Issy; Silvino, Alexandre Magno Dias; Sarmet, Maurício Miranda. (2005). Ergonomia, Cognição e Trabalho Informatizado. *Psicologia: Teoria e Pesquisa, Brasília*, v. 21, n. 2, p. 163–171.

Abrantes, José. (2011). A Ergonomia Cognitiva e as Inteligências Múltiplas. In: *XIII Simpósio de excelência e gestão em tecnologia, 8., 2011. Rio de Janeiro: Gestão em Tecnologia.* v. 1, p. 1–14.

Agrawal, Ajay; Gans, Joshua S.; Goldfarb, Avi. (2019). Artificial intelligence: the ambiguous labor market impact of automating prediction. *Journal of Economic Perspectives*, v. 33, n. 2, p. 31–50.

Almeida, Ivanna Brasileiro de Sá; Oliveira, Francisco Ariclene; Silva Junior, Francisco Raimundo; Oliveira, Luis Adriano Freitas; Batista, Iany Tâmilla Pereira; Sampaio, Rodrigo Castro; Melo, Dayanna Cynthia Moura; Silva, Iandra Maria Sousa da. (2024). Intervenções preventivas de impacto redutivo em absenteísmo e afastamento por estresse ocupacional. *Journal Of Media Critiques, [S.L.]*, v. 10, n. 26, p. 1–20. South Florida Publishing LLC. https://doi.org/10.17349/jmcv10n26-005.

Andrade, Elias Sebastião de. (2020). Desenvolvimento sustentável e sociedade 5.0: rumo à felicidade e ao bem-estar. *Revista Humanitaris*, [S.I.], v. 2, n. 2, p. 1–20.

Araujo, Eduardo Wagner Santos de; Tacconi, Marli de Fátima Ferraz da Silva. (2023). A ergonomia cognitiva na vida do trabalhador. *Revista Estudo & Debate*, [S.L.], v. 30, n. 4, p. 180–203. Editora Univates. https://doi.org/10.22410/issn.1983-036x.v30i4a2023.3511.

Araujo, Fábia Melo de. (2020). A inteligência artificial e os seus impactos no mundo do trabalho. Tese (Doutorado).

Araujo, Fábia Melo de; Rayol, Rayane Araújo Castelo Branco. (2024). A inteligência artificial e os seus impactos no mundo do trabalho. Tst, [S. L.], v. 90, n. 3, p. 183–210.

Areias, Maria Elenice Quelho; Comandule, Alexandre Quelho. (2020). Qualidade de Vida, Estresse no Trabalho e Síndrome de Burnout. *Unicamp*, [S.I.], v. 1, n. 1, p. 183–202.

Ashrafi, Saeed; Majidi, Babak; Akhtarkavan, Ehsan; Hajiagha, Seyed Hossein Razavi. (2023). Efficient resume based re-education for career recommendation in rapidly evolving job markets. IEEE Access, IEEE.

Belo-Silva, Ariádne Elisa. (2020). Impacto do Estresse Crônico Imprevisível Sobre o Eixo HPA e o Comportamento do Tipo Ansioso em Camundongos Enriquecidos *[Dissertação (Mestrado em Farmacologia)]*. São Paulo: Instituto de Ciências Biomédicas, Universidade de São Paulo. *81 p.*

Berg, Christian. (2022). What is Industry 5.0? (In-depth guide with examples). Acesso em 01 fevereiro 2025. Disponível em https://www.clarify.io/learn/industry-5-0.

Blanco, Marília Bazan; Souza, Azair Liane Matos do Canto de. (2018) Ansiedade, memória e transtorno de estresse pós-traumático. *CES Psicol, Medellín*, v. 2, p. 53–65, dezembro de 2018. https://doi.org/10.21615/cesp.11.2.5.

Blečić, Ivan; Cecchini, Arnaldo. (2019). Antifragile planning. *Planning Theory*, [S.L.], v. 19, n. 2, p. 172–192. SAGE Publications. https://doi.org/10.1177/1473095219873365.

Brunoro, Claudio M.; Bolis, Ivan; Sznelwar, Laerte I. (2017). Contribuições da ergonomia da atividade e da psicodinâmica do trabalho para um trabalhar em uma perspectiva de sustentabilidade. *Revista Ação Ergonômica*, [S.L.], v. 12, n. 1, p. 9–17. Semestral. Editora Cubo. https://doi.org/10.4322/rae.v12n1.e201702.

Chagas, Maria Inês Orsoni. (2010). O estresse na reabilitação: a síndrome da adaptação geral e a adaptação do indivíduo à realidade da deficiência. *Acta Fisiátrica*, [S.L.], v. 17, n. 4, p. 193–199. *Universidade de Sao Paulo, Agência USP de Gestão da Informação Acadêmica (AGUIA)*. https://doi.org/10.11606/issn.2317-0190.v17i4a103393.

Chapanis, Alphonse. (1965). Man-Machine Engineering. London: Division Of Wadsworth Publishing Company, Inc. 132 p.

Chapanis, Alphonse. (1972) A engenharia e o relacionamento homem-máquina. São Paulo: Atlas. 153 p.

Coimbra, Marli Aparecida Reis; Ferreira, Lúcia Aparecida; Araðjo, Ana Paula Alves. (2020). Impactos do estresse na exposição ocupacional de bombeiros: revisão integrativa. *Revista Enfermagem Uerj*, [S.L.], v. 28, p. 1–9. Universidade de Estado do Rio de Janeiro. https://doi.org/10.12957/reuerj.2020.52825.

Comissão europeia. (2011). A avaliação do desenvolvimento socioeconómico - O Guia.Agência paraoDesenvolvimentoeCoesão.Acessoem05fevereiro2025.Disponívelemhttps://www.adcoesao.pt/wpcontent/uploads/guia_avaliacao_desenvolvimento_socioeconomico.pdf.

Cortez, Célia Martins; Silva, Dilson. (2007). Implicações do estresse sobre a saúde e a doença mental. *Arquivos Catarinenses de Medicina*, 36(4), 96–108.

Cruz, Élissa Jôse Erhadt Rollemberg; Souza, Norma Valéria Dantas de Oliveira; Amorim, Luanna Klaren de Azevedo; Pires, Ariane da Silva; Gonçalves, Francisco Gleidson de Azevedo; Cunha, Lidiane Passos. (2018). Resilience as an object of study of occupational health: narrative review / resiliência como objeto de estudo da saúde do trabalhador. *Revista de Pesquisa Cuidado É Fundamental Online, [S.L.]*, v. 10, n. 1, p. 283–288. Universidade Federal do Estado do Rio de Janeiro UNIRIO. https://doi.org/10.9789/2175-5361.2018.v10i1.283-288.

Cunha, Ícaro Guilherme Félix da; Baêta, Laís Fortes; Moreira, Rayane Henriques; Barbosa, Aline Gabriela da Silva. (2017). Ergonomia cognitiva: uma análise de seu impacto no desempenho humano. In: *XXIV SIMPEP, 24. Baurú. Anais. Baurú: Simpep.* v. 1, p. 1–14.

European Commission. (2021). Industry 5.0: Towards a sustainable, human-centric and resilient European industry. Brussels: European Commission.

Falzon, P.; Mollo, V. (2009). Para uma ergonomia construtiva: as condições para um trabalho capacitante. *Laboreal*, v. 5, n. 1, p. 61–69.

Fava, Mariana Caramore; Rocha, Aline de Oliveira; Bittar, Cléria Maria Lobo; Tonello, Maria Georgina Marques. (2019). Percepções sobre uma prática de yoga: impactos no estresse diário e na saúde do trabalhador. *Aletheia, [S.I.]*, v. 52, n. 1, p. 37–49.

Fernandes, Priscila Rodrigues; Anacleto, Juarez; Batiz, Eduardo Concepción; Hurtado, Ana Lúcia Berretta. (2014). O papel da cultura organizacional e a conscientização na gestão de Ergonomia com foco em Resiliência: ricardo de la caridad montero martínez. In: *3 Congresso De Engenharia de Produção da Região Sul,* 3. Joinville: Conepro-Sul. v. 1, p. 1–14.

Fernandes, Priscila Rodrigues; Bini, Edmundo; Valentina, Luiz Veriano Oliveira dalla; Batiz, Eduardo Concepcion. (2014). Aprovação de processos pelo viés ergonômico com foco em engenharia de resiliência. *In: XXXIV Encontro Nacional De Engenharia De Producao*, 34, 2014, Curitiba: Enegep. v. 1, p. 1–13.

Freitas, Vinicius Gomes de. (2017). Efeito do treinamento de força sobre os escores de depressão e níveis plasmáticos de BDNF em idosos institucionalizados: estudos de casos. 2017. 85 f. Dissertação (Mestrado) - *Curso de Programa de Pós-Graduação em Ciências do Esporta, Escola de Educação Física, Fisioterapia e Terapia Ocupacional, Universidade Federal de Minas* Gerais, Belo Horizonte.

Gąsiewski, Tomasz Adam. (2017). Influence of physical activity and cardiorespiratory fitness on brain structure and functioning: a review. *Physical Activity Review, [S.L.]*, v. 5, p. 19–28. Uniwersytet Humanistyczno-Przyrodniczy im. Jana Dlugosza w Czestochowie. https://doi.org/10.16926/par.2017.05.04.

Girardello, Ruy José Rueda. (2004). A relação entre o cortisol sanguíneo e o estresse pré-competitivo em lutadores de caratê de alto rendimento. *Dissertação (Doutorado) - Curso de Ciências Biológicas, Departamento de Educação Física, Universidade Federal do Paraná,* Curitiba.

Gonçalves, Juliana Machion. (2014). Ação Ergonômica e Estratégias de Operações: proposta de integração na prática (Tese de doutorado). Universidade Federal de São Carlos, São Carlos.

Hendrick, Hal; Kleiner, Brian. (2006). Macroergonomia: uma introdução aos projetos de sistemas de trabalho. Rio de Janeiro: Virtual Científica.

Hernández, Aneyquel Ricardo Gaitán; Xavier, Antônio Augusto de Paula; Nunes, Bianka Brekailo Silveira. (2019). Ergonomia verde e sustentabilidade: revisão de literatura. *GEPROS. Gestão da Produção, Operações e Sistemas*, v. 14, n. 2, p. 1–16.

Hespanhol, Alberto. (2005). Burnout e stress ocupacional. *Revista Portuguesa de Psicossomática, [S. L]*, v. 7, n. 12, p. 153–162.

Hirschle, Ana Lucia Teixeira; Gondim, Sônia Maria Guedes. (2020). Estresse e bem-estar no trabalho: uma revisão de literatura. *Ciência & Saúde Coletiva, [S.L.]*, v. 25, n. 7, p. 2721–2736. FapUNIFESP (SciELO). https://doi.org/10.1590/1413-81232020257.27902017.

Iida, Itiro; Buarque, Lia. (2016). Ergonomia: projeto e produção. 2. ed. São Paulo: Blucher. 850 p.

James, Derbyshire; George, Wright. (2014). Preparing for the future: Development of an 'antifragile' methodology that complements scenario planning by omitting causation, *Technological Forecasting and Social Change*, v 82, p 215–225, ISSN 0040-1625, https://doi.org/10.1016/j.techfore.2013.07.001.

Jones, Kennie H. (2014). Engineering antifragile systems: A change in design philosophy. *Procedia Computer Science*, [S.L.], v. 32, p. 870–875. Elsevier BV. https://doi.org/10.1016/j.procs.2014.05.504.

Karwowski W. The discipline of ergonomics and human factors: definition. In: Karwowski W, editor. *International encyclopedia of ergonomics and human factors*. 2nd ed. CRC Press; 2006. https://doi.org/10.1201/9780849375477-36.

Khogali, Hisham O.; Mekid, Samir. (2023). The blended future of automation and ai: Examining some long-term societal and ethical impact features. *Technology in Society*, v. 73, p. 102232, Elsevier.

Lanna, Antônio Bahury. (2018). Os impactos socio-econômicos da inteligência artificial. Contextura, Belo Horizonte, v. 1, n. 12, p. 21–30.

Latam, Totvs. (2021). Indústria 5.0: O que há de novo e quais são seus impactos? Acesso em 20 fevereiro 2025. Disponível em: https://es.totvs.com/blog/industria-4-0/industria-5-0-que-hay-de-nuevo-cuales-son-sus-impactos/.

Leandro, Carol Góis; Castro, Raul Manhães de; Nascimento, Elizabeth; Pithon-Curi, Tânia Cristina; Curi, Rui. (2007). Mecanismos adaptativos do sistema imunológico em resposta ao treinamento físico. *Revista Brasileira de Medicina do Esporte, [S. L]*, v. 13, n. 5, p. 343–348.

Lima, Daiciany Espírito Santo; Marques, Sandra Godinho Lagaron; Nilson, Gabriela; Pereira, Karen Pagliani. (2021). Aspectos favoráveis do exercício físico no controle das patofisiologias provocadas pelo estresse crônico. *Revista Didática Sistêmica, Rio Grande do Sul*, v. 1, n. 1, p. 79–94.

Lipp, Marilda E. Novaes. (1984). Stress e suas implicações. *Estudos de Psicologia, [S. L]*, v. 1, n. 34, p. 1–15.

Londoño, Laura Arciniegas; Ussa, German Darío Corzo. (2020). Contextualización de la cuarta revolución industrial, Industria 4.0, Industria 5.0 y tecnología 5G con el sector Defensa y Seguridad. *Perspectivas en Inteligencia*, v. 12, n. 21, p. 245–258.

Louro, Viviane; Louro, Fabiana dos Santos; Duarte, Plinio Gladstone. (2020). O estresse gerado pela pandemia como risco para adoecimento mental e físico do músico a partir das neurociências cognitivas. *Revista Música*, [S.L.], v. 20, n. 2, p. 379–396. *Universidade de Sao Paulo, Agência USP de Gestão da Informação Acadêmica (AGUIA).* https://doi.org/10.11606/rm.v20i2.178817.

Maffia, Lyovan Neves; Pereira, Luciano Zille. (2014). Estresse no trabalho: estudo com gestores públicos do estado de minas gerais. *Read, Porto Alegre*, v. 3, n. 79, p. 658–680.

Mainardi, Marco; Garbo, Angelo di; Caleo, Matteo; Berardi, Nicoletta; Sale, Alessandro; Maffei, Lamberto. (2014). Environmental enrichment strengthens corticocortical interactions and reduces amyloid-β oligomers in aged mice. *Aging Neuroscience, Roma*, v. 6, p. 1–11.

Malvestiti, Rosane; Maas, Larissa; Gontijo, Leila Amaral. (2017). Sobrecarga mental de trabalho como um fator desencadeador do estresse ocupacional em educadores infantis. *Espacios, Florianopólis*, v. 38, n. 45, p. 1–16.

Moraes, Anamaria de; Mont'Alvão, Claudia. (2012). Ergonomia: conceitos e aplicações. 3. ed. *Rio de Janeiro*: 2Ab. 120 p.

Oliveira, Edson Alves de. (2020). Delimitando o conceito de stress. Psicologia Clínica: PUC-Campinas, Campinas, v. 1, n. 1, p. 1–8.

Østergaard EH. Welcome to Industry 5.0. *InTech Magazine* [Internet]. 2018 [cited 2023 Sep 28]. Available from: https://www.isa.org/intech-home/2018/march-april/features/welcome-to-industry-5-0

Paschoal, Tatiane; Tamayo, Álvaro. (2005). Impacto dos valores laborais e da interferência familiar: trabalho sem estresse ocupacional. *Psicologia: Teoria e Pesquisa, [S. L.],* v. 2, p. 173–180. FapUNIFESP (SciELO). https://doi.org/10.1590/s0102-37722005000200007.

Passos, Danielle Sandler dos. (2018). Antifragilidade: como a ciência da complexidade pode auxiliar as organizações. 118 f. Tese (Doutorado) - Curso de Tecnologias e Arquitetura, Ciências e Tecnologias da Informação, Instituto Universitário de Lisboa, São Paulo.

Pena, Felipe Montes; de Araujo Faria, Magda Guimarães. (2024). Reflexões sobre o uso da inteligência artificial em ambientes de profissionais de saúde: impactos atuais e futuros. *Revista Contemporânea*, v. 4, n. 1, p. 2833–2846.

Pheasant, Stephen (2005). Bodyspace: Anthropometry, Ergonomics and the Design of Work. 3rd ed, Boca Raton, FL: CRC Press.

Pimenta, Marcio Luiz Reis e; Rosa, Bruno Leonardo; Pimenta, Paula Cristina Reis e. (2023). Perspectivas futuras para a gestão de projetos e sua relação com a sustentabilidade das organizações. *Revista De Gestão E Projetos*, v. 14, n. 3, p. 87–110. https://doi.org/10.5585/gep.v14i3.25016.

Pinto, Andréa Gonçalves; Tereso, Mauro José Andrade; Abrahão, Roberto Funes. (2018). Práticas ergonômicas em indústrias: natureza, gestão e atores envolvidos. *Revista Ação Ergonômica*, [S.L.], v. 13, n. 1, p. 278–285, 2018. Mensal. Editora Cubo. https://doi.org/10.4322/rae.v13e201823.

Régis, Josiana Florêncio Vieira; Campos, Ana Célia Cavalcanti Fernandes; Torres, Elvia Florêncio; Barbosa, Everton Rodrigues; Desidério, Vanessa. (2008). Ergonomia cognitiva do trabalho: um estudo baseado na gestão do conhecimento e no capital intelectual. In: *XV SIMPEP, Baurú,* p. 1–12.

Sakib, Nazmus; Anik, Fahim Islam; Li, Lei. (2023). Chatgpt in it education ecosystem: Unraveling long-term impacts on job market, student learning, and ethical practices. In: Proceedings of the 24th Annual Conference on Information Technology Education. [S.l.: s.n.], p. 73–78.

Sarmento, Luis Davi da Silva. (2024). Uma revisão sistemática do impacto da inteligência artificial no mercado de trabalho. *Curso de Bacharelado em Tecnologia da Informação, Centro Multidisciplinar de Pau dos Ferros, Universidade Federal Rural do Semi-Árido, Pau dos Ferros*. 36 f.

Silva, Ariádne Elisa Belo da. (2020). Impacto do Estresse Crônico Imprevisível Sobre o Eixo HPA e o Comportamento do Tipo Ansioso em Camundongos Enriquecidos. Dissertação (Mestrado) - Curso de Programa de Pós-Graduação em Farmacologia, Instituto de Ciências Biomédicas, Universidade de São Paulo, São Paulo. 40 f.

Silva, Edvânia da Glória e. (2023). O Efeito da Indústria 5.0 na Produção. *Dissertação- Curso de Engenharia Mecânica, Departamento de Engenharia Mecânica, Instituto Superior de Engenharia de* Lisboa, Lisboa. 187 f.

Silva EC. Efeitos do estresse crônico em áreas do cérebro. (2015). *Revista Eletrônica da Estácio Recife* [Internet], v. 1, n. 1, p. 1–10. Available from: https://reer.emnuvens.com.br/reer/article/view/12

Silva, Jorge Luiz Lima da; Dias, André Campos; Teixeira, Liliane Reis. (2012). Discussão sobre as causas da Síndrome de Burnout e suas implicações à saúde do profissional de enfermagem. *Aquichan*, v. 12, n. 2, p. 144–159.

Tipan, Andrea Valeria Travez; Garzon, Cristian Moises Villafuerte. (2023). Industria 5.0, revisión del pasado y futuro de la producción y la industria. *Ciencia Latina Revista Científica Multidisciplinar*, [S.L.], v. 7, n. 1, p. 1059–1070. Asociacion Latinoamericana para el Avance de la Ciencia. https://doi.org/10.37811/cl_rcm.v7i1.4457.

Tseitlin, Ariel. (2013). The antifragile organization. *Communications of the Acm, [S.L.],* v. 56, n. 8, p. 40–44. Association for Computing Machinery (ACM). https://doi.org/10.1145/2492007.2492022.

Veiga, Rui; Pires, Cristina Cadete. (2018). Percepção do impacto da inteligência artificial em contexto ocupacional. *Isla Multidisplinary, Porto*, v. 1, n. 1, p. 47–60.

Winkelmann, Caroline; Mager, Gabriela Botelho. (2019). Fatores humanos e comunicação de saúde: a relação entre ergonomia cognitiva e design da informação. In: 9° CIDI E 9° CONGIC, Belo Horizonte. Proceedings of the 9th CIDI and 9th CONGIC. Belo Horizonte: Sbdi. p. 1–12.

Witzel, Morgen. (2005). *50 Grandes Estrategistas de Administração. Tradução Paulo Cezar Castanheira*. São Paulo: Contexto.

11 Ergonomics of Gamified Systems in 5.0 Workplaces

Alexandra Nemery and Eric Brangier

11.1 INTRODUCTION

In a logistics warehouse, operators are tasked with processing a set number of parcels each day, earning symbolic rewards along the way. To learn about safety, employees engage in augmented reality (AR) simulations, navigating disaster scenarios to accumulate competence points. Maintenance technicians receive continuous feedback on their progress, with metrics based on customer satisfaction, resolved breakdowns, and pending issues. Fiction? No, these are the emerging realities of contemporary industry.

Welcome to Industry 5.0, where the gamification of work environments is rapidly gaining ground. Gamification presents a powerful opportunity to enhance employee engagement and performance, particularly in contexts where advanced technologies play a pivotal role. When thoughtfully designed, it can turn complex or repetitive tasks into motivating experiences, fostering a seamless integration of human and machine collaboration. However, the effectiveness of gamification hinges on ergonomically sound implementations tailored to meet the needs of operators and industry objectives.

The objectives of this chapter are to (1) explore the key concepts of gamification and their application in the workplace, (2) analyze the impact of Industry 5.0 on new forms of human-machine interaction and collaboration, and (3) reflect on the implications, opportunities, and challenges of gamification in evolving work environments.

11.2 BACKGROUND TO INDUSTRY 5.0 AND GAMIFICATION

The global industrial landscape is undergoing a profound transformation with the emergence of Industry 5.0 (Leng et al., 2022). This new paradigm marks a shift away from the predominantly technology-centric approach of Industry 4.0, placing human well-being at the core of manufacturing systems (Hassan et al., 2024). Industry 4.0—often referred to as the fourth industrial revolution—is defined by the integration of advanced digital technologies such as the Internet of Things (IoT), artificial intelligence (AI), cyber-physical systems, cloud computing, and big data analytics into industrial and manufacturing processes. This transformation has enabled the development of "smart factories," where machines, systems, and humans interact in real

DOI: 10.1201/9781003559399-13

time to optimize automation, predictive maintenance, and supply chain efficiency. By bridging the physical and digital worlds, Industry 4.0 enhances productivity, customization, and sustainability, while simultaneously challenging traditional workflows and demanding new skill sets from the workforce (Ghobakhloo, 2020).

Industry 5.0 aims to achieve social goals beyond mere employment and growth, striving for prosperity and sustainable development for all humanity through characteristics such as human-centricity, sustainability, and resiliency. This paradigm shift, driven by the European Commission's white paper, recognizes the importance of human collaboration, creativity, and overall well-being in establishing a stronger sense of connection and purpose for employees.

In this context, gamification has emerged as a powerful tool to support the humanization of the workplace (Cónegoa et al., 2024). By incorporating game design elements and mechanics into non-game settings, gamification presents a transformative approach to enhancing employee engagement, boosting motivation and productivity, promoting learning and skill development, encouraging collaboration, and ultimately contributing to organizational success in the era of Industry 5.0 (Paschek et al., 2022).

Although the term "gamification" gained prominence in the digital media sector around 2010, the concept of applying game-like elements in non-game contexts dates back further, with the first documented use appearing in 2008 (Krath et al., 2021). This concept builds upon earlier research in human-computer interaction and game studies related to serious games, playful design, and motivational affordances. Deterding et al. (2011) proposed a foundational definition of gamification as "the use of game design elements in non-game contexts." This definition highlights the application of elements found in games, such as points, badges, levels, leaderboards, challenges, and rewards, to motivate and influence user behavior in various domains beyond pure entertainment. The underlying principles of gamification often involve tapping into intrinsic human motivations like competition, achievement, collaboration, and a sense of progress.

Examples of gamification in work environments are becoming increasingly prevalent, particularly within the context of manufacturing and logistics, which are central to Industry 5.0 (Baroroh et al., 2024). Organizations are employing gamification for various purposes, such as employee onboarding through virtual tours and simulations, enhancing productivity and job satisfaction in warehouse operations using AR and game mechanics, improving work efficiency and quality, motivating knowledge transfer and skill development, increasing engagement in learning tasks, and even redesigning specific manufacturing tasks like bolt tightening on an assembly line using goal-based gamification. Furthermore, gamification is being explored to support human monitoring activities in highly automated manufacturing and to personalize assistance systems for operators by adapting information and feedback through gamified elements. The effectiveness of these gamified systems, however, is closely linked to the context of implementation and the characteristics of the users (Arumugam and Tyagi, 2023).

By considering the human element as a central pillar, Industry 5.0 calls for a careful analysis of how gamified solutions affect worker well-being, safety, and overall system performance—core concerns of ergonomics. Understanding the ergonomic implications of designing and implementing gamified solutions is crucial for ensuring that these systems truly contribute to a human-centric and sustainable industrial revolution (Breque et al., 2021).

11.2.1 Aims of Gamification in Industry 5.0

The integration of gamification into Industry 5.0 pursues multiple objectives focused on improving worker experience and well-being. A key goal is to prevent workplace accidents by making safety training and protocols more interactive and engaging. By using gamified simulations, particularly through AR or virtual reality (VR), employees can get up close and personal with risky tasks in a safe environment, strengthening their understanding of potential hazards and best practices for avoiding them. This approach aligns with the ergonomic principles of Industry 5.0, which emphasize creating comfortable, safe, and satisfying work systems for operators (Sophan et al., 2024).

Furthermore, gamification in Industry 5.0 is designed to improve decision-making and minimize errors. By presenting complex information and data in a playful and visually appealing way, workers can better understand situations and make more informed decisions. Gamified systems can also provide immediate performance feedback, facilitating real-time error detection and correction. The increased engagement generated by gamification can also lead to increased focus and alertness during tasks, particularly reducing the likelihood of human errors in production and logistics processes (Şenol and Onay, 2023).

Although gamification is not intended as a "mental health assistant," it helps improve employees' psychological well-being (Fleming et al., 2023). By improving job satisfaction, motivation, engagement, and enjoyment, gamification can make tasks less monotonous and reduce stress. Furthermore, gamification aims to improve worker capabilities by incentivizing the acquisition of new skills through challenges and rewards. The potential integration of AI into gamified systems allows for personalized assistance and learning paths, tailoring support to employees' individual needs and abilities. Mutual cognitive collaboration between human intelligence and AI in machines can also enhance each other's complementary strengths.

Finally, a fundamental objective of gamification in Industry 5.0 is to make tasks more engaging and enjoyable to use. By incorporating game elements such as points, badges, leaderboards, and challenges, gamification can boost employee motivation, making work less repetitive and fostering a sense of accomplishment (Foroughi et al., 2023). This process, however, must be guided by ethical considerations (Figure 11.1), as gamification operates at the intersection of three foundational pillars: game mechanics, psychology, and technology. These elements provide the structural base of effective gamified systems, while ethics stands at the apex of the framework—reminding designers and organizations to consider autonomy, fairness, transparency, and inclusivity in every layer of implementation. Creating a more pleasant work environment can not only increase employee satisfaction and reduce stress, but also encourage collaboration and knowledge sharing within the organization. By putting human well-being at the center, gamification contributes significantly to achieving the vision of a truly human-centric Industry 5.0.

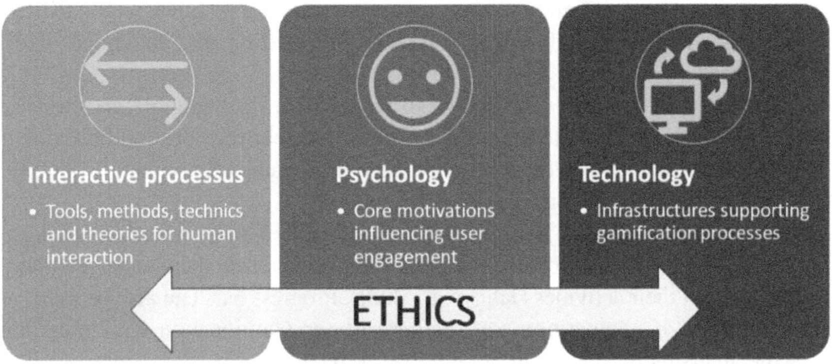

FIGURE 11.1 Fundamental structure of gamification in digital environments.

- Interactive Processus focuses on the tools, methods, techniques, and theories that facilitate human interaction.
- Psychology addresses the core motivations driving user engagement.
- Technology refers to the infrastructures that enable and support gamification processes.

At the core of these pillars lies Ethics, symbolized by the central arrow. It serves as a guiding principle, ensuring that the implementation of gamified systems respects user autonomy, privacy, and fairness across all dimensions.

11.2.2 Gamification in Workplaces

Gamification in workplaces leverages game-inspired elements—such as points, badges, leaderboards, and challenges—to enhance employee motivation, productivity, and engagement. But gamification has many dimensions. We will outline the systematic steps for integrating game mechanics into organizational settings, distinguish between structural and content-based approaches, and analyze real-world applications across industries. Dissecting these components provides a comprehensive understanding of how gamification functions as a strategic tool in modern workplaces.

11.2.3 List of Gamification Processes

Gamification—applying game-design elements in non-game contexts—plays a crucial role in improving ergonomics, reducing physical strain, and enhancing productivity (Marache-Francisco and Brangier, 2013b). Below is an exhaustive list of gamification processes tailored to ergonomic optimization:

- *Use of point systems:* this game element is often used to track progress and provide a sense of accomplishment, level systems to mark advancement, badge systems as visual symbols of achievements, and leaderboards to display progress and create a sense of competition (Lewis et al., 2016). Other elements include virtual currency, storytelling, and challenges with rewards.

- *Provision of rewards:* Rewards can be tangible or intangible and aim to motivate individuals to undertake actions and reinforce desired behaviors. This can include prizes, discounts, access to exclusive content, or simply recognition through badges (Richter et al., 2015).
- *Progress tracking and feedback*: Gamification often involves tracking participants' progress and providing real-time feedback to help them understand the quality of their activities (Jahn et al., 2021). Progress bars can also be used.
- *Creation of competition and collaboration:* Gamification can integrate mechanisms for competition between individuals or teams to increase the quality and speed of work. It can also encourage collaboration and teamwork to achieve common goals (Sailer and Homner, 2020).
- *Personalization:* Personalized gamification aims to offer an intimate experience to users to increase the effectiveness of the application in motivating specific behaviors (Rodrigues et al., 2020).
- *Narrative*: Narration in gamification weaves engaging storylines into tasks, transforming basic workflows into immersive experiences. By framing challenges as quests and progress as a journey, it boosts motivation (Palomino et al., 2019).
- *Use of digital platforms and tools:* Gamification is frequently implemented through the development of custom gamified applications, integration of game mechanics into existing software, or the use of specialized platforms (Jia et al., 2024). Game development tools can be used to create tailored interactive elements.
- *Adaptation to objectives and context:* Successful implementation of gamification requires aligning gamification elements with specific organizational goals and considering the financial and practical considerations related to equipment and devices (Trinidad et al., 2021).
- *User-centered approach and participation in design:* Employee satisfaction with a gamification strategy is positively related to their participation in the design process (Chen, 2019). Considering users when designing gamified solutions is important to respond to their needs.

These processes illustrate that gamification is a flexible and adaptable approach that uses game design elements to achieve various objectives in multiple contexts. By integrating these gamification processes, Industry 5.0 fosters ergonomic resilience, blending human well-being with technological efficiency. Future advancements may incorporate biometric AI adaptivity, further personalizing gamified ergonomic interventions.

11.2.4 Forms of Gamification

Gamification presents dual potential: it can significantly enhance productivity and engagement, yet demands careful design to avoid unintended consequences. Its implementation varies widely depending on specific goals and organizational contexts. Below is a structured overview of the main forms of gamification, categorized by their primary functions within industrial and organizational environments.

- *Learning and training support*: Serious games are interactive simulations designed to develop practical skills through immersive experiences (Armstrong and Landers, 2018). For example, VR welding training in manufacturing allows workers to practice techniques in a risk-free virtual environment. Microlearning quizzes break down complex topics into bite-sized challenges, such as safety protocol tests with instant feedback and point rewards. Branching Scenarios create decision-based narratives where choices directly impact outcomes, like troubleshooting equipment failures in a simulated factory setting.
- *Skills and performance enhancement:* Gamified Simulations replicate real work environments using AR or VR, such as an AR-guided assembly line that helps workers refine their precision (Salman et al., 2024). Mastery Paths structure skill progression through incremental unlocks, like earning a "Level 3 CNC Operator" badge after mastering advanced tasks. Real-Time Feedback Systems use sensors or AI to evaluate performance, such as posture trackers that award points for ergonomic movements (Basit et al., 2021).
- *Motivation and engagement:* Point Systems incentivize task completion by offering redeemable rewards, like productivity points traded for extra breaks. Badges and Achievements visually recognize milestones, such as a "Safety Champion" badge for 30 accident-free days. Progress Bars provide a clear visualization of advancement, like a "Daily Target Meter" showing warehouse pickers how close they are to their goals (Oke et al., 2023).
- *Social and collaborative engagement:* Team Challenges encourage collective problem-solving, such as reducing a department's energy use to unlock group bonuses. Peer Recognition features, like digital "kudos," allow coworkers to celebrate each other's contributions (Lithoxoidou et al., 2017). Leaderboards rank performance metrics, such as displaying the top 10 maintenance technicians based on efficiency scores.
- *Competition and risk management:* Time-Based Races add urgency to tasks while maintaining quality standards, like rewarding the fastest defect-free assembly. Risk/Reward Mechanics offer bonus points for adhering to safety protocols under pressure. Dynamic Difficulty adjusts task complexity based on user skill to prevent frustration or burnout (Moreta et al., 2016).
- *Behavioral and ergonomic improvement:* Habit-building games use streaks to reinforce positive actions, such as daily stretching routines tracked via an app (Lin et al., 2016). Biometric Feedback Games integrate wearables to turn health metrics into gameplay, like earning "health points" for reducing muscle strain.
- *Narrative and immersive experiences:* in industrial gamification, use storytelling and role-playing to transform tasks into engaging missions, enhancing motivation and learning (Ismagilova et al., 2023). Story-driven quests and role-based avatars can align closely with Industry 5.0's human-centric vision. Immersive technologies like VR and AR further deepen engagement. However, narratives must mirror real workflows to avoid diverting focus from operational priorities.

11.2.5 Tangible Implementation of Gamification

Gamification merges play with purpose, and tangible examples include modern board games and IoT. Industry 5.0 emphasizes human-machine collaboration, sustainability, and resilience, making experiential learning crucial. Board games are emerging as effective training tools by using tactile interaction. They simulate real-world scenarios, fostering problem-solving, teamwork, and decision-making in a low-risk environment. Serious games offer the possibility to teach resource optimization, waste reduction, and sustainable supply chain strategies. And role-playing games have been used to expose employees to crisis and resilience training. Scenario-based games prepare teams for disruptions (e.g., supply chain failures) through adaptive storytelling.

IoT plays a pivotal role by enhancing engagement, training, and operational efficiency. Industrial Internet of Things enables manufacturing measurement by combining domains and easing the interpretation of sensor data. This technology succeeded to improve interpretation from different users and enhance their performance (Annanth et al., 2021). Other initiatives mentioned IoT-based smart grid as a way to merge AI and technological devices. Samsung was able to help farmers build sustainability and improve food security (Watts et al., 2024).

Whatever the field, gamification cannot be efficiently implemented if it is not designed based on a good understanding of the users and their context of use. Thus, seven core principles are needed:

1. Analyze the work organization to determine if gamification would be relevant: Is it a technological gadget or does it really lead to a beneficial evolution of work organization (Marache-Francisco and Brangier, 2015).
2. Give the possibility to choose whether to work with a gamified system or not (Marache-Francisco and Brangier, 2013b, 2014): giving the user the freedom to exercise the user's own will, for example being allowed to disable functionalities, or allowed to opt out of the gamified system;
3. Develop shared benefits (Deterding et al., 2011): The influences of gamification must be relevant both to the system owners, who expect positive consequences, and to the end users themselves. It's about seeing how gamification would contribute to human, technological, and social performance. It means anticipating the advantages and disadvantages.
4. Promote a personalized experience (Nicholson, 2012): Different user profiles can lead to several different designs. This is where the added value of gamification comes from, through triggers adapted to each user profile.
5. Consider maintaining long-term interaction (Marache-Francisco and Brangier, 2013b, 2015): Design keeping in mind the evolution of interaction, especially concerning motivational elements, without which motivation will suffocate due to lack of new stimulation.
6. Anticipate unwanted side effects: Unwanted effects may include stress induced by pressures related to efficiency requirements (Apter, 1991), loss

of the feeling of privacy and credibility, circumvention of system rules, or focus on quantity rather than quality to obtain a reward.

7. Ensure compliance with legal and ethical issues (Werbach and Hunter, 2020): Take into account the existing legal context, for example regarding data and privacy, as well as the interests of end users.

11.2.6 Ethical Considerations in Gamification for Industry 5.0

Gamified interfaces present both ethical risks and benefits in that they encourage certain behaviors, suppress others, and shape cognitions, emotions, and more broadly opinions. Omitting this aspect of gamification could have serious repercussions on organizational behaviors. It is clear that gamification is capable of causing harm to human beings unless prevented by adding an ethical component in the interfaces themselves, in design principles, and in organizational operations.

From this perspective, ethics informs us about how individuals behave when faced with dilemmas. Imagine yourself as a taxi driver or delivery person. The end of the day is approaching and you want to go home and spend time with your children. But a message kindly suggests that you work a few more minutes: "Do you really want to quit your application?", topped with "dollars" and followed by a small text saying there is money to be made … What will you do: work 15 more minutes or spend 15 more minutes with your family? You are therefore facing a dilemma that relies on a gamified interface that can strongly suggest one option over another. Gamification thus goes beyond nudges by accentuating a contextual element (dollars), activating the shortcut (earning more money), which in turn is likely to trigger the expected behavior from the worker (I work a little longer than I had planned). Gamification is then a magic recipe for bringing a worker in the desired direction, without exerting either positive or negative pressure, by asking them to choose between two options.

Gamification constitutes a gentle manipulation that subtly leads toward a profitable, reasonable, or responsible option, without prohibiting other possible choices. Thus, formatted by algorithms and gamification, the management of professional activity becomes an engaging interaction between a worker and persuasive prewiring where interactive design plays a role as important as the algorithm that optimizes work.

Gamification dresses up algorithms. More consistent and convincing than nudges (Thaler and Sunstein, 2008), gamification goes beyond a friendly encouragement to change behavior in a way that makes it more positive. It offers subtle, fine, and implicit incentives to orient behavior toward an external expectation. Gamification is often more effective than direct instruction or a reminder of guidelines. As a result, human fears manifest: being led to perform tasks through interactive manipulations stemming from an ethical insufficiency of these interfaces and their designers. With this new form of interactive servitude, can behaviors contrary to ethics be promoted?

Ethics must become the keystone of design in order to clarify how people might behave in ethical dilemmas. Werbach and Hunter (2012, 2020) have highlighted the importance of gamification for the development of organizations. They have already outlined the risks of manipulation, the limits of physical fatigue, and the ethical issues involved. Let's develop these aspects further.

11.2.7 The Benefits of Gamification in Industry and Beyond

Gamification has revolutionized industries by integrating game-like elements into everyday tasks, creating engaging and motivating experiences. By leveraging tools such as points, badges, leaderboards, and immersive storytelling, it transforms mundane activities into dynamic challenges (Polyanska et al., 2022). Research shows that this approach significantly enhances learning, productivity, and collaboration while fostering a more enjoyable work environment.

One of the most significant advantages of gamification lies in its capacity to enhance learning and skill development. Interactive approaches—such as serious games and simulations—have demonstrated considerably higher effectiveness than traditional training methods, with some studies reporting up to 90% greater knowledge retention (Redelinghuys, 2021). For example, VR-based welding training in manufacturing allows workers to practice in a risk-free virtual space, while microlearning quizzes reinforce safety protocols through instant feedback. Mastery paths, such as earning "Level 3 Operator" badges, further encourage continuous improvement by providing clear milestones for skill progression.

Beyond training, gamification boosts motivation and engagement by tapping into intrinsic psychological rewards. Reward systems like points and badges create a sense of achievement, turning repetitive tasks into fulfilling challenges. Progress bars and quest-like objectives, such as completing "30 accident-free days," give employees tangible goals, reducing burnout and increasing participation. Platforms like Duolingo demonstrate this principle effectively, where gamified language learning leads to higher user retention and success rates.

The impact of gamification on performance and productivity is equally compelling. Real-time feedback mechanisms, such as posture trackers that score ergonomic movements, help workers refine their techniques while minimizing strain (Landers et al., 2017). Competitive features like leaderboards and timed challenges encourage employees to optimize their efficiency without sacrificing quality—for instance, assembly line workers striving for the fastest defect-free production. These gamified systems create measurable improvements in output while maintaining high standards.

In addition, gamification reinforces collaboration and teamwork by fostering a sense of shared purpose. Team-based challenges, such as departmental energy-saving goals, unite employees toward common objectives, while peer recognition tools like "kudos" systems celebrate individual contributions. Particularly in remote or physically isolated work environments, these social features play a vital role in maintaining morale and cohesion, ensuring that individual efforts align with collective success (Vegt et al., 2015).

Another critical benefit is the promotion of healthier workplace behaviors. Habit-building games, such as daily stretch reminders, encourage employees to adopt ergonomic practices, reducing the risk of injury. Wearable devices that gamify health metrics—like earning "health points" for proper posture—turn well-being into an engaging activity rather than a chore. Continuous improvement could be obtained through gamified feedback loops (Lin et al., 2016).

Gamification significantly lowers training and onboarding expenses by streamlining the learning process and reducing reliance on traditional, resource-intensive methods. Interactive simulations and microlearning modules allow employees to train at their own pace, minimizing downtime and the need for in-person instructors. VR and AR tools, for example, enable hands-on practice without the costs associated with physical equipment or materials (Haoming and Wei, 2024). Additionally, gamified systems often require less time to achieve competency, accelerating the integration of new hires and reducing the financial burden of extended training periods. By making learning more efficient and scalable, organizations can achieve faster ROI while maintaining high engagement and retention rates.

Finally, gamification provides valuable data-driven insights. By tracking performance trends, skill gaps, and engagement levels, organizations can identify areas for improvement and tailor interventions more effectively. This analytical approach ensures that gamification strategies remain aligned with real-world objectives, maximizing their impact.

Overall, gamification is far more than a trend—it's a powerful tool that enhances learning, motivation, productivity, and well-being across industries. As highlighted by researchers, its success depends on thoughtful design that balances entertainment with functionality, ensuring that the benefits translate into tangible real-world outcomes. Whether in manufacturing, healthcare, or education, gamification proves that work and play can coexist to drive meaningful progress.

11.2.8 The Limitations and Risks of Gamification

While gamification can be a powerful engagement tool, it is not universally applicable and carries several risks if poorly implemented. A fundamental limitation is that gamification is not suitable for all tasks or contexts (Behl et al., 2024). Complex, creative, or highly specialized work often resists gamification, as reducing such activities to points or levels may oversimplify their value.

One significant concern is that gamification tends to prioritize surface skills—measurable, quantifiable behaviors—while neglecting deeper competencies like critical thinking or emotional intelligence. This can create a mismatch between rewarded behaviors and actual job requirements. Additionally, an excessive focus on game mechanics may detour from core task goals, as employees might prioritize "winning" the game over achieving meaningful outcomes (Hamari et al., 2014).

Psychological risks are equally pressing. Poorly balanced reward systems can demotivate employees, especially if they perceive goals as unattainable or comparisons as unfair. Leaderboards, for instance, may discourage those consistently at the bottom (Krath et al., 2021). There's also a risk of infantilization, where adult professionals feel patronized by simplistic game elements that trivialize their expertise.

Although gamification can boost engagement for some learners, it risks alienating those who prefer traditional methods or perceive game-like mechanics as trivializing. Systems that reward only active participation (Wang et al., 2025) may inadvertently marginalize reflective or self-paced learners. Critically, poorly designed gamification can infantilize users through overly simplistic rewards or cartoonish interfaces (Marache-Francisco and Brangier, 2020), eroding trust and triggering rejection. Rapid gratification loops—such as points or badges—may distract users from core objectives, causing them to prioritize "winning" over understanding the business context. This undermines the system's perceived usefulness, as users question its relevance to real-world tasks. Additionally, excessive gamification can introduce operational complexity, slowing workflows and reducing perceived efficiency. When users interpret these elements as patronizing, resistance intensifies, compounding disengagement. To avoid backlash, designers must balance game mechanics with transparency, aligning incentives with meaningful skill development and organizational goals.

Gamification succeeds only when carefully tailored to context, audience, and purpose. Without ethical design and genuine alignment with organizational values, it risks alienating employees or reducing work to a hollow game.

11.2.9 Ethical Perspectives

Broadly speaking, gamification ethics encompasses the various ways in which game mechanics influence human behavior—ranging from subtle persuasion to more unconscious forms of manipulation. While gamification is often employed to enhance motivation, performance, and task engagement, it can also introduce unintended negative consequences, including:

- Encouraging over-performance, workaholism to the detriment of employee well-being,
- Excluding learners who don't like computer games, as traditional teaching methods remain valid for them,
- Reducing transparency, as users are often unaware of the true objectives and intentions behind gamified systems,
- Compromising fairness, a critical aspect of gamification design. Poorly calibrated reward mechanisms may unintentionally favor certain groups over others, leading to discrimination or exclusion.

These ethical risks highlight the need to design gamified experiences that enhance user engagement without manipulation or exploitation. Gamification should remain a supportive and equitable tool—one that respects the autonomy and dignity of its users.

In the workplace, the expansion of gamified technologies raises a host of social, political, and moral questions. Employees are increasingly expressing concerns about the ethical implications of these systems. By integrating playful elements into non-playful contexts (work, education, health, etc.), gamification can be a powerful lever for professional motivation and commitment. However, its use in work situations poses ethical dilemmas, particularly in terms of manipulation, voluntary servitude, critical thinking, well-being, and organizational transparency ... in short, a whole range of ethical issues that ergonomics needs to address. To navigate these challenges, an ethical guide to gamification design practices can serve as a valuable reference (see Figure 11.2). This guide is structured around five key principles:.

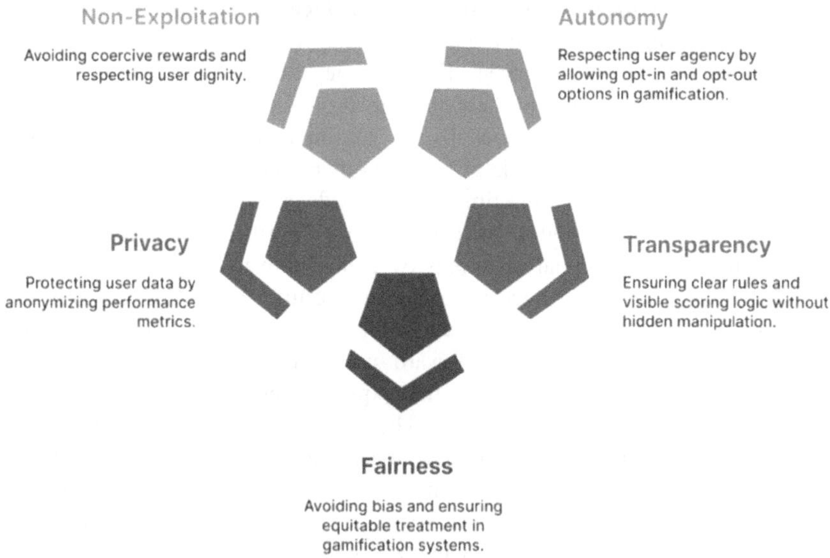

FIGURE 11.2 Ethical principles in the design of gamified systems.

1. Non-exploitation: Avoiding the instrumentalization of users solely for productivity gains;
2. Autonomy: Ensuring that participation is voluntary and supports individual agency;
3. Transparency: Clearly communicating the goals, mechanics, and data use of gamified systems;
4. Fairness: Designing reward systems that are inclusive and equitable;
5. Privacy: Safeguarding user data and respecting personal boundaries.

Together, these principles serve as a framework to guide ethical decision-making and foster trust in gamified environments.

Each of these pillars aims to ensure that gamified systems respect users' rights, protect their data, foster trust, and promote agency and equity, rather than merely maximizing engagement at all costs (Figure 11.2). By aligning gamification strategies with ethical design practices, organizations can leverage their benefits without compromising on integrity or well-being.

Gamification leads to abuses, while raising fundamental ethical questions about the legitimacy of gamifying interactions to manipulate organizational behavior. In this respect, existing ethical theories can provide a suitable framework for, depending on professional contexts, proposing the resolution of practical moral questions.

11.2.10 Ethical Regulation of Gamification

The ethical regulation of gamification comprises three distinct and complementary elements:

- Integrating **ethical criteria into the interfaces** themselves: this involves implementing possible and impossible uses of gamification. Here, the interface becomes an ethical agent. For example, some people's scores may be made invisible because of their previous professional experience, so as not to scare off beginners. Here, it's the gamified interface that is endowed with the resources to regulate itself. Another example is cross-referencing diary items (e.g., spending time with my kids, rather than wanting to earn more money by continuing my cab business). Family dilemmas (being with family) vs. professional dilemmas (working more for more money) could be adjusted and prioritized according to the information gathered on the employee's activity. Thus, the ethics of gamified systems aim to ensure that the behavior of machines toward human users is ethically acceptable, hence the need for technologies that explicitly represent ethical principles (Anderson and Anderson, 2007). The ethical sophistication of gamified interfaces should develop our willingness to prevent technologies from doing harm, as well as prevent humans from unfairly avoiding responsibility for their actions. This induces the idea that the moral competence of technical systems (Malle and Scheutz, 2014) could increase the moral competence of humans. However, we still need to seriously identify the moral expectations of machines.
- The ethics of **the designers who design gamified interfaces**: what do they expect from ethics in gamifying interfaces? Utilitarianism plays a central role here, as it comes into potential conflict with ethics. Is there a good reason for the interface to engage the user without explicitly admitting it? Is it useful and ethical for the individual or the organization to adopt a behavior without his or her knowledge? On the other hand, a professional activity is inseparable from its raison d'être; it is finalized, and respect for the employee's autonomy is implicitly based on implicit consent, in compliance with the law and without interference from unethical behavior.
- The **ethics of organizations** working with and developing gamified interfaces: why and how do they move toward a globalized ludo-capitalism? Gamification also corresponds to an ideological process of transformation of work and its organization. It produces uses that go beyond a simple modus operandi and can reshape users' cognition and emotions. In this respect, gamification is far from neutral. Games "illustrate the moral and intellectual values of a culture" (Caillois, 1967, p. 76). The current characteristics of gamification (measurement of behavior by managers, real or virtual monetization of work and leisure, pitting employees against each other) make us wonder about its underlying value. With gamification, we're a long way from the reserved, protected, and friendly world of childhood games. Games, as Triclot (2011) argues, are political by nature. In this sense, gamification becomes a mechanism of translation, turning diverse activities into performance indicators to be tracked, optimized, and rationalized. It embodies what Joseph (2013) refers to as Ludocapitalism: a system that entangles contradictory logics—persuasion and control, productivity and pleasure, motivation and subordination. While gamification can foster engagement, satisfaction, and desire, it

> also risks deepening forms of psychological alienation. It encourages workers to internalize performance imperatives, turning self-improvement into a continuous game loop. Yet, despite generating these psychological effects, gamification often fails to confront the structural sources of alienation embedded in contemporary work environments. In this light, ethical reflection on gamification must go beyond questions of design mechanics or user well-being. It should critically examine the socio-political logics it serves and the kind of world of work it contributes to shaping.

Gamification is, therefore, also an ideology of productive work. It evacuates the question of alienation at work by setting up lures, i.e., seductive and stimulating interfaces. In tomorrow's company, ergonomics will have a crucial role to play in defusing the riskiest trends and ensuring that we benefit from this redefinition of our relationship with work.

11.2.11 Gamification and the Organization of Ethics in Professional Conduct

Far from being monolithic, the ethics of gamification explores various moral perspectives on the world, particularly in the corporate sphere. Rooted in personal, philosophical, religious, normative, financial, legal, or political positions, this moral framework takes diverse forms that aim, from very different perspectives, to:

- Defend privacy
- Preserve professional lives
- Promote institutional transparency
- Develop democracy in social and professional life

On the most activist side, the moral vision of the "cypherpunks" advocates for the use of strong digital cryptography to protect individuals' privacy and promote institutional transparency in the digital age (Anderson, 2022). Supporters of Julian Assange, this movement champions an ethics that fiercely defends against surveillance, protects privacy, supports whistleblowing, upholds democracy, fights censorship, safeguards journalism, and questions intellectual property and power structures. Conversely, state and corporate surveillance promote a morality of control based on a digital panopticon, justified by security narratives and an ideology of mass personal data collection—including through games and gamified applications. Certain government agencies exemplify this approach. Other moral orientations are driven by a form of neo-paternalism that seeks to regulate data and associated behaviors. This includes advocates of a centralized and regulated Internet to protect users, prevent misinformation, and combat cybercrime. Finally, the moral framework of some companies relies on a centralized digital identity system that grants access to technical systems, where usage is recorded, analyzed, and sometimes controlled—such as in certain remote work situations. Although these systems are justified by security and corporate efficiency, they conflict with the cypherpunk vision of decentralization and absolute privacy, even in professional contexts.

11.2.12 The Role of the Ergonomist: Gamification Experiences as Ethical Collaboration

In reality, labor laws in many countries grant significant power to employers to monitor their employees' digital activities. As a result, companies, by gamifying interfaces, often position themselves as defenders of a centralized, monitored, and controlled Internet—where user identity is known and privacy is subordinated to financial, economic, security, technical, social, and managerial imperatives.

For the ergonomist, the primary advantage of ethical considerations is the ability to organize a controversy, a debate, or a critical discussion about work, its meaning, and the methods for accomplishing it. This controversy, conducted with practitioners from various disciplines, challenges simplistic intuitions and weak convictions, encouraging a close analysis of work and the construction of shared—even consensual—work experiences. Integrating ethics into ergonomic intervention thus gives moral considerations a practical value for designing future work systems that align with organizational culture. Ethics becomes risks and benefits (Tab. 11.1).

Does this make the ergonomist a "game designer"? In part, yes—ergonomists increasingly take on responsibilities that align with game design principles, particularly in the context of gamification. Their role may involve tasks traditionally associated with gamification specialists, such as:

- Identifying situations and domains where gamification can improve motivation and productivity,
- Establishing an effective gamification development team, including project managers, designers, programmers, engineers, etc.
- Avoiding counterproductive incentives,
- Ensuring the security and ethical use of user data.

However, in other aspects, the tasks remain within the scope of ergonomics:

- Analyzing work situations to be gamified,
- Designing gamified systems that align with real work processes,
- Managing dialogue and coordination with various industry stakeholders,
- Safeguarding and promoting operator health, well-being, and sustainable performance,

TABLE 11.1
Gamification mechanics with ethical implications

Mechanic	Ethical risk	Ethical benefit
Leaderboards	Social pressure, shame for low performers	Healthy competition, recognition
Points/badges	Over justification effect (↓intrinsic motivation)	Skill validation, progress tracking
Real-time feedback	Surveillance concerns	Ergonomic improvement, safety
Narratives/quests	Trivialization of serious tasks	Enhanced engagement, contextual learning
Personalization	Privacy invasion (data collection)	Tailored support, inclusivity

- Crafting user experiences that are both engaging and human-centered,
- Respecting and enforcing ethical standards and advocating for transparency and fairness in workplace practices.

New professional responsibilities are enriching the scope of ergonomics and the practice of ergonomists. This reinforces a more comprehensive approach that aims for a balanced and respectful gamification—one that values both the nature of work and the humanity of the worker.

11.3 CONCLUSION

This chapter offers an in-depth examination of the integration of gamification into Industry 5.0 work environments, bridging theoretical foundations with practical applications, while addressing the associated challenges and ethical implications. To navigate these complexities, we propose a gamification framework (Figure 11.3) structured around four key dimensions: self-determination theory, user understanding, ethics, and game design. This multidimensional approach is designed to ensure that gamified systems are not only engaging and user-centered but also ethically responsible and structurally robust.

- **User Understanding**: Focuses on identifying user personas, needs, and pain points to design more relevant and engaging gamified experiences.
- **Self-Determination Theory**: Enhances user motivation by supporting autonomy, competence, and relatedness.
- **Ethic**: Emphasizes fairness, user well-being, and the avoidance of manipulative practices.

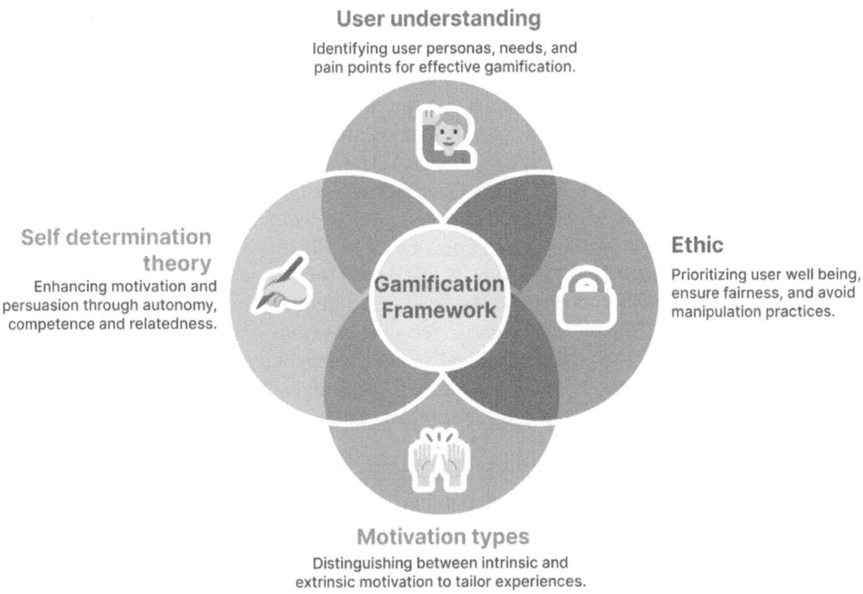

FIGURE 11.3 Conceptual framework of gamification.

- **Game Design**: Involves the practical and structural tools used to implement gamification mechanics effectively.

Together, these components create a balanced approach to gamification that is both **engaging and ethically grounded**, ensuring a user-centered experience that fosters trust and motivation.

Beneath the surface, we witness a shift from an ergonomics focused on effort and work constraints to one centered on pleasure and playful interactions in the workplace—an evolution from ergonomics to hedonics. This marks a new transition in ergonomics and human factors engineering, moving from the prevention of pain to the promotion of pleasure and well-being (Oron-Gilad and Hancock, 2017). But is this truly progress, or just another form of alienation?

Gamification does not fundamentally change the nature of work; rather, it **reframes how work is experienced** by introducing aesthetic, motivational, and socio-technical overlays. Points, trophies, badges, and social recognition hold value only within the gamified system and rarely translate into actual career advancement. As such, gamification may deepen engagement while simultaneously distorting the worker's relationship to their labor—encouraging emotional investment without tangible rewards.

This raises essential questions about the ergonomics of gamification—not only concerning interface design or the effectiveness of gamification teams but also its broader impact on organizational dynamics as a whole.

REFERENCES

Anderson, M., & Anderson, S. L. (2007). Machine ethics: Creating an ethical intelligent agent. *AI Magazine*, 28(4), 15–26.

Anderson, P. D. (2022). *Radical Ethics for the Digital Age*. Imprint Routledge. doi:10.4324/9781003220534

Annanth, V. K., Abinash, M., & Rao, L. B. (2021, July). Intelligent manufacturing in the context of industry 4.0: A case study of siemens industry. *Journal of Physics: Conference Series*, *1969*(1), 012019. IOP Publishing.

Apter, M. J. (1991). A structural-phenomenology of play. In H. Kerr & M. J. Apter (Eds.), *Adult Play. A Reversal Theory Approach* (pp. 13–22). Swets & Zeitlinger Amsterdam.

Armstrong, M. B., & Landers, R. N. (2018). Gamification of employee training and development. *International Journal of Training and Development*, *22*(2), 162–169.

Arumugam, S. K., & Tyagi, A. K. (2023). Gamification for Industry 5.0 at the core of Society 5.0. In *Concepts, technologies, challenges, and the future of Web 3* (pp. 307–336). IGI Global Scientific Publishing. https://doi.org/10.4018/978-1-6684-9919-1.ch016.

Baroroh, D. K., Santoso, H. B., & Van, N. T. H. (2024). Gamification for Manufacturing (GfM) Towards Era Industry 5.0. In *2024 21st International Conference on electrical engineering/Electronics, Computer, Telecommunications and Information Technology (ECTI-CON)* (pp. 1–5). IEEE Thailand Section, Bangkok.

Basit, A., Hassan, Z., Omar, N., Sethumadavan, S., & Jhanjhi, N. Z. (2021). Gamification: A Tool To Enhance Employee Engagement And Performance. *Turkish Online Journal of Qualitative Inquiry*, *12*(5), 3251–3269.

Behl, A., Jayawardena, N., Bhardwaj, S., Pereira, V., del Giudice, M., & Zhang, J. (2024). Examining the failure of gamification in implementing innovation from the perspective of problematization in the retail sectors of emerging economies. *Technovation*, *129*, 102902.

Brangier, E., Marache-Francisco, C. (2020). Measure of the Lived and Functional Effects of Gamification: An Experimental Study in a Professional Context. In: Rebelo, F., Soares, M. (eds) *Advances in Ergonomics in Design. AHFE 2019. Advances in*

Intelligent Systems and Computing, (Vol. 955, pp. 242–253). Springer, Cham. https://doi.org/10.1007/978-3-030-20227-9_22

Breque, M., De Nul, L., & Petridis, A. (2021). *Industry 5.0: Towards a sustainable, human-centric and resilient European industry* (p. 46). *Luxembourg, LU: European Commission, Directorate-General for Research and Innovation, e-document.* https://research-and-innovation.ec.europa.eu/knowledge-publications-tools-and-data/publications/all-publications/industry-50-towards-sustainable-human-centric-and-resilient-european-industry_en#details

Caillois, R. (1967). *Les jeux et les hommes*. Paris: Gallimard.

Chen, Y. (2019). Exploring design guidelines of using user-centered design in gamification development: A Delphi study. *International Journal of Human–Computer Interaction*, *35*(13), 1170–1181.

Cónegoa, L., Pinto, R., Pinto, J., & Gonçalves, G. (2024). Leveraging gamification in Industry 5.0: Tailored solutions for workplace' employees. *Procedia Computer Science*, *232*, 1769–1778.

Deterding, S., Dixon, D., Khaled, R., & Nacke, L. (2011a). From game design elements to gamefulness: defining "gamification". In *Proceedings of the 15th International academic MindTrek conference: Envisioning future media environments* (pp. 9–15). ACM Digital Library. https://doi.org/10.1145/2181037.2181040

Deterding, S., Sicart, M., Nacke, L., O'Hara, K., & Dixon, D. (2011b). Gamification. Using game-design elements in non-gaming contexts. In *CHI '11 Extended Abstracts on Human Factors in Computing Systems* (pp. 2425–2428). ACM Digital Library. https://doi.org/10.1145/1979742.1979575

Fleming, T., Poppelaars, M., & Thabrew, H. (2023). The role of gamification in digital mental health. *World Psychiatry*, *22*(1), 46.

Foroughi, B., Iranmanesh, M., Kuppusamy, M., Ganesan, Y., Ghobakhloo, M., & Senali, M. G. (2023). Determinants of continuance intention to use gamification applications for task management: An extension of technology continuance theory. *The Electronic Library*, *41*(2/3), 286–307.

Ghobakhloo, M. (2020). Industry 4.0, digitization, and opportunities for sustainability. *Journal of cleaner production*, *252*, 119869.

Hamari, J., Koivisto, J., & Sarsa, H. (2014). Does gamification work? -- A literature review of empirical studies on gamification. In *2014 47th Hawaii international conference on system sciences* (pp. 3025–3034). IEEE Xplore. DOI: 10.1109/HICSS.2014.377.

Haoming, L., & Wei, W. (2024). A systematic review on vocabulary learning in AR and VR gamification context. *Computers & Education: X Reality*, *4*, 100057.

Hassan, M. A., Zardari, S., Farooq, M. U., Alansari, M. M., & Nagro, S. A. (2024). Systematic analysis of risks in industry 5.0 architecture. *Applied Sciences*, *14*(4), 1466.

Ismagilova, G., Lysenko, E., & Bozheskov, A. (2023). Gamification in industry: Simulation-game modeling of production processes. In *International conference on professional culture of the specialist of the future* (pp. 231–245). Cham: Springer Nature Switzerland.

Jahn, K., Kordyaka, B., Machulska, A., Eiler, T. J., Gruenewald, A., Klucken, T., & Niehaves, B. (2021). Individualized gamification elements: The impact of avatar and feedback design on reuse intention. *Computers in Human Behavior*, *119*, 106702.

Jia, F., Bao, X., & Yu, J. (2024). Gamification on digital platform: A meta-analysis of affordance on behavior from value perspective. *Electronic Commerce Research and Applications*, (pp. 1–15). Elsevier Science Publishers. https://doi.org/10.1016/j.elerap.2024.

Joseph, D. (2013). A new philosophy of play: The entanglement of gamification and work. In *Proceedings of DiGRA 2013: Defragging Game Studies*, Atlanta, USA.

Krath, J., Schürmann, L., & Von Korflesch, H. F. (2021). Revealing the theoretical basis of gamification: A systematic review and analysis of theory in research on gamification, serious games and game-based learning. *Computers in human behavior*, *125*, 106963.

Landers, R. N., Bauer, K. N., & Callan, R. C. (2017). Gamification of task performance with leaderboards: A goal setting experiment. *Computers in Human Behavior*, *71*, 508–515.

Leng, J., Sha, W., Wang, B., Zheng, P., Zhuang, C., Liu, Q., … Wang, L. (2022). Industry 5.0: Prospect and retrospect. *Journal of Manufacturing Systems*, *65*, 279–295.

Lewis, Z. H., Swartz, M. C., & Lyons, E. J. (2016). What's the point?: A review of reward systems implemented in gamification interventions. *Games for Health Journal*, *5*(2), 93–99.

Lin, K., Spindell, H., Cambo, S., Kim, Y., & Zhang, H. (2016). Habitsourcing: Sensing the environment through immersive, habit-building experiences. In *Proceedings of the 29th annual symposium on user interface software and technology* (pp. 639–650). Association for Computing Machinery New York NY United States.

Lithoxoidou, E. E., Doumpoulakis, S., Tsakiris, A., Krinidis, S., Ioannidis, D., Votis, K., & Tzovaras, D. (2017). Improvement of the workers' satisfaction and collaborative spirit through gamification. In *Internet science: 4th International Conference, INSCI 2017, Thessaloniki, Greece, November 22–24, 2017, Proceedings 4* (pp. 184–191). Springer International Publishing.

Malle, B. F., & Scheutz, M. (2014). Moral competence in social robots. In *2014 IEEE International Symposium on ethics in science, technology and engineering*, Chicago, IL, USA (pp. 1–6). doi:10.1109/ETHICS.2014.6893446

Marache-Francisco, C., & Brangier, E. (2013b, October 27–November 1). Process for gamification: From the decision of gamification to its practical implementation. In *CENTRIC 2013: The sixth International Conference on advances in human-oriented and personalized mechanisms, technologies, and services* , Venice, Italy (pp. 126–131). ISSN: 2308-3492; ISBN: 978-1-61208-306-3.

Marache-Francisco, C; Brangier, E. (2014). The Gamification Experience: UXD with a Gamification Background. In. Blashki, K; Isaias, P. *Emerging research and trends in interactivity and the human-computer interface*. Advances in Human and Social Aspects of Technology. Advances in Computer Human Interaction ACHI. Iaria Press. (pp. 205–223).

Marache-Francisco, C., & Brangier, E. (2015). Gamification and human-computer interaction: a synthesis. *Le travail humain*, *78*(2), 165–189. https://www.cairn.info/revue-le-travail-humain-2015-2-page-165.htm

Moreta, L. L., Gamboa, A. C., & Palacios, M. G. (2016). Implementing a gamified application for a risk management course. In *2016 IEEE Ecuador Technical Chapters Meeting (ETCM)* (pp. 1–6). IEEE Guayaquil, Ecuador.

Nicholson, S. (2012). A User-Centered Theoretical Framework for Meaningful Gamification", *Proceedings of Games+Learning+Society 8.0, GLS 8.0* Conference Proceedings Madison, WI. https://doi.org/10.17613/d36z4-pmx76

Oke, A. E., Aliu, J., Mwanaumo, E. M. U. M., Abayomi, T., & Kahanji, C. (2023). Leveraging gamification to enhance productivity and employee engagement in the Nigerian construction industry. *Built Environment Project and Asset Management*, *13*(6), 813–829.

Oron-Gilad, T., & Hancock, P.A. (2017). Chapter 7- From ergonomics to hedonomics: Trends in human factors and technology—The role of hedonomics revisited. In M. Jeon (Ed.), *Emotions and affect in human factors and human-computer interaction* (pp. 185–194). Academic Press. doi:10.1016/B978-0-12-801851-4.00007-0

Palomino, P. T., Toda, A. M., Oliveira, W., Cristea, A. I., & Isotani, S. (2019). Narrative for gamification in education: why should you care?. In *2019 IEEE 19th International Conference on Advanced Learning Technologies (ICALT)* (Vol. 2161, pp. 97–99). IEEE Computer Society.

Paschek, D., Luminosu, C. T., & Ocakci, E. (2022). Industry 5.0 challenges and perspectives for manufacturing systems in the society 5.0. *Sustainability and Innovation in Manufacturing Enterprises: Indicators, Models and Assessment for Industry 5.0*, Springer-Advances in Sustainability Science and Technology (ASST) 17–63.

Polyanska, A., Andriiovych, M., Generowicz, N., Kulczycka, J., & Psyuk, V. (2022). Gamification as an improvement tool for HR management in the energy industry—A case study of the Ukrainian market. *Energies*, *15*(4), 1344.

Redelinghuys, Z. (2021). *Gamification and simulation teaching--A system created to improve the depth of knowledge and knowledge retention of engineering students* (Doctoral dissertation). Stellenbosch: Stellenbosch University.

Richter, G., Raban, D. R., & Rafaeli, S. (2015). *Studying Gamification: The Effect of Rewards and Incentives on Motivation.* In: Reiners, T., Wood, L. (eds) Gamification in Education and Business. Springer, Cham. https://doi.org/10.1007/978-3-319-10208-5_2.

Rodrigues, L., Toda, A. M., Palomino, P. T., Oliveira, W., & Isotani, S. (2020). Personalized gamification: A literature review of outcomes, experiments, and approaches. In *Eighth International Conference on technological ecosystems for enhancing multiculturality* (pp. 699–706). Association for Computing Machinery, New York, NY, United States https://doi.org/10.1145/3434780

Sailer, M., & Homner, L. (2020). The gamification of learning: A meta-analysis. *Educational Psychology Review*, *32*(1), 77–112.

Salman, O. K., Khasawneh, Y. J. A., Alqudah, H., Alwaely, S. A. K., & Khasawneh, M. A. S. (2024). Tailoring gamification to individual learners: A study on personalization variables for skill enhancement. International Journal of Data and Network Science 8 (2024) 789–796. doi: 10.5267/j.ijdns.2023.12.025.

Şenol, D., & Onay, C. (2023). Impact of gamification on mitigating behavioral biases of investors. *Journal of Behavioral and Experimental Finance*, *37*, 100772.

Sophan, K., Granger, J., Chokprajakchad, M., & Jongaramraung, J. (2024). Prevention of risky motorcycle driving behaviors using gamification among senior secondary school students in Bangkok: A Quasi-experimental Study. *Pacific Rim International Journal of Nursing Research*, *28*(4), 825–837.

Thaler, R.H., & Sunstein, C. R., (2008) *Nudge. Comment inspirer la bonne décision.* Presses Pocket- Paris.

Triclot, M. (2011). *Philosophie des jeux vidéo.* Paris: La Découverte.

Trinidad, M., Calderón, A., & Ruiz, M. (2021). GoRace: A multi-context and narrative-based gamification suite to overcome gamification technological challenges. *IEEE Access*, *9*, 65882–65905.

Vegt, NJH., Visch, VT., de Ridder, H., & Vermeeren, APOS. (2015). Designing gamification to guide competitive and cooperative behavior in teamwork. In T. Reiners, & L. Wood (Eds.), *Gamification in education and business* (pp. 513–533). Springer. https://doi.org/10.1007/978-3-319-10208-5_26.

Wang, C.-C., Chang, S.-C., Yu, Y. H. (2025). Using gamification to enhance learning: A college course case study, *Entertainment Computing*. doi:10.1016/j.entcom.2025.100942

Watts, S., Hoa, N. T. T., Martens, W., Doan, D. T., & Guzman, A. (2024). An examination of internet of things in the South Korean agricultural industry: The case of Samsung. *World Review of Entrepreneurship, Management and Sustainable Development*, *20*(3), 374–396.

Werbach, K., & Hunter, D. (2012, 2020). *For the win, revised and updated edition: The power of gamification and game thinking in business, education, government, and social impact.* Wharton School Press (University of Pennsylvania).

12 Validation of Kinebot
Pioneering AI-Driven Ergonomics for the 5.0 Era

Alison Alfred Klein and Maria Lucia Okimoto

12.1 INTRODUCTION

The assessment of human interaction with the work environment is a central function of contemporary ergonomic practice. To this end, tools are used to measure worker exposure to various occupational conditions. However, these instruments are mostly task-focused rather than tailored to the actual needs of ergonomists, contradicting the principles of human-centered design (Maguire, 2001).

These tools are often narrow in scope and specific to certain tasks, failing to encompass the entirety of work activities and showing a high dependence on the analyst's judgment. Consequently, Ergonomic Work Analyses (AET) and Preliminary Ergonomic Assessments (AEP) become susceptible to subjectivity, undermining the reliability of their outcomes (Klein & Okimoto, 2023).

The diversity of interpretations among evaluators, influenced by factors such as educational background, experience, and personal preferences, compromises the standardization of results and the effectiveness of ergonomic interventions. Reducing such variability is thus essential for ensuring consistent and sustainable assessments.

In this context, the use of emerging technologies, such as computer vision and artificial intelligence (AI), has emerged as a promising alternative. These approaches allow for the automated analysis of human movement, with the potential to reduce interpretative bias and enhance measurement precision.

This chapter aims to validate the use of computer vision for angular measurements in ergonomics by comparing the Kinebot platform with the Vicon system—an international benchmark in motion capture. The goal is to contribute to the development of more objective and replicable assessment methods in ergonomic practice.

12.2 LITERATURE REVIEW

Ergonomic assessment of work has historically relied on observational methods and postural analysis tools such as RULA, OWAS, and REBA. Although widely adopted, these tools exhibit significant limitations in terms of precision, subjectivity, and applicability in real-world occupational settings (Klein & Okimoto, 2023). Many of these analyses are heavily dependent on the evaluator's experience and the

DOI: 10.1201/9781003559399-14

manual placement of markers or sensors, making the process error-prone and difficult to replicate.

Motion capture has become a fundamental resource in biomechanical and ergonomic analysis, with applications in fields such as rehabilitation, sports science, and occupational evaluation (Monnier, 2004). However, systems based on inertial sensors, optical markers, or magnetic devices face serious challenges in occupational environments. These systems require calibration, electromagnetic interference-free zones, and unrestricted movement—conditions not always feasible in real-world scenarios (Vignais et al., 2013; Battini et al., 2014; Plantard et al., 2017)

Faced with these limitations, video recording has gained traction as a viable alternative, enabling task segmentation and practical postural and temporal analyses. However, video usage demands careful attention to camera placement, lighting, and angles and often still relies on visual interpretation by the analyst (Roebuck, 1993).

With technological advancements, new approaches based on computer vision and AI are being explored to address these gaps. Among them, Human Pose Estimation (HPE) stands out, enabling the identification of anatomical keypoints from 2D (RGB) images and generating articulated skeletons that represent human movement (Klein et al., 2023; Xuzhong et al., 2017). These skeletal models typically include 15–30 joints and allow the calculation of biomechanical variables such as flexion, extension, and rotation of body segments (Plantard et al., 2017).

HPE leverages deep learning algorithms such as Convolutional Neural Networks to detect keypoints and establish their interconnections. This structure comprises two main elements: Confidence Maps, which indicate the likelihood of a joint being at a specific location in the image, and Part Affinity Fields, which are vectors encoding the direction and association between adjacent joints (Cao et al., 2021).

These representations enable the reconstruction of highly accurate 2D skeletal models, allowing automated postural analysis without physical sensors or manual intervention. According to Xuzhong et al. (2017), joint angles extracted from such skeletons can be used as input variables for postural classification algorithms, enabling the automatic identification of ergonomic risk patterns.

Furthermore, the human anatomy can be modeled as an articulated structure comprising rigid segments interconnected by joints with varying degrees of freedom. Biomechanical models, such as those proposed by Terlemez et al. (2014), define angular limits and axes of rotation for each joint, providing essential parameters for computer vision-based ergonomic analyses.

Figure 12.1 illustrates the reference biomechanical model, in which the human body is represented as a rigid structure with joints endowed with specific degrees of freedom. This model provides the basis for the biomechanical limits described in the following section.

Table 12.1 presents the angular limits for each joint and their respective degrees of freedom, as extracted from the same biomechanical model.

AI, particularly through machine learning and deep learning, offers a new paradigm for human movement analysis. By processing large volumes of visual data, these systems are capable of learning complex motion patterns, eliminating the need for direct human intervention and reducing evaluator subjectivity (Klein et al., 2023). Artificial neural networks, inspired by the structure of the human brain, enable the

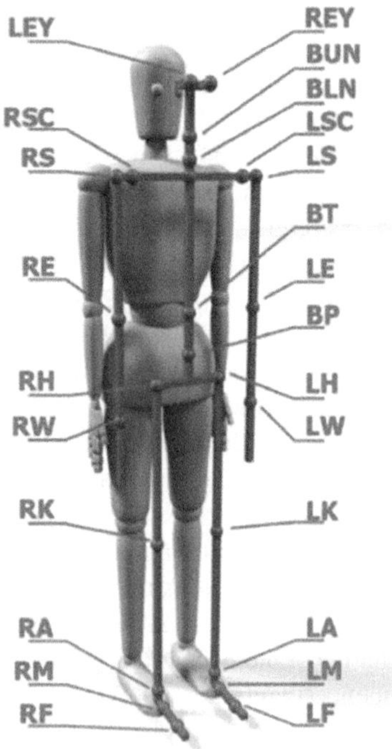

FIGURE 12.1 Kinematics of the reference model.

Source: Adapted from Terlemez et al. (2014).

training of robust predictive models with high generalization capacity—especially useful in dynamic occupational settings.

Accordingly, computer vision techniques combined with AI have shown promise in the automatic detection of postures and the angular measurement of body segments. Their application in ergonomics represents a significant step forward in standardizing, automating, and increasing the reliability of field assessments.

Given this context, it becomes essential to empirically validate these emerging approaches by comparing them with well-established motion capture systems such as Vicon Nexus. Such validation contributes to strengthening the practical application of these tools in occupational ergonomics, supporting more precise and evidence-based interventions.

12.3 METHOD

For the validation proposed in this study, a volunteer was selected and marked with anatomical reference points according to the official protocol of the Vicon system. The subject was then positioned at the center of the capture space at the Kinematics Laboratory of the Pontifical Catholic University of Paraná (PUCPR), where Vicon system cameras were installed alongside five standard smartphone cameras for simultaneous recording.

TABLE 12.1
Joint angular limits and degrees of freedom

Join	Degrees of freedom	X-limits	Z-limits	Y-limits
LF/RF	1+1	[−30°,45°]	–	–
LM/RM	1+1	–	[−30°,45°]	–
LA/RA	3+3	−40°,30°]	[−30°, 30°]	[−20°,20°]
LK/RK	1+1	[−130°,0°]	–	–
LH	3	[−50°,95°]	[−45°,45°]	[−20°,65°]
RH	3	[−50°,95°]	[−45°,45°]	[−65°,20°]
LW	2	[−30°,20°]	[−70°,50°]	–
RW	2	[−30°,20°]	[−50°,70°]	–
LE/RE	2+2	[0°,160°]	[−90°,90°]	–
LS	3	[−70°,190°]	[−70°,60°]	[0°,160°]
RS	3	[−70°,190°]	[−60°,70°]	[−160°,0]
LSC/RSC	2+2	–	[−20°,20°]	[−20°,20°]
LEY/REY	2+2	[−60°,60°]	–	[−60°,60°]
BUN	3	[−20°,30°]	[−20°,20°]	[−15°,15°]
BLN	3	[−45°,15°]	[−15°,15°]	[−20°,20°]
BT	3	[−35°,27°]	[−36°,36°]	[−20°,20°]
BP	3	[−50°,35°]	[−45°,45°]	[−20°,20°]

Source: Adapted from Terlemez et al. (2014).

During the recording session, the volunteer performed movements simulating common load-handling tasks, including arm flexion, extension, and abduction; elbow, neck, and trunk flexion and extension; as well as leg elevation and knee flexion. The goal was to replicate bodily gestures frequently observed in occupational settings.

The cameras were strategically positioned to simulate the main viewpoints typically adopted by ergonomists in the field, with the following capture angles: 135°, 0°, 315°, 180°, and 90°. The devices were configured to record video at a resolution of 1080p at 30 frames per second (FPS), with distances ranging from 250 cm to 300 cm from the volunteer. The camera arrangement is shown in Figure 12.2:

The videos collected by the cameras were submitted to the Kinebot platform, which enabled the extraction of angular data in .CSV format. Each column of the file represents a body segment, totaling 11 columns, with the first corresponding to the frame number and the others to the joint angles, as shown in Figure 12.3, where each column represents a body segment, totaling 11 columns. Thus, the first column serves as an indicator of frames, while the subsequent columns reflect the angles associated with different parts of the body.

Simultaneously, the data from the Vicon system were exported in .XLSX format, organized into 24 separate worksheets, each pair representing the angular axes of a body segment (Figure 12.4).

Before comparing the data sets, a processing approach based on the ETL (Extraction, Transformation, and Loading) method was applied. First, the data were

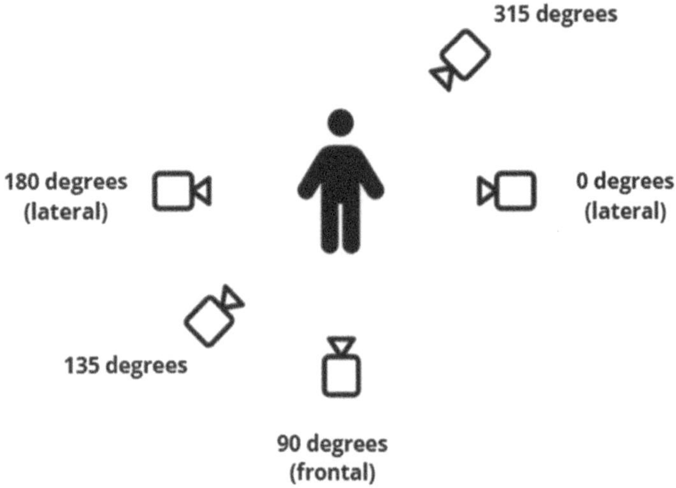

FIGURE 12.2 Camera positioning.

Source: The authors (2024).

Frame	Neck	Trunk	Left LowerA	Right Lower	Left UpperA	Right Upper	Left Hip	Right Hip	Left Knee	Right Knee
1	33.93	2.46	1.65	0.53	92.4	101.45	11.4	7.39	5.96	13.91
2	34.49	1.69	1.34	0.86	93.8	104.04	11.01	8.16	5.52	13.96
3	0	1.67	1.39	7.92	92.49	105.46	10.95	8.39	6.07	15.3
4	0	1.69	0.09	11.33	93.23	107.46	10.8	7.96	4.65	16.28
5	34.15	1.69	1.99	18.17	92.11	100.01	10.8	7.96	4.7	16.44
6	0	2.43	1.34	5.52	95.21	102.51	12.6	6.3	2.67	15.99
7	0	2.43	1.61	7.58	93.8	103.28	12.6	6.48	3.51	17.38
8	34.15	2.46	1.28	8.26	95.16	101.82	12.63	6.45	3.54	17.18
9	35.34	2.46	0.26	7.58	93.23	103.28	12.63	6.45	3.54	15.97
10	34.15	2.43	2.18	4.57	91.93	106.28	12.79	5.29	4.66	17.18
11	35.82	2.43	2.01	1.4	92.8	103.28	13.87	5.46	2.51	17.21
12	35.82	2.46	0.37	7.58	94.41	103.28	12.63	5.1	3.52	17.18
13	34.15	2.43	2.36	16.37	92.57	100.19	12.6	5.13	3.47	17.21
14	34.15	2.43	5.47	1.68	90.87	105.31	12.38	6.09	3.44	15.03

FIGURE 12.3 Data generated by Kinebot.

Source: The authors (2024).

extracted from the raw files. Then, they underwent cleaning and transformation steps to standardize the scales and prepare the variables for analysis. Finally, the data were loaded into dataframes using the Pandas library in Python (Figure 12.5).

During the analysis, a temporal mismatch between the camera data was identified, as the recording on each device started at different times. This resulted in frame desynchronization, as illustrated in Figure 12.6.

To resolve the scale issue and align the data for comparison, the normalization technique was applied, standardizing the values to a common scale between 0 and 1 (Ali et al., 2014). This enabled the variables to be comparable, regardless of unit differences, as shown in Figure 12.7.

Subsequently, the Pearson correlation coefficient was used to measure the linear association between the angles obtained by Kinebot and Vicon (Ly et al., 2018). The

16,5844	-1,64293	-0,74296
16,5306	-1,61245	-0,74062
16,4783	-1,6164	-0,74825
16,4042	-1,58749	-0,77505
16,3262	-1,55049	-0,76703
16,2469	-1,48579	-0,7824
16,2093	-1,46003	-0,76455
16,2425	-1,47503	-0,77152
16,2588	-1,46236	-0,81691
16,294	-1,45716	-0,91718
16,3244	-1,43868	-0,9981
16,3276	-1,44881	-1,03787
16,3367	-1,46458	-1,0634
16,3462	-1,45841	-1,1181
16,3577	-1,43575	-1,15571
16,3844	-1,44431	-1,19111

+ var_angular_pelve_direito var_angular_pe_direito var_angular_pescoco_direito var_angular_espinha_direito

FIGURE 12.4 DATA generated by Vicon.

Source: The authors (2024).

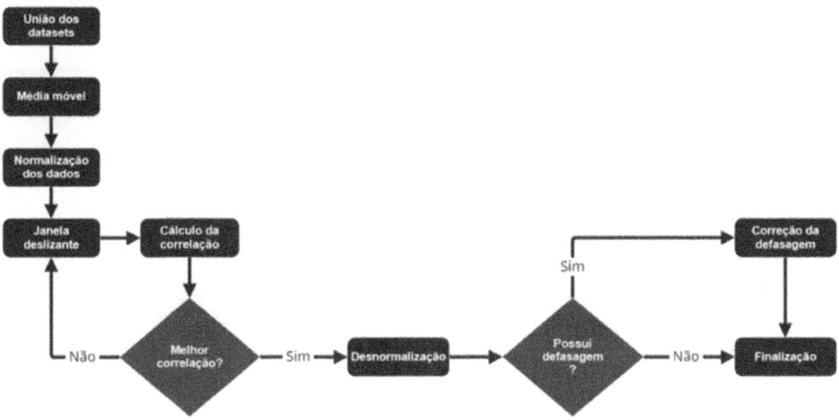

FIGURE 12.5 Data processing flowchart.

Source: The authors (2024).

coefficient values range from −1 to +1, indicating negative, null, or positive correlation. A p-value associated with each calculation was used to assess the statistical significance of the correlations, with $p < 0.001$ considered indicative of a significant correlation.

To maximize the alignment between the temporal signals, a sliding window algorithm was used, which compares successive segments of the data and calculates the correlation at each point, automatically identifying the shift necessary for optimal alignment (Raab et al., 2010).

With the data aligned, a moving average of 30 frames was applied to smooth out sharp fluctuations and eliminate noise resulting from rapid movements or variations

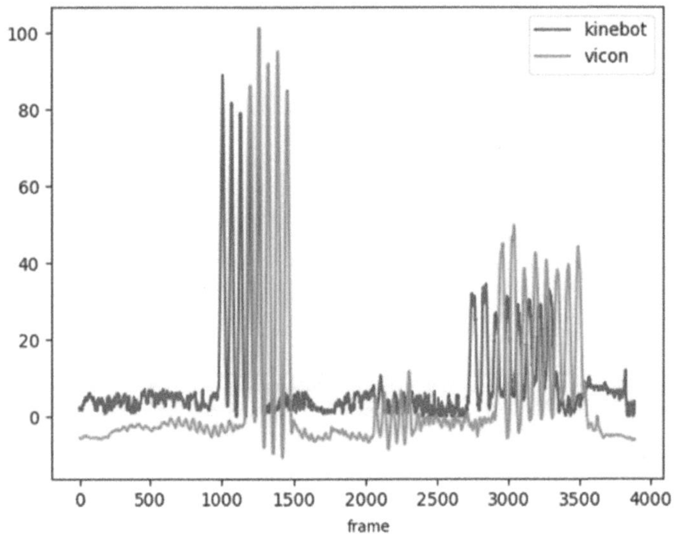

FIGURE 12.6 Example of desynchronized data.

Source: The authors (2024).

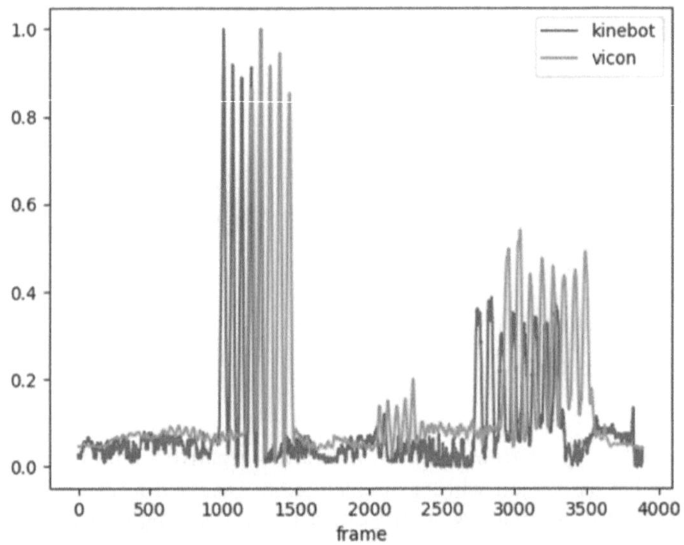

FIGURE 12.7 Normalized data.

Source: The authors (2024).

in the sensitivity of the capture systems. This corresponds to one second of recording, respecting the 30 FPS rate, as shown in Figure 12.8:

Additionally, a systematic difference in the location of the anatomical points was observed between the two systems (see Figure 12.9). Vicon uses physical markers (foam balls) manually placed, while Kinebot applies automatic marking via AI, as shown in Figure 12.10. This reference difference affected three main joints, with the following average angular deviations observed:

- Neck: +30°
- Shoulder: +20°
- Elbow: −10°

Upon completing the extraction, transformation, loading, normalization, and synchronization steps, a unified and validated dataset was obtained for statistical analysis, presented in Section 12.3.1.

12.3.1 Results

The comparative study between the results aimed to evaluate the effectiveness of the assessments conducted by Kinebot. To this end, after data collection and processing

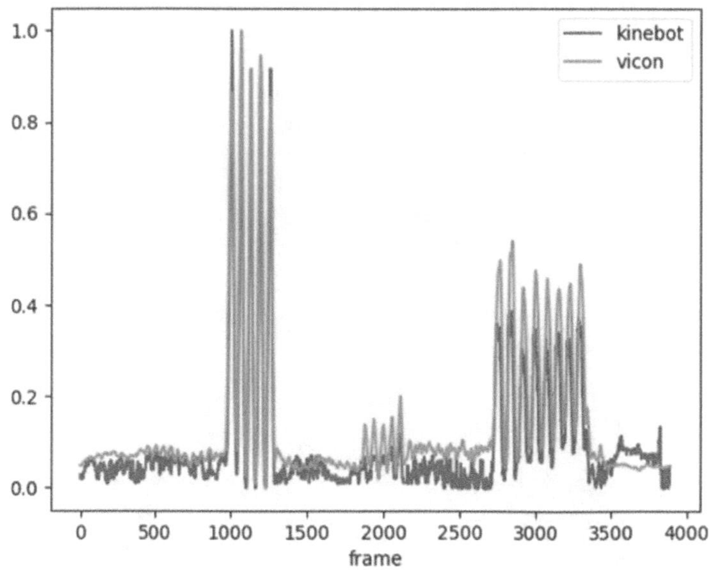

FIGURE 12.8 Synchronized dataframe.

Source: The authors (2024).

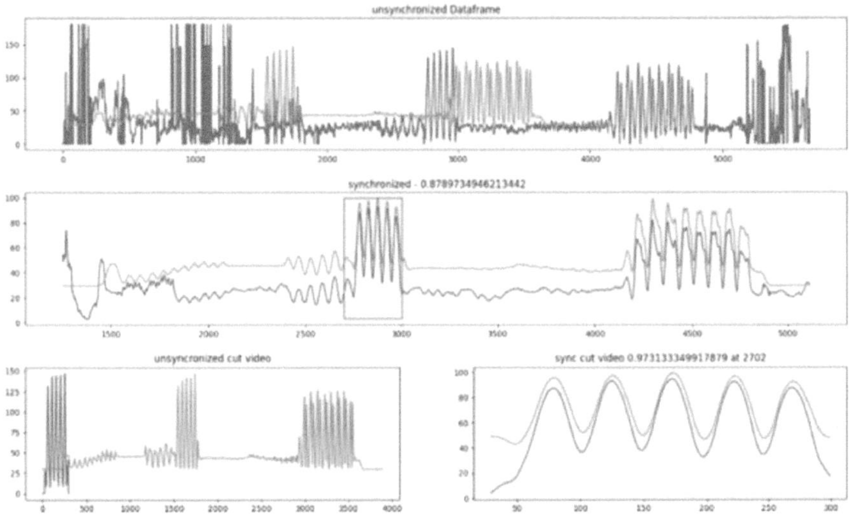

FIGURE 12.9 Summary of right elbow data processing.

Source: The authors (2024).

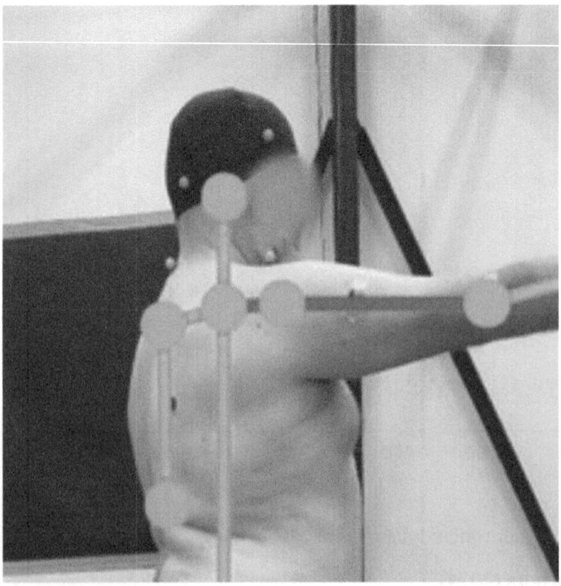

FIGURE 12.10 Data collection cutout.

Source: The authors (2024).

TABLE 12.2
Correlation of sagittal planes

Body part	180 degrees		0 degree	
	Correlation(r)	p-value	Correlation(r)	p-value
Neck	0,983	<0,0001	0,992	<0,0001
Trunk	0,983	<0,0001	0,992	<0,0001
Left elbow	0,982	<0,0001	0,994	<0,0001
Right elbow	0,991	<0,0001	0,968	<0,0001
Left shoulder	0,966	<0,0001	0,990	<0,0001
Right shoulder	0,983	<0,0001	0,985	<0,0001
Left hip	0,914	<0,0001	0,923	<0,0001
Right hip	0,940	<0,0001	0,937	<0,0001
Left knee	0,991	<0,0001	0,994	<0,0001
Right knee	0,990	<0,0001	0,986	<0,0001

Source: The authors (2024).

following the ETL methodology, the correlation indices of the results were calculated, as presented in Table 12.2:

Measurements in the sagittal plane showed very high positive correlations for all body parts, with correlation values exceeding 0.98. The p-values less than 0.0001 provide very strong statistical evidence that the correlation between the variables is significant. Therefore, the collected data indicate that Kinebot has an excellent ability to replicate VICON measurements in this plane, as evidenced by the correlation coefficients presented.

Measurements in the diagonal plane also showed very high positive correlations, with p-values less than 0.0001, indicating a strong linear dependence between the variables. This suggests that the angles captured in the diagonal plane are effective for evaluating body postures and that Kinebot is able to replicate VICON measurements with high accuracy, as demonstrated in Table 12.3.

Measurements in the coronal plane indicated some limitations, mainly for the neck, where measurement could not be performed due to the lack of an adequate reference point in the coronal plane (see Table 12.4). For the other body parts, the correlations were positive, with particular emphasis on the shoulder and hip, which showed very high correlation, while other parts, such as the right elbow, showed moderate correlation. Overall, the results in the coronal plane were not as consistent as in the sagittal and diagonal planes.

Measurements in the coronal plane indicated some limitations, mainly for the neck, where measurement could not be performed due to the lack of an adequate reference point in the coronal plane (see Table 12.4). For the other body parts, the correlations were positive, with particular emphasis on the shoulder and hip, which showed very high correlation, while other parts, such as the right elbow, showed moderate correlation. Overall, the results in the coronal plane were not as consistent as in the sagittal and diagonal planes.

TABLE 12.3
Correlation of diagonal planes

Body part	135 degrees		315 degrees	
	Correlation(r)	p-value	Correlation(r)	p-value
Neck	0,924	<0,0001	0,937	<0,0001
Trunk	0,987	<0,0001	0,983	<0,0001
Left elbow	0,971	<0,0001	0,980	<0,0001
Right elbow	0,895	<0,0001	0,982	<0,0001
Left shoulder	0,990	<0,0001	0,985	<0,0001
Right shoulder	0,938	<0,0001	0,968	<0,0001
Left hip	0,924	<0,0001	0,920	<0,0001
Right hip	0,929	<0,0001	0,637	<0,0001
Left knee	0,987	<0,0001	0,975	<0,0001
Right knee	0,832	<0,0001	0,986	<0,0001

Source: The authors (2024).

TABLE 12.4
Correlation of the coronal plane

Body part	90 degrees	
	Correlation(r)	p-value
Neck	–	–
Trunk	0,783	<0,0001
Left elbow	0,743	<0,0001
Right elbow	0,642	<0,0001
Left shoulder	0,963	<0,0001
Right shoulder	0,973	<0,0001
Left hip	0,924	<0,0001
Right hip	0,939	<0,0001
Left knee	0,746	<0,0001
Right knee	0,883	<0,0001

Source: The authors (2024).

It can be observed that the highest correlations occurred at angles of 0 and 180 degrees, with values of 0.9735 and 0.9753, respectively. This suggests that lateral perspectives are the most effective for capturing body part angles with higher accuracy. On the other hand, the 90-degree angle, with a correlation of 0.8442, showed lower consistency in measurements, indicating that the frontal positioning may be less precise, as summarized in Table 12.5.

It is worth noting that significant variations in correlation occur in certain scenarios, with these discrepancies partly due to the use of different references for selecting the angle calculation points. For example, when capturing neck data, Kinebot uses the ear as a reference point, while Vicon Nexus uses a point located between the ear and the nape.

TABLE 12.5
Overall correlation by camera position

Degrees	Correlation
315 degrees	0,9408
180 degrees	0,9735
135 degrees	0,9325
0 degree	0,9753
90 degrees	0,8442

Source: The authors (2024).

The correlation analysis by camera position revealed that, for most body parts, the right and left cameras showed higher correlations compared to the frontal camera. However, an exception occurred for the hip measurement, which benefited from a frontal capture. In general, the variations between the lateral positions showed no significant differences, but the frontal camera displayed lower correlation for several body parts.

Based on the data provided, synthesized in Table 12.6, it is observable that the most effective approach for capturing angles requires the use of 0-degree and 180-degree positions. In other words, the ideal perspective for recording occurs when positioned laterally to the person

Kinebot demonstrated an overall correlation of 0.935 compared to Vicon Nexus, which is significantly above the expected value of 0.90. This result confirms the reliability of Kinebot as an ergonomic assessment tool. Based on these results, we can assert that Kinebot is a valuable tool for the analysis and prevention of work-related diseases, eliminating human dependency and providing precise measurements with high reliability

TABLE 12.6
Correlation between body parts and camera positioning

Body part	Correlation		
	Right camera	Left camera	Frontal camera
Neck	0,953	0,960	–
Trunk	0,982	0,982	0,783
Left elbow	0,982	0,986	0,742
Right elbow	0,931	0,986	0,642
Left shoulder	0,990	0,975	0,962
Right shoulder	0,977	0,961	0,973
Left hip	0,924	0,917	0,924
Right hip	0,933	0,788	0,939
Left knee	0,990	0,983	0,746
Right knee	0,909	0,988	0,882

Source: The Authors (2024).

12.4 CONCLUSION

The present study aimed to validate the use of computer vision for angular measurements applied to ergonomics, using the Kinebot platform and comparing it with the Vicon system, considered a benchmark in motion capture. The research was conducted in a controlled setting, where a volunteer performed a series of movements representative of occupational tasks, which were simultaneously captured by conventional cameras and the Vicon system.

The method used for data processing and analysis, which involved the steps of ETL, normalization, and synchronization of data, provided a robust approach to handle the differences between the capture systems. The application of statistical analysis techniques, such as the Pearson correlation coefficient, allowed for the assessment of the consistency and reliability of the data obtained from the Kinebot platform in comparison to Vicon.

The results showed a high correlation between the joint angles extracted from Kinebot and Vicon, indicating that the Kinebot platform offers a valid and reliable alternative for postural analysis in ergonomics, without relying on physical markers or complex capture systems. Although a systematic difference was observed in the location of some anatomical points, this discrepancy did not compromise the accuracy of the obtained results. Kinebot demonstrated potential to reduce evaluator bias by automating postural analysis and offering a more practical and scalable solution for ergonomic assessments.

In alignment with the study objectives, this work contributes to the validation of computer vision as a tool for postural analysis in the ergonomic context, demonstrating its applicability in real-world work environments. The Kinebot platform proves to be a promising technological solution capable of facilitating and optimizing ergonomic practice, providing more precise, consistent, and replicable data, while reducing reliance on evaluator subjectivity.

This study paves the way for the development of new methodologies and tools that can be integrated into the field of ergonomics, focusing on automation, precision, and accessibility. The use of AI and computer vision could be the next step for the evolution of ergonomic analysis, promoting more efficient, evidence-based interventions, with a direct impact on workers' health and well-being.

12.5 ETHICAL CONSIDERATIONS IN THE RESEARCH

The research was submitted to and approved by the Research Ethics Committee of the Federal University of Paraná (UFPR), as indicated by the Certificate of Presentation of Ethical Appreciation (CAAE) 71138223.9.0000.0102, and received the Substantive Opinion No. 6.480.725, dated November 1, 2023.

Participation in the study was entirely voluntary, with no form of remuneration provided to the participants. The selection of participants was conducted through a convenience sampling method. In accordance with the ethical principles of research, all personal data and photographic records were anonymized to ensure the privacy and confidentiality of the participants.

This ethical approval and the measures taken ensured that the rights, dignity, and privacy of the participants were respected throughout the study. The research maintained a commitment to ethical standards, ensuring the safety and well-being of all involved.

REFERENCES

Ali, P. J. M. et al. *Data normalization and standardization: A technical report*. Koya University, Iraque: Machine Learning Lab, 2014. DOI:10.13140/RG.2.2.28948.04489. Accessed in: 16 jul. 2024.

Battini, D.; Persona, A.; Sgarbossa, F. Innovative real-time system to integrate ergonomic evaluations into warehouse design and management. *Computers & Industrial Engineering*, [S.l.], v. 77, p. 1–10, 2014. DOI:10.1016/j.cie.2014.08.018. Accessed in: 08 jun. 2024.

Cao, Z.; Hidalgo, G.; Simon, T.; Wei, S.; Sheikh, Y. OpenPose: Realtime multi-person 2D pose estimation using part affinity fields. *IEEE Transactions on Pattern Analysis & Machine Intelligence*, [S.l.], v. 43, n. 1, p. 7291–7299, 2021.

Klein, A. A.; Lima, R. A. F.; Grandi, L. F.; Marinho, C. *Kinebot - A Inteligência Artificial na Ergonomia*. São Paulo: Câmara Brasileira do Livro, 2023. Ebook. Disponível em: kinebot.com.br/wp-content/uploads/2023/05/Kinebot-a-inteligencia-artificial-na-ergonomia.pdf. Accessed in: 17 jul. 2024.

Klein, A. A.; Okimoto, M. L. L. R. Mapping of ergonomic assessment tools. *DAT Journal*, [S. l.], v. 8, n. 2, p. 136–159, 2023. DOI:10.29147/datjournal.v8i2.720. Accessed in: 14 mar. 2024.

Ly, A.; Marsman, M.; Wagenmakers, E.-J. Analytic posteriors for pearson's correlation coefficient. *Statistica Neerlandica*, [S. l.], v. 72, n. 1, p. 4–13, 2018. DOI:10.1111/stan.12111. Accessed in: 17 jul. 2024.

Maguire, M. Methods to support human-centred design. *International Journal of Human-Computer Studies, Leicestershire, Inglaterra*, v. 55, p. 587–634, 2001. DOI:10.1006/ijhc.2001.0503. Accessed in: 17 jul. 2024.

Monnier, G. Simulation de mouvements humains complexes et prediction de l´inconfort associé - Applicarion à l'évaluation ergonomique du bouclage de la ceinture de sécurité. *Tese (doutorado em ciências aplicadas). Instituto Nacional de Ciências Aplicadas de Lyon* - França. Lyon, 2004.

Plantard, P.; Shum, H. P. H.; Le Pierres, A. S.; Multon, F. Validation of an ergonomic assessment method using Kinect data in real workplace conditions, *Applied Ergonomics*, [S. l.], v. 65, p. 562–569, 2017. DOI:10.1016/j.apergo.2016.10.015. Accessed in: 17 jul. 2024.

Raab, D.; Graf, M.; Notka, F.; Schödl, T.; Wagner, R. The geneoptimizer algorithm: Using a sliding window approach to cope with the vast sequence space in multiparameter dna sequence optimization. *Systems and synthetic biology*, [S. l.], v. 4, p. 215–225, 2010. DOI:10.1007/s11693-010-9062-3. Accessed in: 17 jul. 2024.

Roebuck, J. A. *Anthropometric methods: Designing to fit the human body*. Santa Monica: Human Factors end Ergonomics Society, 1993.

Terlemez, Ö. et al. Master Motor Map (MMM) - Framework and toolkit for capturing, representing, and reproducing human motion on humanoid robots. In *IEEE-RAS International Conference on Humanoid Robots*, Madrid, Spain, 2014, p. 894–901, 2014. DOI:10.1109/HUMANOIDS.2014.7041470. Accessed in: 17 jul. 2024.

Vignais, N.; Miezal, M.; Bleser, G.; Mura, K.; Gorecky, D.; Marin, F. Innovative system for real-time ergonomic feedback in industrial manufacturing. *Applied Ergonomics*, [S. l.], v. 44, n. 4, p. 566–574, 2013. DOI:10.1016/j.apergo.2012.11.008. Accessed in: 17 jul. 2024.

Xuzhong, Y. et al. Development of ergonomic posture recognition technique based on 2D ordinary camera for construction hazard prevention through view-invariant features in 2D skeleton motion. *Advanced Engineering Informatics*, [S. l.], v. 34, p. 152–163, 2017. DOI:10.1016/j.aei.2017.11.001. Accessed in: 17 jul. 2024.

13 Technological Devices for Ergonomic Intervention in the Automotive Industry

Maria Victoria Cabrera Aguilera,
Nilo Antonio de Souza Sampaio,
Robson José de Moura Barbosa
and Bernardo Bastos da Fonseca

13.1 INTRODUCTION

Manipulators are technological devices used in industrial activities to assist in the assembly, positioning and movement of materials, components, and products (Dwivedy and Eberhard, 2006). These devices use technology to perform tasks with human intervention or no human intervention that can be manual, automatic, and semi-automatic depending on the characteristics and automation level of the manufacturing line (Romero et al., 2016). When used in activities with human intervention, it is necessary to understand the principles of human-centered manufacturing and its correlation with Ergonomics 5.0 (Trstenjak et al., 2025) to ensure that the interaction between operator and technology is efficient and effective, healthy and safe, and above all, sustainable (Santos, 2024).

Manipulators are widely used to support monotonous, dangerous, and heavy jobs. Most manipulators are designed and built in a manner to maximize stiffness in an attempt to minimize the vibration of the end-effector to achieve a good position accuracy (Yin et al., 2025). Nevertheless, there are some conflicts between project requirements such as high speed and high accuracy, productivity demand, and settling time required.

Manipulators are used in various sectors, such as the electronics sector for precision activities in the manufacture of circuit boards, chips, and equipment; the food sector in food packaging and preparation processes; the pharmaceutical sector in the process of packaging medicines; and the automotive sector for assembling, painting, and welding vehicles. In the automotive sector, the most commonly used manipulators are manual, pneumatic, hydraulic, electric, vacuum, magnetic, and robotic.

The automotive sector uses human labor intensively, although some stages are automated. This requires workers to perform repetitive assembly tasks and use

 DOI: 10.1201/9781003559399-15

excessive force (Lee and Jung, 2020; Sales et al., 2022). There are manual tasks that can be automated, but many of them have a certain complexity because they require human skills of precision, decision-making, flexibility, and the ability to move. These working conditions can affect workers' occupational health and safety. In this context, manipulators can be implemented to enable precise movements safely and efficiently, reduce physical effort, the burden of moving materials and the presence of people during activities that present risks.

This chapter presents a work situation in the automotive industry in which manipulators were implemented on the basis of an ergonomic work analysis, with the aim of increasing production efficiency, reducing effort and improving working conditions. In the first situation, the device focused on handling doors on a car body production line. In the second work situation, the device was used to mount vehicle tires on the production line. The situation involves the automotive company's process engineering, industrial engineering, ergonomics, occupational safety, maintenance, and production areas.

13.2 ERGONOMIC INTERVENTION PROCESS

The methodology adopted in the study presented in this chapter consists of five stages (Figure 13.1). The first stage was the Ergonomic Assessment to identify work constraints relating to posture, effort, adjustment, frequency, and duration, based on the principles of the Muscle Fatigue Assessment method, also known as the functional work assessment technique (Rodgers, 2005), to characterize the discomfort described by workers on car assembly lines and manufacturing tasks. Discomfort or risks can arise, especially as restrictions accumulate: repetitive work, poor posture and great effort, unfavorable environment, and psychological factors. Ergonomic assessment tools are used to score risk factors based on observed posture and strength ratings. All ergonomic assessment tools use the same biomechanical parameters. The Ergonomic Assessment was carried out by collecting data on-site at the assembly line and from operator reports. Based on the assessment, it was possible to identify the level of ergonomic risk to which the operator is exposed at the workplace and direct the scope of the project through the Ergonomic Design Requirements to define which Technological Solution meets the requirements.

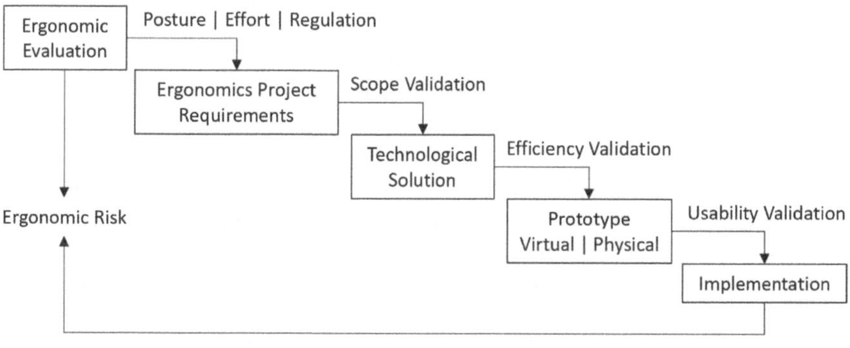

FIGURE 13.1 Ergonomic intervention process with a technological solution.

For the work situation presented in this chapter, the choice was made to implement semi-automatic manipulators that went through the validation process, from the project proposal to installation and use. In order to validate the scope of the project, the efficiency of the proposed technological solution, and the usability of the equipment, it was necessary to develop infographic diagrams and prototypes, both virtual and physical, with the participation of different actors in the company: ergonomics, process engineering, industrial engineering, manufacturing, work safety, and maintenance. The participation of these actors was essential throughout the process, as the ergonomic intervention, through a technological solution, required several joint actions, such as: balancing the production line to meet the cycle time, defining a new standard operating procedure with the use of the manipulator, reconfiguring the layout, new supply routes, trained staff to carry out maintenance on the new equipment, a training schedule for operators to carry out job allocation, updating documents and complying with safety standards related to the workstation and the new equipment.

After full implementation of the manipulator, with the operator using 100% of the equipment while performing the work activity, a new Ergonomic Assessment was carried out to define the level of ergonomic risk present in the new work activity.

13.3 ERGONOMIC EVALUATION

The study took place in the door production sector of the automotive industry's Body Shop. This sector is made up of two production lines, as shown in Figure 13.2: the Hemming Line, with six workstations where parts are joined and flanges are bent (graphing), and the Metal Line, with three workstations where moving parts are fixed to the body. The moving parts correspond to the front and rear doors on the right and left side of the vehicle, as well as the back door, trunk lid, and hood. Each part is produced in an automated cell on the Hemming Line and, after the engraving process, they are deposited on a karakuri (gravity conveyor) and transported to the Metal Line.

In the Hemming Line for the front and rear doors on both sides of the vehicle, the production volume corresponds to 32 cars per hour with 1.63 minutes of cycle time (TCY), 99% engagement and 89% Overall Equipment Effectiveness. The Hemming line operating process involves the manual handling and transportation of parts by the operator, who carries out the activities throughout the cycle time (Figure 13.3):

1. Take the inner door plate from the dolly parts;
2. Position the inner door plate on the turntable;
3. Apply sealer and mastic to the inner door plate;
4. Take the outer door plate, stiffener, and hinges from the dolly parts;
5. Hang the outer door plate on the hanger at the entrance to the robotic engraving cell and support the hinges on the intermediate support;
6. Attach the stiffener to the outer door plate;
7. Position the outer plate on the hemming die (engraving cradle inside the robotic cell);
8. Position the inner door plate greased with sealer and mastic over the outer plate on the hemming die;

9. Press the button on the robotic cell to carry out the embossing process;
10. Remove the door from the hemming die;
11. Position the door on the karakuri (gravity conveyor);
12. Attach the hinges to the door.

Table 13.1 shows the results of the ergonomic analysis carried out at the workplace based on data collected on site and records obtained from the automotive industry's medical clinic. A total of eight constraints were identified relating to aspects of posture, effort, and regulation based on the parameters of the principles of ergonomics, using as a reference postures adopted up to 100 times per hour over a long cycle time, manual efforts that exceed biomechanical and comfort limits, and production factors such as the frequency of 32 cars per hour. The regulation aspect identified corresponds to the combination of postures or efforts associated with the time available to the operator to carry out his operation with dexterity, agility, and precision, avoiding production errors. According to production parameters, each workstation must work at a maximum of 95% engagement in order to have 5% recovery between the start and end of each TCY.

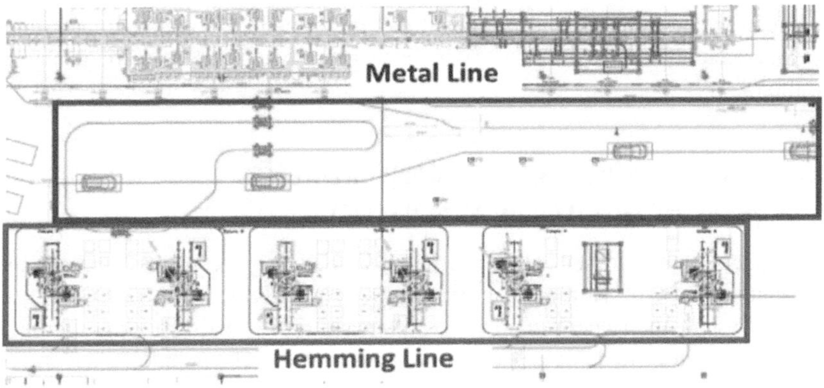

FIGURE 13.2 Door production sector at the Body Shop.

FIGURE 13.3 Door assembly process on the Hemming Line.

TABLE 13.1
Ergonomic assessment of the Body Shop door production line.

#	Aspect	Description
1	Posture	Trunk Flexion >30°
2	Posture	Arms outstretched upwards or shoulder raised
3	Effort	Tension in the wrists/hands/fingers.
4	Effort	Difficulty in handling, picking up, or placing parts.
5	Effort	Manual handling of pieces > 10kg.
6	Effort	Combination of poor posture with effort of manual handling.
7	Effort	Walking > 14 m/min.
8	Regulation	Accumulation of restrictions (1 posture or effort + Engagement > 95%.
Medical Complaints		3–lumbar spine
		3–Fingers, hands, and fists
		3–Shoulder

This result, combined with the medical data, shows that the level of risk to which the operators are exposed is high, which calls for immediate ergonomic intervention to avoid accumulating work constraints concerning posture, effort, frequency, and duration.

13.4 ERGONOMICS PROJECT REQUIREMENTS AND TECHNOLOGICAL SOLUTION

The aspects identified served as the basis for defining the project requirements in conjunction with Human Factors Engineering (Czaja and Nair, 2012) and Human-System Interface guidelines (Karwowski, 2012) and Brazilian regulatory standards, in order to achieve a technological solution. The technological solution defined was the implementation of zero-gravity assistance for handling doors in Hemming, with the aim of eliminating the postures and physical strain caused by work activities.

The scope of the project involved the design, construction, and installation of door handlers with zero gravity. In addition to the regulatory requirements, it also had to meet production requirements, such as maintenance, quality, and having a flexible system that would allow it to absorb new car models and meet the operation cycle time, i.e., the handling equipment must not encumber the cycle time prior to its installation.

To develop the project, ergonomic, design, safety, pneumatic system and standardization premises were taken into account, as described below:

13.4.1 Ergonomic Assumptions

- Ensure a neutral/comfortable posture for gripping, moving, and positioning doors (Table 13.2).

TABLE 13.2
Posture and effort criteria

	Excellent	Recommended
Arm I Hand		
Finger I Hand I Wrist		
Trunk		
Maximum force to start and stop the manipulator	x < 5 kg	–
Maximum force to pull and push the manipulator	x < 3 kg	–
Maximum force to lower and raise the manipulator	x < 2 kg	–

TABLE 13.3
Sizing for the manipulator handles

Diameter and length of the handle	Distance between two handles
Length ≥ 130 mm 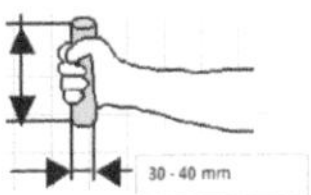	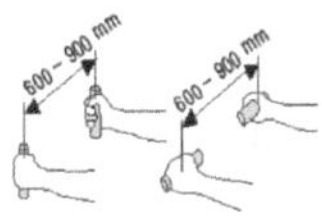

- Make the assistance absolutely essential to carrying out the task.
- Package the parts in such a way that it is essential to use the assistance.
- Involve end-users in the design approach.
- Offer simplicity in use.
- Follow the cycle time: operating time with assistance ≤ manual operating time.

13.4.2 Design Assumptions

- The manipulator must have two handles for better control and correct balancing with the center of gravity.
- Handling handles must be made of a flexible, non-metallic material. Table 13.3 shows the dimensions for the manipulator handles.

13.4.3 Standardization Assumptions

- The project took into account the specific standardized colors for the panel's control buttons, pushbuttons, emergency button, reset switch, indicator lamps, equipment color, among others.

- Pneumatic, electromagnetic, and limit switch valves must be identified by number and name.
- All identification of electrical panels, documentation, cabling, and safety must be in Portuguese.
- Pneumatic components and controllers must come from manufacturers already used by the automotive company.
- The piping for the air preparation system must follow the pneumatic installation model already used by the company.

13.4.4 Security Assumptions

- The design of the manipulator must comply with Brazilian regulatory standards 10, 11, 12, and 17.
- Have a safety system in the event of a failure in the compressed air distribution system, to maintain the integrity of the operator in the event of a fall from the equipment (air tank).
- Have a presence confirmation sensor for triggering.
- Part cannot fall off if compressed air is cut off.
- It must have an emergency button and a signal light in the event of a trigger. In addition, in the event of a trigger, it must communicate with the main line panel with a trigger message.
- No burrs, sharp points, unprotected pistons, or crushing points for hands.

13.4.5 Pneumatic System Assumptions

- The manipulator must have a pressure display that is easy for the operator to see.
- The residual pressure relief valve in an easily accessible area.
- The load capacity of the pneumatic cylinder must be at least two times greater than the handling load.
- The system must contain an accumulator.

13.5 PROTOTYPE VIRTUAL AND PHYSICAL

The process of validating the scope of the project involved making virtual and physical prototypes of the manipulator and the layout of the equipment's area of operation. Validation made it possible to verify the effectiveness of the door manipulator project in relation to the aspects identified in the ergonomic assessment, by analyzing usability issues in the Human-System interface and having an amplified assessment of the work area, providing for the installation of other devices in addition to the door manipulators throughout the Hemming Line and Metal Line area, as seen in Figure 13.4.

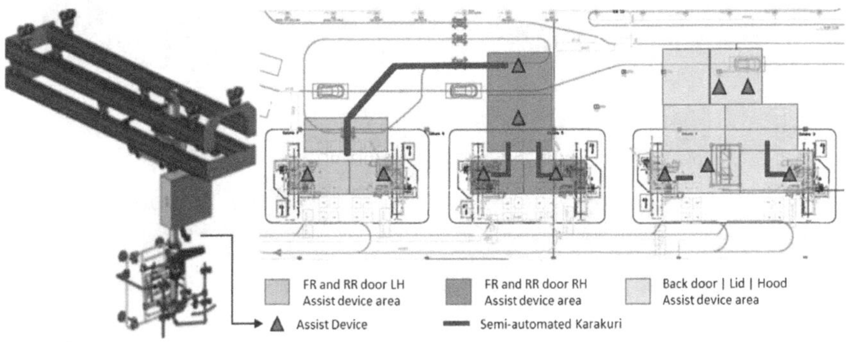

FIGURE 13.4 Project prototype and area of operation of the Hemming and Metal Line door manipulators.

13.6 IMPLEMENTATION

The implementation of the door handler (Figure 13.5) took into account the aspects related to postures and efforts identified in the ergonomic assessment, which were considered harmful to the operator's health when carrying out work activities. The door handler eliminated the most physically demanding postures and efforts made by the operator, especially when removing the door from the hemming die and moving the door to the karakuri.

The design of the front and rear door handlers made it possible to develop and implement the hood and back door handlers for the Hemming Line and the Metal Line. This was possible because both production lines have similar operational characteristics and ergonomic risks with physical overload from the work activity. Figure 13.6 shows two different workstations, where the manipulator was implemented to remove the back door and hood from inside the robotic cell after the engraving process and position it in the karakuri easily and without physical overload.

Figure 13.7(a) shows the working conditions prior to the implementation of the project, where operators were forced to carry out their activities manually and were exposed to occupational risks. Figure 13.7(b) shows the back door manipulator implemented to move the parts from the karakuri to the truck body.

FIGURE 13.5 Implementation of the door manipulator in the manufacturing process.

FIGURE 13.6 Implementation of the back door manipulator (a) and the hood manipulator (b) on the Hemming Line.

FIGURE 13.7 Work situation before and after implementation of the back door manipulator in Metal Line.

FIGURE 13.8 Work situation before and after implementation of the hood manipulator in Metal Line.

The work situation before the hood manipulator project was implemented, as seen in Figure 13.8, also involved manually loading and moving the part from the karakuri to the body, as well as the process of attaching the back door. This process was improved with the implementation of the semi-automatic manipulator with zero gravity, allowing the activity to be carried out safely and without any physical effort, ensuring that the operator can move and fix the part in the cycle time set to produce 32 cars per hour. A total of seven manipulators were implemented on the Hemming Line and four on the Metal Line.

After implementing this equipment, a new ergonomic assessment was carried out, where it was possible to verify the positive impacts generated by the new working conditions without ergonomic risks related to physical overload. In this respect, it is essential that a new ergonomic assessment is always carried out when there is any change in the workplace, whether it is balancing a line, implementing a new piece of equipment or tool, changing the layout, new production volume, etc.

13.7 CONCLUSION

Aiming to achieve better manufacturing goals in automotive Industry 5.0, a new generation of production models requires placing humans at the center of the production system, interacting with intelligent and efficient machines. Industry 5.0 denotes a human-centered, resilient, and sustainable system with a broader application of a socio-technical (manufacturing) system to enterprise automation and production efficiency. In this sense, humans are back at the center, and humans' relations and communication with machines is the principal challenge overcoming the concept of Industry 4.0, which focused only on technology, automation, digitalization, and robotic manipulators while neglecting the human role and perspective. A new generation of production models is about combining human intelligence and creativity with advanced technologies to create a more sustainable, human-centric, and ethical industrial environment (Ordieres-Meré et al., 2023).

The implementation of a technological device in order to reduce occupational risks, improve the working conditions, and optimize productivity in the automotive production line was possible due to a human-centered approach along the ergonomic interventions. The technology offers the potential to level the playing field by transforming old tasks with high physical effort in decent working conditions, equipped with the right tools to boost diversity and inclusion, support a sustainable green economy, and shape the future of work in the automotive industry.

The participation of different actors and sectors of the company to implement successfully the manipulators was an important aspect in order to better understand the dynamic relationship at the workstation (operator, activity, equipments, robots, and so on), defining the design requirements, to defining the type of equipment and the technology involved. This coordination and involvement made it possible for managers to become aware of the working conditions on the production line that needed positive interventions and changes in order to provide suitable working conditions for production and assembly operators.

The ergonomic intervention was imperative to address the adaptation of the automotive workstations through the implementation of a technological device, placing the human as the primary driver of the production system and involving the necessary support from those responsible for managing production systems.

The combination of organizational principles and technologies to design and manage operations and supply chains as resilient, sustainable, and human-centric systems the synergy of Industry 5.0 emphasizing the importance of multidisciplinary technologies and actors (Ivanov, 2023).

The automotive industry is an industrial and economic force that contributes to the global economy and growth and development around the world, driving innovation and generating billions of dollars in investments and millions of jobs and livelihoods (ILO, 2020). It is an essential sector for achieving the 2030 Agenda for Sustainable Development, in particular for achieving Goal 8 of the Sustainable Development Goal, which aims to promote sustained, inclusive and sustainable economic growth, full and productive employment and decent work for all (ILO, 2021).

REFERENCES

Czaja, S. J., Nair, S. N. (2012). Human factors engineering and systems design. Chapter 2 in *Handbook of human factors and ergonomics*, edited by Gavriel Salvendy, 4th ed. John Wiley & Sons, Inc., Hoboken, NJ., pp. 38–56.

Dwivedy, S. K., Eberhard, P. (2006). Dynamic analysis of flexible manipulators, a literature review. *Mechanism and Machine Theory*, 41, pp. 749–777. https://doi:10.1016/j.mechmachtheory.2006.01.014

ILO. (2020). *The future of work in the automotive industry: The need to invest in people's capabilities and decent and sustainable work*, Issues paper for the Technical Meeting on the Future of Work in the Automotive Industry (Geneva, 15–19 February 2021), International Labour Office, Sectoral Policies Department, Geneva.

ILO. (2021). *The future of work in the automotive industry, technical meeting on the future of work in the automotive industry* (Geneva, 15–19 February 2021), Note on the proceedings, International Labour Office, Sectoral Policies Department, Geneva.

Ivanov, D. (2023). The industry 5.0 framework: Viability-based integration of the resilience, sustainability, and human-centricity perspectives. *International Journal of Production Research*, 61(5), pp. 1683–1695. https://doi.org/10.1080/00207543.2022.2118892

Karwowski, W. (2012). The discipline of human factors and ergonomics. Chapter 1 in *Handbook of human factors and ergonomics*, edited by Gavriel Salvendy, 4th ed. John Wiley & Sons, Inc., Hoboken, NJ., pp. 03–37

Lee, K., Jung, M. (2020). Investigation of common insertion hand postures and directions and determination of the voluntarily maximal and preferred insertion forces in automotive assembly tasks. *Human Factors and Ergonomics in Manufacturing & Service Industries*, 30(2), pp. 93–102. https://doi.org/10.1002/hfm.20825

Ordieres-Meré, J., Gutierrez, M., Villalba-Díez, J. (2023). Toward the industry 5.0 paradigm: Increasing value creation through the robust integration of humans and machines. Computers in Industry, 150, p. 103947.

Rodgers, S. H. (2005). Muscle fatigue assessment: Functional Work Analysis Technique. *Chapter* 12 in *The handbook of human factors and ergonomics methods*, edited by Neville Stanton, et al. CRC Press, Boca Raton, pp. 104–113.

Romero, D., Stahre, J., Wuest, T., Noran, O., Gorecky, D. (2016). Towards an operator 4.0 typology: A human-centric perspective on the fourth industrial revolution technologies. Tianjin.

Sales, J. P. d., Reis, J. S. d. M., Barros, J. G. M. d., Fonseca, B. B. d., Junior, A. H. d. A., Almeida, M. d. G. D. d., Barbosa, L. C. F. M., Santos, G., Sampaio, N.A. d. S. (2022). Quality management in the contours of continuous product improvement. *International Journal for Quality Research*, 16(3), pp. 689–702. https://doi.org/10.24874/IJQR16.03-02

Santos, E. F. (2024). O que é Ergonomia 5.0? Blog do Professor Ferro [online].

Trstenjak, M., Benešova, A., Opetuk, T., Cajner, H. (2025). Human factors and ergonomics in Industry 5.0—A systematic literature review. *Applied Sciences*, 15(4), p. 2123. https://doi.org/10.3390/app15042123

Yin, X., Yang, Y., Li, X., Wu, R., Guo, A., Zhao, Q. (2025). Research on the balancing problem of human–robot collaborative assembly line in SMEs considering ergonomic risk and cost. *Computers & Industrial Engineering*, 204, p. 111091. https://doi.org/10.1016/j.cie.2025.111091

14 A Human-Centric Assessment of Musculoskeletal Risks to Facilitate a Human-Centric Workplace for Train Drivers

Paul Thompson and David Hitchcock

14.1 INTRODUCTION

This chapter presents a blow-by-blow account of the development of a human-centric assessment tool to reduce dissatisfaction, discomfort and injury – to contribute to the provision of a human-centric workplace. Its intention is to inspire others to develop similar assessments for their own applications. The contention is that by deploying user-friendly and meaningful assessments, it is possible to identify and prioritise measures to create, or improve existing, workplaces.

14.2 THE IDENTIFICATION OF A NEED

Initially, a Loughborough University postgraduate student, Paul Thompson, chose to look at the driver's cab problem as part of his postgraduate degree in Occupational Health & Safety, on the appreciation of the problems of musculoskeletal discomfort experienced by drivers associated with driving certain trains in Great Britain. He recognised that the ideal driving cab environment may be compromised by the seat, available space and the position of primary controls.

Train operators had seen a rise in complaints and sickness absence regarding alleged dissatisfaction, discomfort and in some cases the onset of injury whilst driving trains.

At the time, if an employee reported experiencing musculoskeletal symptoms that they believed were related to the work they do, an employer would be expected

DOI: 10.1201/9781003559399-16

to modify that employee's working practices. To facilitate these, operators followed their own methods of assessment for each complaint or alleged injury. This typically involved making subjective judgements on the significance of hazards, which undoubtedly led to widely differing opinions and inconsistent treatment about such risks.

There wasn't enough approved ergonomic design data for manufacturers to consistently model suitable proportions in driving cabs to be used in Great Britain. No industry-specific design criteria, approach, or guidelines existed for evaluating design elements that might otherwise compromise or restrict some drivers' usage of their workspace.

A standard on the requirements for driving cabs of railway vehicles was issued on 01.08.1995 (GMRT 2161, Issue 1). Despite the lack of essential human body measures for cab design and physical dimensions, the standard did establish a general need for ergonomic layout to accommodate drivers. It referred to the publication: Pheasant, S. *Anthropometrics: An Introduction*, BSI Publication PP7310 as a starting point.

The design approach for the posture, positioning, and use of the controls should include an ergonomic assessment. Risk assessments must be performed by train operators to safeguard the health and safety of workers, such as train drivers. In accordance with the Management of Health and Safety at Work Regulations 1999 (MHSWR) it is necessary to conduct a risk assessment that considers all risks to which drivers may be exposed.

According to Regulation 4 of the Provision and Use of Work Equipment 1998 (PUWER), employers must take ergonomic risks into account when selecting appropriate equipment and educate and train employees on how to use it properly (Regulations 8 and 9, respectively).

14.3 LEGAL CASES

There were three successful civil cases in connection with musculoskeletal injuries where a properly conducted risk assessment would have recognised that the risk of injury was foreseeable.

14.3.1 Allison -v- London Underground Ltd, 25 January 2007

Damages were claimed for developing Tenosynovitis believed to be caused by repeated action and the way the driver held the brake handle against the spring loading, and resting against the chamfered end of the handle. The judgement concluded that there was inadequate risk assessment (MHSWR, Regulation 3) and training (PUWER, Regulation 9) following the redesign of the safety brake (known as the dead man's handle).

14.3.2 Eaglen -v- Arriva Trains Northern Ltd, 13 July 2007

Damages were claimed for the aggravation of a pre-existing back condition due to the shape and design of the driver's seat. The Judge recognised a link between postural problems and aggravation of back injuries. The judgement concluded that there was inadequate risk assessment (MHSWR, Regulation 3) and failure to select suitable work equipment (PUWER, Regulation 4).

14.3.3 Thomas, Studholme & Rogan -v- Arriva Trains Wales Ltd, 30 November 2009

Damages were claimed by three train drivers for developing Carpal Tunnel Syndrome and Chronic Compartment Syndrome caused by driving Class 142, 143, 150, 153 and 158 trains. The judgement concluded that there was inadequate risk assessment (MHSWR, Regulation 3), failure to select suitable work equipment, and inform and train the drivers (PUWER, Regulations 4, 8 and 9).

14.4 THE CHALLENGE

Because there are many aspects to consider in order to determine how well a person fits their environment, conducting ergonomic risk assessments can be challenging.

It was not uncommon to have risk assessments for musculoskeletal health that were too general to understand risk in a particular workplace context, such as the driver's cab. For example, specific information about the existence of risks, cab physical characteristics, individual (driver) physical characteristics, the job/tasks being performed, the source-path-discomfort mechanism (ligament, muscle, tendon) and subjective assessments of the overall level of risk.

Occupational health staff, the individual driver, their line manager, and even doctor could all be involved in the process for addressing individual risk factors.

For example, the occupational health specialist would advise whether the driver was fit or unfit to return to work, subject to a risk assessment by the line manager. And here arose a major obstacle. Assessing the risk was considered complex, not least because diagnosis of the driver's musculoskeletal condition(s) may be too general or lacking in specific detail. It was not uncommon for a specialist, such as an ergonomist, to be called in to help resolve things.

The aim of Paul's study was to determine key dimensions and ranges, using anthropometric data to accommodate 5th and 95th percentile males and females to evaluate the person and cab fit.

The primary objective of the study was to develop a pragmatic cab (workplace) assessment tool to facilitate a common method of assessment to evaluate design risk factors associated with awkward postures and movements, which may cause, contribute to or exacerbate musculoskeletal discomfort or pain and provide valid recommendations to improve and manage the risks. In doing so, this human-centric assessment would provide the evidence to facilitate improvements to create a human-centric workplace for the train driver.

14.5 THE APPROACH

The approach was based on established ergonomics triangulation, considering the agreement or otherwise of three sources: task analysis, literature review and observations (including measurements). For example, if the task analysis identified key activities that had the potential to result in awkward postures and movements, and the literature review confirmed that such postures and movements could result in discomfort and injury, and the observations confirmed such postures and movements were happening, the level of agreement would indicate a musculoskeletal risk factor worthy of consideration for inclusion in the assessment tool.

14.5.1 Task Analysis

Context was particularly important, given the need to provide a specific tool that advanced on the existing vague, generalist assessments, so it made sense to clarify the context through task analysis.

A driving cab working group (DCWG) including trade union representatives was established to aid with the study. A number of task analysis studies in train driving were evaluated in relation to the work environment and functional activities to identify key tasks performed by the sedentary driver, with the following six key tasks being identified:

1. Observing internal (instrumental) and external (window) information.
2. Operating the brake controller with a push-and-pull handle to increase and decrease braking.
3. Operating the Driver Safety Device (DSD) by maintaining pressure on the DSD foot pedal operated by a driver's lower limb. For vigilance, this could also require the release and reset of the pedal.
4. Operating the power controller with a push and pull handle to decrease and increase acceleration.
5. Operating the warning horn with a push and pull horn stick to sound a two-tone horn.

FIGURE 14.1 Class 158 train cab: key functional activities and driver controls.

6. Resetting the Automatic Warning System (AWS) – Push button to remind driver when approaching a cautionary signal, temporary speed restrictions, emergency speed restrictions and certain permanent speed restrictions (Figure 14.1).

14.5.2 Literature Review

Following the task analysis, a literature review, primarily to identify and clarify the issue of potential discomfort caused by posture and movement, was conducted, which centred on the key tasks identified. This revealed probable levels of driver discomfort associated with four risk variables:

1. Awkward posture relates to the extent of deviation of the joint and/or upper limb (or part of the limb) from a neutral position. There is strong evidence from a large range of studies showing an association between awkward postures and inappropriate dimensions of workplace design. Where a dimensional match of the person and design is suboptimal, posture is often adjusted to compensate, affecting the distribution of body weight around a centre. The individual voluntarily sitting comfortably rather than sitting correctly is also a contributory factor leading to an awkward posture. Whilst musculoskeletal discomfort and musculoskeletal injury both share a change in physiological conditions, the former does not necessarily lead to the latter. The extent of discomfort depends upon the centre of gravity and the extent to which posture is from neutral, the person's extremes of the anthropometric range and the person's body mass.
2. Duration of the task can result in risk when the joints or muscles are overused due to the period they are in contact with the exposure. Injury occurs as the tissues do not have adequate time to recover or replenish materials.
3. Forceful movements can result in injury when loading on the joints (wrist, shoulder, elbow, neck and lower back) or the muscles exceed the capacity of tendons and tissues. Research strongly suggests a combination of other factors such as posture and repetition is necessary to be strongly linked with force to create a musculoskeletal risk.
4. Repeated movements have been shown to increase the degree of risk, with strong evidence positively correlating with an increasing number of repetitions. Components of the musculoskeletal system can degrade, and physiological changes such as inflammation can occur when the joints, muscles and tissues are overused and do not have adequate time to recover or replenish.

14.5.3 Observations and Measurements

Field observations to assess the presence of awkward postures and movements (or otherwise) were conducted using drivers from Northern Rail's DCWG. Key findings from these observations were:

- Compromised clearance between the side or underside of the desk to footplate.
- Inadequate depth of the seat pan to accommodate support of the thighs and buttocks.
- Inadequate horizontal adjustment of the seat pan or the length of the driver's upper thigh limiting forward seating position.
- Inadequate vertical adjustment of the seat pan.
- Prolonged leaning forward and bending and/or rotation of the neck.
- Working with unsupported shoulders or raised shoulders due to lack of armrests, limited armrest adjustment or inappropriate height of fixed armrests.

Key findings from the observations regarding the duration of the task were:

- Duration is considered a risk factor for lower limbs from DSD operation for those who have an awkward posture. DSD operation requires a constant load on the pedal. Average journey times were around 40 minutes, making an average duration of 20 minutes for each leg (which is not in every case) and 50% exertion if each leg equally shares the task throughout the shift.
- Duration is considered high risk if a task is held for longer than one minute, or if it constitutes more than 50% of the duty cycle (a period of continuous activity). No risk is considered if the duration is less than 10% of the duty cycle.
- Duration is not considered a risk factor in upper limb disorders from task operation. The duration of tasks is short for the brake controller, power controller, AWS and warning horn at about one second (e.g., 2.6% duty cycle for brake controller) for each separate movement.

Using an ISOTEST® IST-300D Series digital force gauge, the controls of 11 different Classes of train used by Northern Rail were measured:

- The AWS pushing average of 32.1N – medium risk.
- The brake controller 'Emergency brake' pushing average was 43.1N, and pulling 41.2N – an intentionally high force movement by design to avoid inadvertent emergency brake applications (which are rare).
- The DSD foot pedal pushing average of was 83.3N, considered a high force risk.
- The forces measured for the operation of power, brake, warning horn and AWS tasks are unlikely to present a foreseeable risk of injury.
- The power controller force movement results were obtained for each step taken separately through notch 'zero' to 'step 7'. The pushing average of 23.7N and the pulling average of 13.9N rendered both medium risks.
- The pushing average of the brake controller was 11N which is marginally over the 'low' threshold making it a medium risk. The pulling average was 9.5N which is marginally under the 'medium' threshold making it a low movement force.
- The warning horn pushing average was 16.6N, and the pulling average was 11.5N – both medium force risks.

Repeated movements were determined using downloads of the 'On Train Monitoring Recorder' driving operation data for three distinct service levels, intensive, rural and urban. Typically, Northern Rail drivers would be exposed to 20% intensive, 40% rural and 40% urban.

- Of the six key tasks identified in the task analysis, the highest repeated movements occurred for the brake operation, averaging 1/38.5s, and power controller operation, averaging 1/44.9s.
- The closest to highly repetitive according to Silverstein's definition of, 'one movement every 30 seconds' is the brake operation during intensive route. However, at Northern Rail no driver was exposed to 100% of intensive service level routes over the typical maximum driving time of six hours.
- Although there were no highly repetitive tasks, Northern Rail did not operate a combined power/ brake controller, but such a design would likely breach the repetition threshold as it combines both the brake and power control activities in one task.

14.5.4 Triangulation

The design and construction of the train driver's cab can have an impact on how well the train driver performs, including safety-related tasks.

It was evident that risk factors associated with upper-body repetitive tasks and forceful movements could be considered low risk in train driving. Operation of the hand controls was not sufficient to give rise to concern as there were regular and relatively long periods when there was no significant activity by the driver.

It was not the same for the lower limbs. Operating the DSD foot pedal required a continuous and high static loading using a single foot located centre to the pelvis if the seat is centrally aligned to the DSD and window. Drivers at the anthropometric extremes (particularly small or large) would be more likely to adopt awkward postures and, therefore, an increased risk of musculoskeletal discomfort and injury.

There is strong evidence from a large range of studies showing an association between awkward postures and the dimensions of workplace design, equipment and users' understanding of correct posture and operation. The layout of the primary controls in relation to the train seat, available space, location of the DSD and fixed armrests are key areas producing an undesirable posture. When a driver's size or other physical characteristics require them to adopt an awkward posture whilst doing their tasks, they are likely to exacerbate the situation. In support of this, it was also evident that drivers need to be trained on how to correctly adjust their seat and armrests as part of the cab set-up process, aiming, whenever possible, to attain and maintain a near-neutral driving posture.

14.5.4.1 Anthropometry Assessment

So, having gathered and analysed the evidence, the next stage of the development of a meaningful and effective workplace assessment clearly needed to be based on the suitability of the anthropometric provision of the cab to determine if it would support, or otherwise, the adoption of neutral operating posture and movements.

The evidence indicated that 11 Key anthropometric dimensions were of particular importance:

- Buttock-Knee Length
- Buttock-Popliteal Length
- Forward Grip Reach
- Hip Breadth
- Popliteal Height
- Sitting Elbow Height
- Sitting Eye Height
- Sitting Knee Height
- Sitting Shoulder Height
- Stature
- Thigh Depth.

In accordance with design practice and standards, using the 5th percentile female and 95th percentile male static anthropometry measurements, these 11 dimensions would be used to compare body proportions with associated cab design parameters. The anthropometry software 'PeopleSize' (2008) was the data source used throughout this work.

A form was created to record the body dimensions of an individual driver and the corresponding dimensions of the driver's cab.

14.5.5 The Driver

A Harpenden anthropometer was purchased for taking driver body measurements. A training handbook was developed for using the Harpenden anthropometer and a tape measure to measure the proportions of body parts in the standardised static postures. Although the Harpenden anthropometer was a more precise method for measuring, it was obvious that the occupational health team doing the measurements would have preferred using a regular tape measure.

14.5.6 The Cab

Cab measurements were captured using a tape measure. Interestingly, the dimensions of the same Class (type) of driving cab varied when direct measurements were obtained across the 11 Classes of Northern Rail trains. This was largely due to the imprecise way the seat had been fixed to the floor or bulkhead. The average was determined using the range of measurement data gathered to obtain the base data for the 11 anthropometric dimensions in each type of cab.

14.5.7 ECAT

A spreadsheet tool – ECAT (Ergonomic Cab Assessment Tool) - was subsequently developed for recording measurements against each of the 11 dimensions and the train cab.

The ECAT tool had a data input table comparing the size of the drivers with the 5th and 95th percentile range. This data is then automatically fed into other tables to compare them with the dimensions and available space in the driving cab.

ECAT was then able to report on six comparative dimensions to evaluate how well the cab and individual fit with the 5th and 95th percentile range.

To obtain software expertise and ergonomic specialist assistance for implementing Paul's findings for Northern Rail, Paul's draft study was presented in late 2009 to the Health and Well Being Team of the Rail Safety and Standards Board (RSSB), which was led by Daryll Hopper. It was decided that assistance would be given under Martin Stout's direction, an expert in ergonomics. Working in conjunction with Martin Stout the ECAT spreadsheet tool was initially completed in October 2010.

In June 2012, ECAT underwent an independent review which indicated that ECAT was an important and useful tool for swiftly establishing driver-cab fit, that the 11 dimensions were appropriate and sufficient for achieving the aims without being excessive, and that the method was clear and easy to follow (Figures 14.2–14.5).

1. Place anthropometric measurements in table below

Anthropometric Dimensions	Male mm	Based on Peoplesize data			Female mm	Based on Peoplesize data		
		Actual percentile	5th percentile mm	95th percentile mm		Actual percentile	5th percentile mm	95th percentile mm
1 Stature	1803	83%	1644	1875			1518	1729
2 Sitting eye height	801	44%	748	865			650	800
3 Sitting elbow height	249	64%	208	276			198	264
4 Thigh thickness	177	80%	138	194			132	202
5 Buttock-knee length	612	43%	557	674			537	645
6 Buttock-popliteal	506	52%	456	555			437	534
7 Popliteal height			411	501			370	456
8 Sitting Knee height			500	592			458	544
9 Hip breadth			340	433			347	469
10 Forward grip reach			693	812			623	748
11 Sitting shoulder height			556	658			520	619

Note: Dimensions 1, 7 and 8 includes a correction of 25mm to take account of footwear, not adjusted in 5th / 95th references

FIGURE 14.2 Ergonomic cab assessment tool (ECAT) input table for anthropometric comparison.

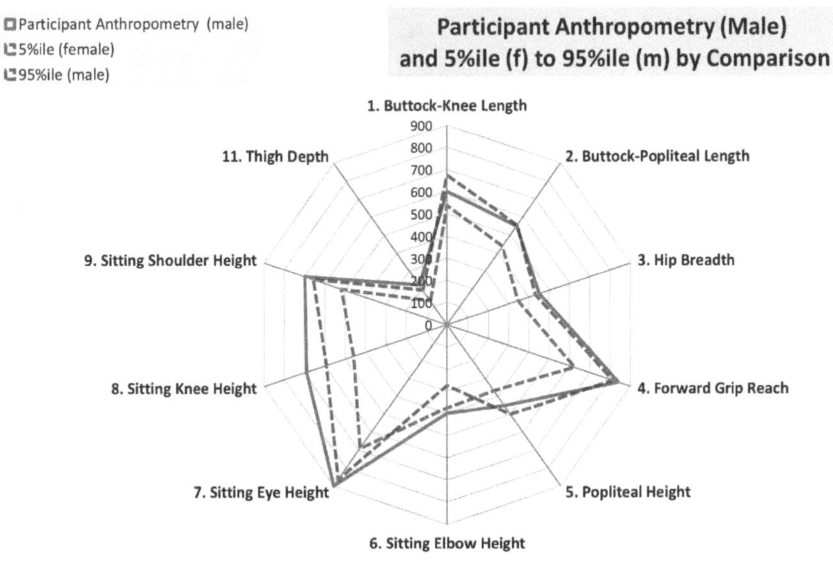

FIGURE 14.3 Comparative analysis of participant anthropometry and anthropometric percentile limits.

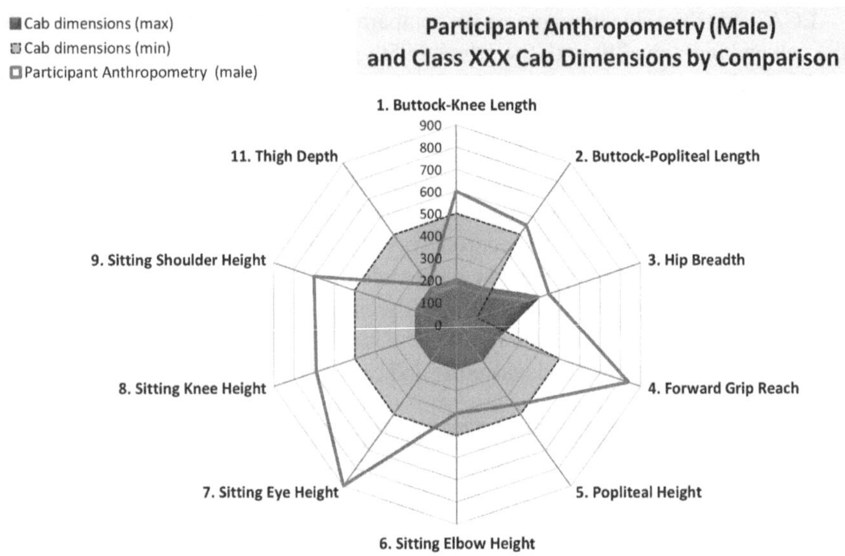

FIGURE 14.4 Comparative analysis of participant anthropometry and train cab dimensions.

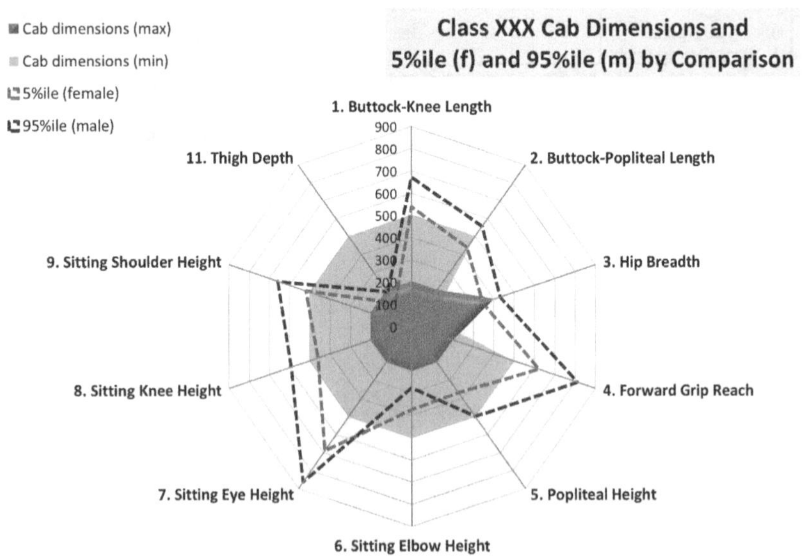

FIGURE 14.5 Comparative analysis of driving cab dimensions with 5th and 95th percentile anthropometric limits.

14.5.8 MAT

A joint presentation was subsequently conducted with RSSB sharing best practice on Northern Rail's ECAT tool and the implementation plans, which featured a joint Ergonomic Cab Assessment Process procedure (OHSMP 03) established with an occupational health provider.

At this point, pre-employment medicals and other associated health referrals now measured the crucial elements required as input for this tool, and the instrument's output indicated any potential educational and training requirements for MSD risk for certain employees.

The recent Thomas, Studholme & Rogan -v- Arriva Trains Wales Ltd court case was the topic of discussions in the industry headed by Peter Lovegrove, Operations Resilience Manager at the Association of Train Operating Companies (ATOC). The case was seen by the rail sector as one that might serve as a model for similar cases and further civil court action. The case highlighted the potential for improvements in the way that MSD risks associated with train driving were managed.

The issue at hand was how to ensure that train operators conducted a suitable and sufficient risk assessment that fully conformed with the requirements of the civil courts.

The Operations Focus Group (OFG) subsequently asked RSSB to evaluate the currently available tools and, if necessary, create such a tool for assessing and managing these risks.

A research project (T940) was prepared by Anita Scott (RSSB) and approved by RSSB steering committee. Paul was part of the steering group that RSSB established after giving the project's approval.

The first phase of the research was to review the scientific literature, laws, resources, and advice that were already available on MSDs. RSSB commissioned Sharon Allaway of Marum Healthcare to undertake the review. Several areas of Paul's study were referenced in her report dated September 2010. The review concluded that there was no procedure or resource that train operators might use to consider all train operator risk factors required to control MSDs in train drivers.

The project steering group asked RSSB to commission a second phase of work to close this gap. This involved either further improving the Northern Rail ECAT or creating a new RSSB tool that met the needs of all train operators including that of developing guidelines on how to evaluate MSD risk, measure cab dimensions, and conduct anthropometric measurements using the tool. A cost-benefit analysis for undertaking the additional work on the Northern Rail ECAT tool was favoured over creating a tool from scratch.

Through the steering group, RSSB developed the Northern Rail ECAT tool and renamed it MSD risk Assessment for Train Drivers (MAT). An improved user interface and significant functionality were added with detailed guidance to support understanding of the risks. The tool was launched in Autumn 2012.

MAT adopted a couple of important features which made it a usable and meaningful tool. First – and arguably the most important. It distilled a complex and highly multifactorial assessment into a prioritised and manageable set of factors. Second – and also helpful in identifying the key risks to address it presented the assessment in a 'traffic light' format, which to all intents and purposes, inferred that red risks ought to, at least, be reduced to amber risks, and ideally amber risks improved to be green.

An article was created, presented and submitted to the National Rail Awards panel at the request of the rail industry. Paul's study, titled 'Identifying and Managing the Risk of MSD in Train Driving' was summarised in a written submission about collaborative working with RSSB and Industry stakeholders leading to the development of the Industry MAT tool.

In July 2012, a 20-minute presentation followed by a 20-minute Q&A was given on 12th July 2012, at 13.30. Hosted by the British Transport Police, the panel of Judges comprised of Alan Cooksey, Clive Fletcher-Wood and Andrew Trotter who shortlisted the project to the forthcoming September National Rail Awards.

Today, train cabs in Great Britain are required to adhere to the most recent Railway Group Standards document GMRT 2161, Issue 2.1, March 2022. The coding 'GM' falls into the Trains (rolling stock) category and 'RT' defines mandatory requirements. Although GMRT 2161 has undergone several revisions, the content itself has not significantly changed in terms of ergonomics.

A supplementary standard RIS-2761-RST, (Issue 1.1, March 2022) is a Good Practice Guide entitled 'Rail Industry Standard for Driving Cabs' which was released for the first time in March 2020. To support engineering and operational concerns, it specifies additional requirements which may be used regarding the design, layout, and functional equipment in driving cabs.

14.6 BEYOND MAT

14.6.1 The Train Cab Workplace

The MAT tool has been applied in a proactive manner in the assessment and adjustment of new vehicles before implementation, and in a reactive manner when drivers have reported problems, discomfort and injury. Hitchcock, Didio and Rees (2016) reported on its application.

Although not mandatory, Section G 2.1.7 of this Rail Industry Standard now offers best practice recommendations by utilising the RSSB MAT tool to assess the driver and cab fit.

14.6.2 Other Activities and Workplaces

The Health & Safety Executive has successfully exploited similarly distilled and simplified 'traffic light' human-centric assessments for repetitive work (ART – Assessment of Repetitive Tasks), lifting and carrying (MAC – Manual handling Assessment Chart) and (RAPP – Risk Assessment of Pushing and Pulling). Like the MAT tool, these also recognise that to facilitate human-centric workplaces, the hierarchy of risk controls should be used, with the preference being to eliminate risks through the selection and design of facilities, equipment, and processes, followed by workforce education and training.

ACKNOWLEDGEMENTS

In the development of ECAT:

- Deborah Walker (Loughborough University), for general research advice
- Gary Stewart (Northern Rail)
- Martin Stout (RSSB)
- Roy Cochran (BUPA)

In the development of MAT:

- Anita Scott; Daryll Hopper & Matin Stout
- Chris Didio, Paul Bell & Simon Groves (Arriva Trains Wales)
- Colin Clifton (Southeastern)
- Colin Fish & Robert Gardner (Freightliner)
- Dr David Shackleton
- Neil Whisler & Richard Burnham (AMEY)
- Peter Lovegrove (ATOC)
- Philip Howells (First Group)
- Sarah Hext (Northern Rail)
- Sharon Allaway (Marum Healthcare).

REFERENCE

Hitchcock, D., Didio, C., & Rees, T. (2016). *Help or Hindrance? A Case Study of Pragmatic Anthropometry*, Contemporary Ergonomics and Human Factors 2016. Eds. Rebecca Charles and John Wilkinson. CIEHF.

15 The Importance of the Passive Skeleton for Industry 5.0

The Case of the Track Maintenance Activity of a Mining Industry

Márcia Raissa A.F. Pereira,
Raimundo Lopes Diniz, and Marcelo M. Soares

15.1 INTRODUCTION

The Brazilian railway industry, responsible for transporting large volumes of ore, requires intensive maintenance of the permanent way, such as rail grinding, due to the increasing load and operational speed (ABIFER 2023; Selig & Waters, 1994). In mining, ergonomic challenges are significant, requiring strategies to mitigate musculoskeletal risks (Andrade, 2018). This study investigated grinding activities from an ergonomic perspective, applying the Macroeconomic Analysis of Work—AMT method (Guimarães, 1999) and analyzing the use of passive exoskeletons. There is also a transition in these sectors toward Industry 5.0, with the incorporation of human-centered technologies aimed at personalization, ergonomics, and sustainability (Nahavandi, 2019; Breque et al., 2021).

This chapter describes a case study related to the mining industry, considering the relevance of using passive exoskeletons—a piece of equipment that incorporates technology into its structure to enhance the performance of activities that require physical effort. This study examines an ergonomic intervention in the permanent way sector of a mining industry, specifically in the grinding of rails. The AMT method (Guimarães, 1999) was applied up to the ergonomic diagnosis phase. Initially, the ergonomic assessment phase was necessary to survey and map ergonomic demand items (EDIs) and assess the ranking of EDIs related to physical load. Subsequently, we looked into the level of physical load and its relationship to the use of a passive exoskeleton during the grinding activity.

 DOI: 10.1201/9781003559399-17

15.1.1 Mining Industry and Rail Transportation: A Brief Description

Mining is one of Brazil's main economic activities, essential for job creation and mineral exports. According to the Brazilian Mining Institute (IBRAM), the sector employs more than 176,000 workers directly, with each direct job generating an estimated 3.5 jobs in the mineral transformation chain. It is an extractive activity that includes the research, extraction, and processing of mineral resources found underground. It is one of the foundations of economic and technological development in many countries. However, in addition to extraction itself, modern mining is a complex industry that involves integrated logistics, processing, automation, and data management operations. It can therefore be considered an integral part of basic industry.

The mining industry encompasses the planning, operation, and management of mines, as well as the processing, transportation, and export of minerals. In this chain, the strategic role of rail transportation is particularly notable, especially in countries with extensive territorial distances, such as Brazil. This model is highly efficient for transporting minerals on a large scale, contributing to the sector's logistical sustainability and international competitiveness (IPEA, 2018).

The adoption of digital technologies, autonomous systems, and ergonomic solutions has transformed mining into an increasingly connected, intelligent, and human-centered industry, characteristics aligned with the concept of Industry 5.0. This new approach proposes the integration of automation and the valorization of human work, through technologies such as passive exoskeletons, which optimize ergonomics and reduce occupational risks in critical activities, such as the maintenance of railway wagons used to transport ore (European Commission, 2021).

In this context, mining is no longer viewed merely as an extractive activity, but as a strategic and innovative industry that combines productivity, sustainability, and workers' well-being—fundamental pillars of the transition to Industry 5.0.

15.1.2 Permanent Way Maintenance and Rail Grinding Activity

Maintenance of the permanent way is crucial for railway operational safety, as it ensures the integrity of the rails, sleepers, and ballast, thereby maintaining the stability of the railway system. This activity involves continuous interventions to keep the components in good condition, especially the rails, which cannot fail without compromising the operation (Neto, 2012). Rail wear caused by various factors requires periodic replacement (Azevedo, 2017). Grinding is a fundamental technique for removing microcracks and ensuring that joints are level, reducing impacts, and increasing the useful life of the track.

Performing these tasks requires considerable physical effort and exposes individuals to inappropriate postures, repetitive motions, and pressure, which can lead to musculoskeletal disorders (Massad et al., 2011). Working conditions vary according to the environment, tools, and materials used (Dias et al., 2021). Despite technical training and safety measures, workers are still exposed to significant ergonomic risks, especially in activities under the wagon, where there is a lack of data on health impacts (Sousa et al., 2015; Kunda et al., 2013; Reardon et al., 2014).

15.1.3 Exoskeleton: History, Types, and Industrial Applications

Exoskeletons initially appeared for medical and military purposes, aiming to facilitate rehabilitation and enhance physical capacity, but they have only recently been adapted for occupational use. Starting in the 1960s, the US Department of Defense began researching the concept of "power armor" to increase the strength of soldiers (Lowe et al., 2019; Schnieders & Stone, 2017). Since then, various models have been developed focusing on different body regions, such as the arms, back, and legs (Bosch et al., 2016).

These devices are capable of physically assisting the worker, reducing musculoskeletal overload and preventing injuries related to forced postures and repetitive movements (Looze et al., 2016; Zhang & Huang, 2018). Considered a form of cooperation between humans and robots, exoskeletons are particularly notable in tasks that cannot be automated due to their inherent variability (Lee et al., 2012; Claus et al., 2019). The most common versions are classified as either active, which utilize motors to amplify movements, or passive, which employ elements such as springs and shock absorbers to redistribute loads (Bosch et al., 2016; Looze et al., 2016).

With the advent of Industry 4.0, technologies such as exoskeletons have been introduced to enhance productivity and ergonomics in industrial environments (Schreiber, 2020; Russo et al., 2021). However, their use still requires technical advances in terms of comfort, weight, and adaptation to the human anatomy.

In Industry 5.0, the focus has shifted to human-machine collaboration, valuing personalization, sustainability, and worker well-being (Demir & Cicibas, 2021; Nahavandi, 2019). In this context, exoskeletons have become fundamental by combining technology and human capabilities, promoting safer and healthier environments (Javaid & Haleem, 2020). Recent studies have contributed to a more efficient application of these devices, adapting them to specific tasks and profiles, which strengthens their acceptance and effectiveness in the occupational scenario.

15.1.4 Ergonomics and Exoskeleton Research

The evolution of ergonomics is closely tied to industrial progress, driving solutions that enhance efficiency, safety, and quality in production operations. In this context, exoskeletons have emerged as a promising innovation. However, their large-scale adoption depends on robust evidence as to the effectiveness of these devices (Crea et al., 2021).

Recent studies have demonstrated physiological and ergonomic benefits. Maurice et al. (2019) reported that the use of a lightweight upper limb exoskeleton reduced the load on the shoulder during aerial tasks, without negatively impacting other regions of the body. Kong et al. (2022) found, through EMG and biomechanical modeling, that lower limb exoskeletons significantly reduce muscle load. In the case of passive lumbar models, laboratory studies indicate reductions in muscle activity in the lumbar region of between 10% and 57% (Frost et al., 2009; Ulrey & Fathallah, 2013; Koopman et al., 2019). Field studies show reductions of up to 25% in activities such as automotive assembly (Graham, 2009; Bosch et al., 2016).

Despite these advances, there are still no specific studies on the use of lumbar exoskeletons in activities such as rail grinding, which require static postures of trunk flexion and twisting. Recent literature highlights the need to expand field studies, which, although more representative of reality, are still limited in terms of sampling and scope. Such studies are essential for mapping potential side effects and informing revisions to ergonomic standards and guidelines (Bogue, 2018; McFarland & Fischer, 2019; Torricelli et al., 2020).

Research, such as that by Veronesi and Aquino (2020), demonstrates the efficiency of various models in industrial contexts, including a 17% increase in lumbar strength with the use of elastic exoskeletons. In addition, Ippolito, Constantinescu, and Riedel (2020) emphasize the importance of methodologies that integrate computer simulations and personalized adaptation to maximize the ergonomic benefits of technology.

The growing incidence of musculoskeletal disorders in the workplace, such as back, neck, and upper limb pain, underscores the importance of exoskeletons as a preventive measure (Parent-Thirion et al., 2017; Bräunig, 2013). To this end, a user-centered approach is recommended, based on ergonomic assessments, usability, and acceptance, ensuring comfort and safety (Ippolito et al., 2020).

15.2 METHODS AND TECHNIQUES

15.2.1 Typification of the Research

This research is of an applied nature, descriptive and exploratory in terms of its objectives and the level of knowledge produced, as well as being a case study in terms of its technical procedures and a qualitative-quantitative approach. This research complied with the ethical principles outlined in Resolution 466/12 or 510/16 of the National Health Council and was conducted after approval by the ethics board under CAAE number 74192523.3.0000.5208.

15.2.2 Macro-Ergonomic Analysis of Work (Guimarães, 1999)

The Macroergonomic Work Analysis method was adopted, as described by Guimarães (1999), which is a participatory approach based on action research. The study comprised phases 0 (launch) and 1 (ergonomic assessment), grounded in the principles of participatory ergonomics (Hasani et al., 2022). In phase 0, the research objectives and schedule were presented to managers and workers in health and safety meetings and dialogues, ensuring free, informed consent and confidentiality. In Phase 1, observations were made, open interviews were conducted with six permanent way workers (30% of the population), and a validated questionnaire was administered to 19 workers (100%). The aim was to identify and prioritize EDIs using the GUT matrix. In phase 2 (diagnosis), specific tools were used to assess the physical load: postural analysis (REBA), Body Segment Diagram (Corlett), and lumbar isometric dynamometry, with and without the use of the passive exoskeleton LAEVO®, Figure 15.1 (Barca, 2020), selected for transferring efforts from the spine to the lower limbs.

The wearable exoskeleton is designed to aid in lower-body movement. The device includes padded thigh supports, a waist belt, adjustable shoulder straps, and a mechanical linkage system with cables to help reduce the physical load on the user's legs and back during activities such as walking, standing, or lifting.

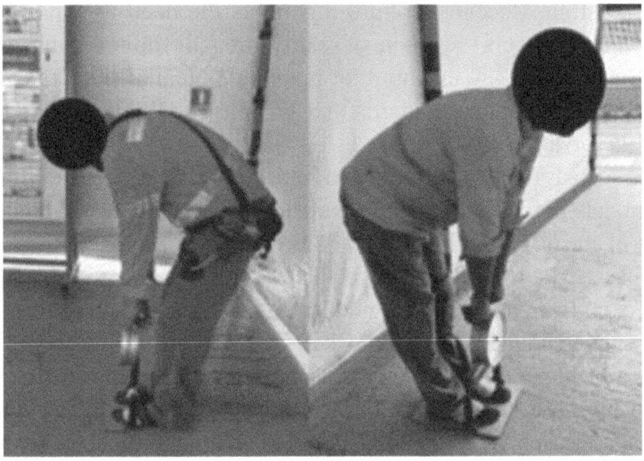

FIGURE 15.1 Dynamometry with and without the exoskeleton.

Source: The authors (2024).

A pre-test was conducted with five workers to assess the feasibility of the measurements, followed by the application of the techniques to 15 participants. The sample size was calculated using G*Power software, employing a two-tailed test with a power of 80% and an α level of 5%. Analyses were conducted before and after rail grinding, with and without the use of the exoskeleton, using a t-test or Wilcoxon test, depending on the data distribution (Shapiro-Wilk, $p < 0.05$), and Cohen's d test for effect size.

Task analysis included systematic observation using the Kinebot system and a descriptive activity protocol. Discomfort was assessed using Corlett's Body Segment Diagram, a widely validated tool in ergonomics, at various times during the activity, both with and without the exoskeleton. The averages per body segment were tabulated and analyzed graphically.

The postural analysis employed the REBA (Rapid Entire Body Assessment) technique, with filming conducted in the sagittal plane at three key moments (beginning, middle, and end of the activity). Body segments were assessed according to the methodology outlined by Hignett and McAtamney (2000). The score was processed using REBA v.1.3, and the data were analyzed in IBM SPSS v.26, employing the Wilcoxon test for comparisons and the Friedman test for repeated measures.

Lumbar isometric muscle strength was assessed using an Oswaldo Filizola dorsal dynamometer (Crown, 200 kgf), according to the protocol by Veronesi (2020), with three repetitions of 3 seconds each and a 1-minute interval. The task consisted of maximum traction in the standardized position, with and without the exoskeleton, before and after the activity (Figure 15.1). The data were analyzed in Excel using the Friedman test.

The image shows two workers performing a lifting task with a dynamometer to measure force exertion. The left side illustrates the subject using a passive exoskeleton, while the right side shows the same task performed without the device. This setup is used to evaluate the biomechanical assistance provided by the exoskeleton in reducing muscular strain during manual handling tasks.

The Spearman coefficient (rho) was used to verify the correlation between the work variables. This measure of linear association utilizes the intervals and order numbers of each group of subjects, comparing these intervals (Santander & Ruiz, 2004).

15.3 METHODOLOGICAL DESIGN

The research was conducted in three phases to analyze the activity of grinding rails and the associated ergonomic constraints. In the initial phase (phase zero), the study was launched. In the first phase (ergonomic assessment), interviews, questionnaires, and systematic observation were used to validate workers' perceptions and describe the occupational posture. In the second phase (ergonomic diagnosis), a qualitative-quantitative approach was employed to test hypotheses, utilizing techniques such as the Corlett Body Segment Diagram, REBA, and dynamometry. The design followed the AMT method, guiding the tools applied at each stage.

15.4 RESULTS AND DISCUSSIONS

In the Ergonomic Assessment stage, the main constraints identified were:

a. Environment:
   - Exposure to the sun.
b. Biomechanics/Workstation:
   - Discomfort in the lumbar spine.
   - The longer they work, the greater the discomfort.
   - Pain in the arms and knees.
   - Repetitive movements.
   - Inadequate posture.
   - Grinding the side of the rail is more difficult.
   - Heavy activity.
   - Positioning the rail on the ground.
c. Organization:
   - Routine activity.
   - There is a rotation of activities.
   - Activity carried out in groups.
d. Company:
   Uses a personal checklist to identify points of attention.
   - Hydration is available.

Inadequate postures were observed during tasks performed during the working day. These postures were particularly noticeable when grinding the rail with a sander and when finishing with a grinder, as the workers bent their spines for extended periods. As the activity is carried out in uncovered areas, workers are exposed to the weather, and sun exposure is frequent, given that the activity cannot be conducted during rainy periods. You can also see from the picture that the floor is uneven (Figure 15.2).

FIGURE 15.2 Examples of inappropriate postures when grinding the rail with a sander.

Source: The authors (2024).

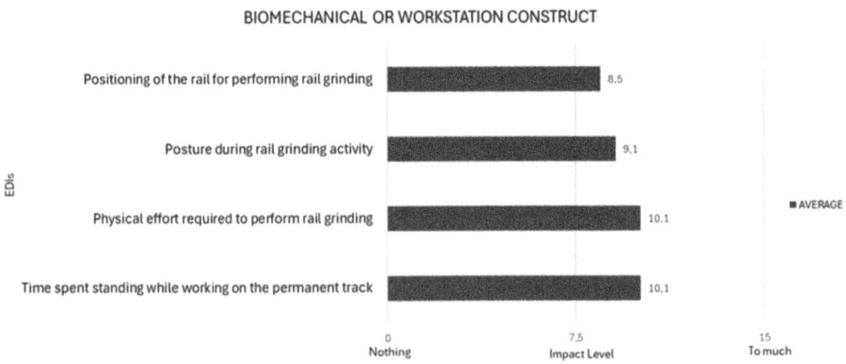

FIGURE 15.3 Level of impact of biomechanical or workstation constructs.

Source: The authors (2024).

The images depict railway workers performing grinding operations on rail tracks while adopting ergonomically unfavorable postures, such as excessive trunk flexion and prolonged kneeling. These positions can contribute to increased risk of musculoskeletal disorders, highlighting the need for ergonomic interventions or assistive devices to reduce physical strain in such tasks.

The following are the results of the main aspects (EDIs) noted by the respondents when the closed questionnaires were administered to the permanent way workers in the two classes on the day shift.

As far as the environment construct is concerned, it showed higher levels of impact overall, with the FDI "temperature in your work environment" standing out (average 12.5—environment), which corroborates the location where the activity is carried out, where workers have no protection from exposure to inclement weather.

In the biomechanical construct or workstation (Figure 15.3), it is evident that the workers are not satisfied with the physical effort and occupational posture they adopt while carrying out their activities, a fact that was also evident during the systematic observations.

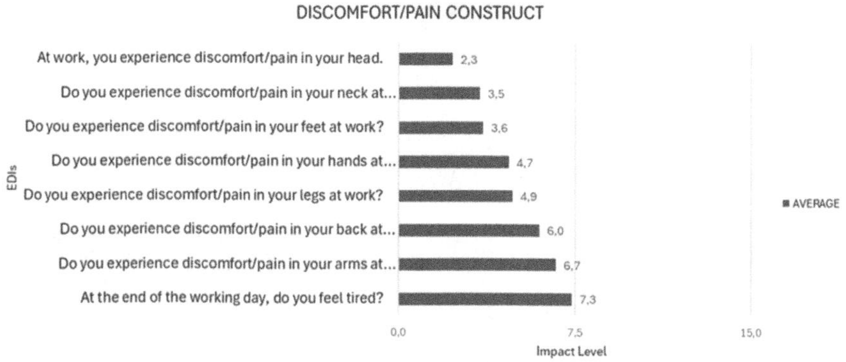

FIGURE 15.4 Impact level of the discomfort/pain construct.

Source: The authors (2024).

The bar chart presents average ratings for four ergonomic discomfort indicators (EDIs) related to rail grinding activities: rail positioning, working posture, physical effort, and time spent standing. All factors were rated on a scale from 0 (Nothing) to 15 (Too much), with physical effort and prolonged standing receiving the highest average impact scores (10.1), followed by posture (9.1) and rail positioning (8.5). These results highlight critical ergonomic challenges faced by workers during rail maintenance operations.

In the organizational construct, the EDIs with the highest impact were "your work involves responsibility" (mean 13.5) and "your work makes you feel valued" (mean, 12.1), while those with the lowest impact were "you feel psychological pressure from your boss" (mean, 2.3) and "your work is monotonous" (mean, 4.5). All the average EDIs had impact scores below 7.5, the highest being related to discomfort in the back (mean, 6.0) and arms (mean, 6.7). It was observed that the postural demands associated with the height of the floor force flexion of the spine and overload of the upper limbs, resulting in pain and discomfort when grinding rails. Body regions involved in dynamic movements, such as the head, neck, and feet, reported less pain, in line with the literature on biomechanical overload and postural discomfort (Diniz, 2003; Serranheira et al., 2003; Pauluk & Michaloski, 2016) (Figure 15.4).

This bar graph presents average ratings for various physical discomfort or pain indicators experienced by workers. It includes body regions such as the head, neck, feet, hands, legs, back, and arms, as well as overall fatigue at the end of the workday. The highest impact levels were reported for fatigue (7.3), arm pain (6.7), and back pain (6.0), indicating significant physical strain associated with work activities.

The Macroergonomic Analysis of Work proposes that, immediately after concluding the data analysis of the appraisal, we return to the company to return the information obtained and discuss the EDIs with the participants. In agreement with Guimarães (1999), a presentation of the data was made, which enabled a better understanding of the situation and, therefore, a more reliable diagnosis.

According to systematic observations and a flowchart, the ergonomic diagnosis, based on task analysis and a kinesiological description of postures, revealed that the

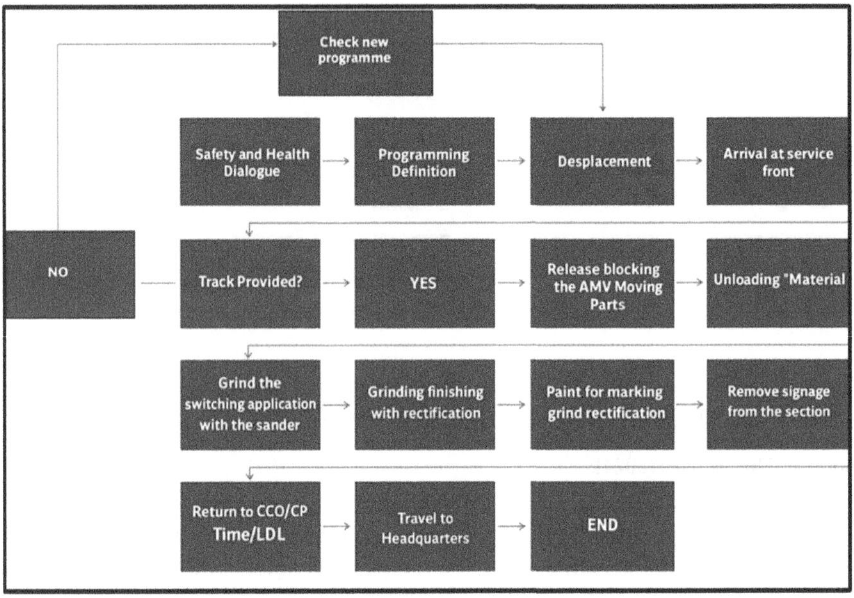

FIGURE 15.5 Flowchart of the activities of permanent way officers.

Source: The authors (2024).

activities of permanent track workers follow a structured workflow. Initially, after the track has been cleared by the Operational Control Center and the Yard Control Center, the track switcher is mechanically blocked. This is followed by unloading the material and setting up the support site. With the tools ready, grinding begins on the track switcher, followed by finishing with a grinder. The site is then painted with white spray paint, and finally, the line is released, including the removal of tools, signaling, and returning the Line Release Time.

The workers alternate standing postures (static and dynamic), with flexions and extensions of the trunk, upper and lower limbs, as well as neck tilts. During grinding (3.5 kg sander) and finishing (2 kg grinder), they hold their palms, flex their elbows and wrists, and tilt their hips and torso to the side. When painting, they adopt a squatting posture for better reach, with the upper limbs extended and the fingers flexed to press the spray.

The execution of the activities is graphically represented in a flowchart to illustrate the conditions and processes carried out by the officers (Figure 15.5).

This flowchart outlines the activities of permanent way officers, detailing the sequence of tasks involved in checking a new program, conducting safety dialogues, programming, displacement, and material handling, as well as the steps for grinding, finishing, painting, and removing signage from a section. The process concludes with returning to CCO/CP Time/LDL and traveling to headquarters.

To analyze the task and provide a kinesiological description of the postures adopted during the activities, we utilized the Kinebot system, a kinesiological assessment tool used in ergonomics, which identifies inappropriate postures based on biomechanical principles. Table 15.1 illustrates the analyzed postures, highlighting the

TABLE 15.1
Table with the postures assumed

Activities		**Postures assumed**	
Material unloading			
Put up road signs			
Sander assembly		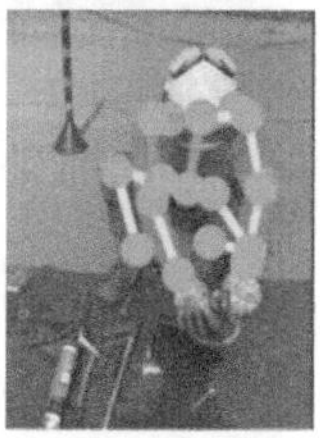	
Grinding of switches and crossings			
Grinding finish with grinding wheel			

(Continued)

TABLE 15.1
(Continued)

Activities	**Postures assumed**		
Painting to mark the grinding area			
Remove road signs from the stretch			
Material load			

Source: The authors, 2024 based on Kinebot (2023).

angles at most significant risk of discomfort or pain in specific colors (red and lilac). The activities of grinding the track changer and finishing with a grinder were identified as the most critical, with the most significant impact on the spine. To deepen the analysis, the REBA method was applied to grinding with a sander, as this is the most physically demanding task among the examined tasks.

This table, titled "Postures Assumed," illustrates various activities performed by workers, such as material unloading, putting up road signs, sander assembly, grinding of switches and crossings, grinding finish with a grinding wheel, painting to mark the grinding area, removing road signs from the stretch, and material load. For each activity, a series of images depicts the different postures adopted by the workers, likely for ergonomic analysis.

15.4.1 Corlett Body Segment Diagram—Analysis of Discomfort/Pain Before and After the Activity, Without and With the Exoskeleton

The results showed that the trunk is one of the most affected regions, as illustrated in Figures 15.6–15.8. The main factors associated with discomfort include inadequate postures, high repetitiveness, and unfavorable environmental conditions, confirming previous findings by Paz (2021).

This triangular graph presents self-reported levels of discomfort or pain in the cervical spine region before and after the rail grinding activity, both with and without the use of an exoskeleton. Each line represents an individual participant (n=15). The data suggest a reduction in discomfort levels when the exoskeleton is used, indicating its potential effectiveness in minimizing cervical spine strain during physically demanding tasks.

This triangular chart illustrates the levels of discomfort or pain reported in the thoracic spine before and after the rail grinding activity, comparing conditions with and without the use of an exoskeleton. Each line represents an individual participant (n=15). The figure shows a noticeable increase in discomfort after the activity without the exoskeleton, while levels remained lower and more consistent when the exoskeleton was used, suggesting its potential in reducing thoracic spine strain during physical tasks.

This triangular graph presents individual participant ratings (n=15) of lumbar spine discomfort or pain at three time points: before the activity, after the activity with an exoskeleton, and after the activity without it. The data reveal higher discomfort levels after the activity when performed without the exoskeleton, while

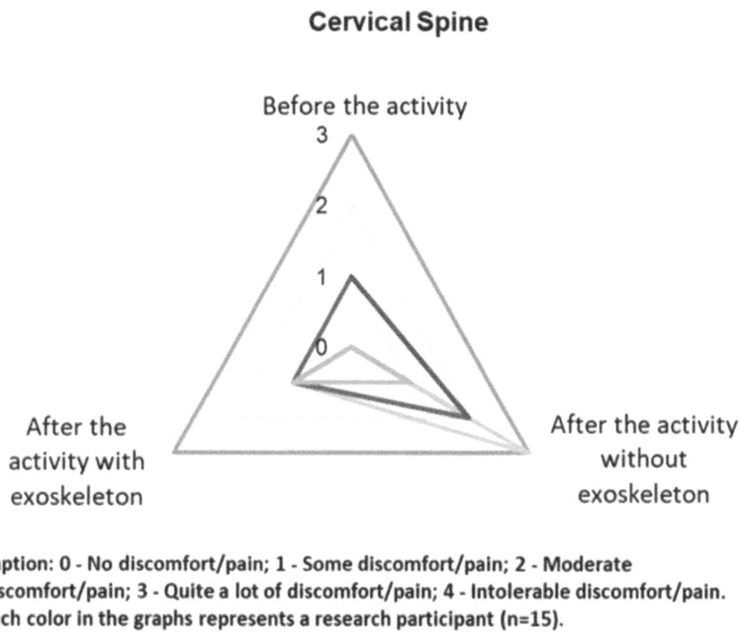

FIGURE 15.6 Results of the discomfort/pain analysis for the cervical spine.

Source: The authors (2024).

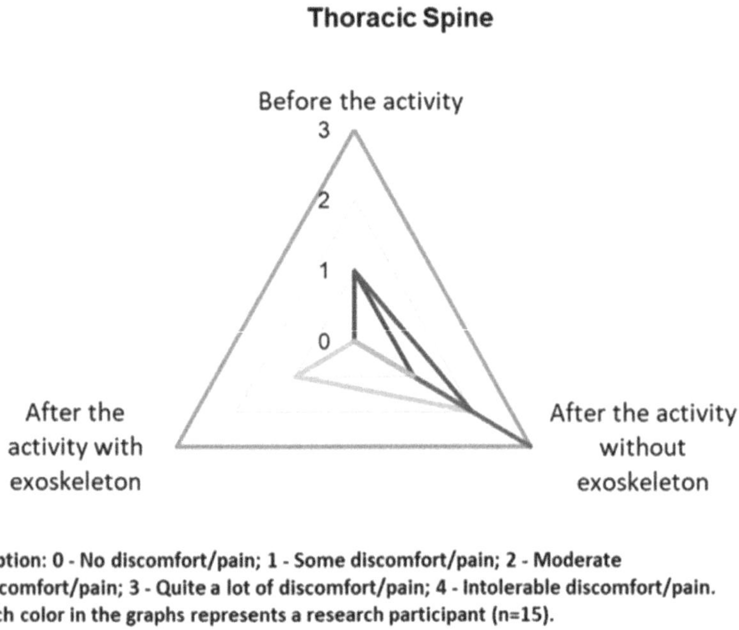

FIGURE 15.7 Results of the discomfort/pain analysis for the thoracic spine.

Source: The authors (2024).

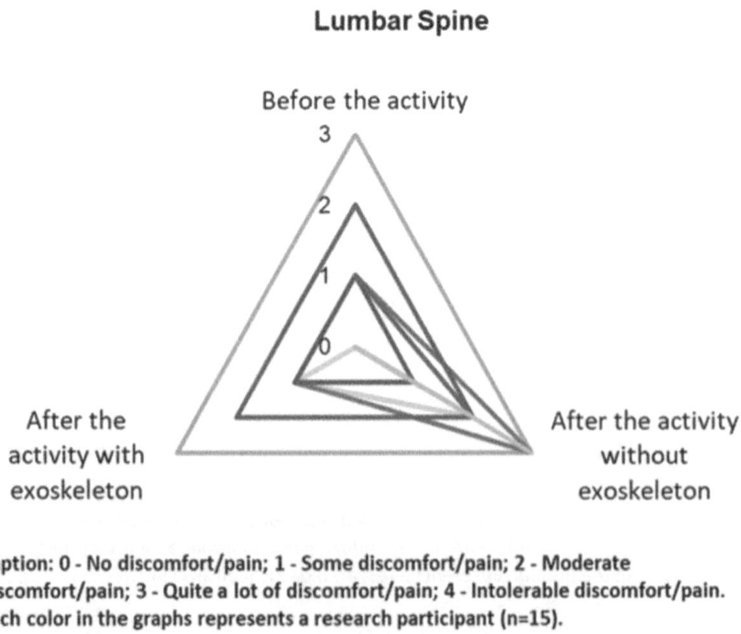

FIGURE 15.8 Results of the discomfort/pain analysis for the lumbar spine.

Source: The authors (2024).

post-activity discomfort was generally lower and more consistent when the exoskeleton was used. These results suggest that the exoskeleton may provide meaningful lumbar support during physically demanding tasks.

The results showed a higher frequency of reports of moderate or intense pain and discomfort (levels 2 and 3) in the trunk region after the activity without the use of the exoskeleton. In some cases, lumbar discomfort remained constant throughout the measurements, confirming Possebom et.al (2017) findings, which relate increased discomfort to performing tasks outside the muscular comfort zone. Comparisons between the right and left sides of the body revealed similar patterns, reinforcing the symmetry in the biomechanical demands of the task.

The shoulders and legs exhibited the highest levels of discomfort after the activity, regardless of whether the exoskeleton was used, reflecting the effort required during grinding and prolonged standing (Zander et al., 2004; Messing et al., 2008; Balasubramanian, 2009). In the other regions, discomfort was more pronounced before the activity and, above all, after performing it without the exoskeleton, with a predominance of moderate or intense reports (levels 2 and 3).

15.4.2 Posture Analysis Using REBA (Hignett & McAtamney, 2000)

Permanent track workers face high physical loads, characterized by repetitive movements and uncomfortable postures, as assessed by the Rapid Entire Body Assessment (REBA) technique (Hignett & McAtamney, 2000). Trunk flexions above 60°, neck rotation, deep knee flexions, arm extension, and wrist flexions above 15° were observed, resulting in a medium risk score of 7.37 without the exoskeleton and 5.95 with the device (Figure 15.9). Although the use of the exoskeleton reduced the score, the risk level remained unchanged, indicating the need for additional interventions.

These findings align with those of Cardoso (2019) and Dahmen and Hefferle (2018), who highlight the limitations of the exoskeleton in the face of environmental and postural factors, such as the need for trunk flexion to achieve visual and motor precision.

This sequence of images illustrates the most frequently observed movements and postures of the trunk, neck, legs, arms, and wrists during permanent way

FIGURE 15.9 Most frequent trunk, neck, leg, arm, and wrist movements among permanent track workers observed.

Source: The authors (2024).

TABLE 15.2
Results of the REBA technique

	REBA		
	Without the exoskeleton	**With the exoskeleton**	**p-valor**
Standard	7,37	5,95	<0,001

Source: The authors (2024).

maintenance activities. The tasks typically require sustained forward bending, squatting, or semi-kneeling positions, and repetitive upper limb exertion, highlighting the biomechanical demands placed on workers during rail track operations (Table 15.2).

15.4.3 Dynamometry

The results indicated strength averages between 116.82 and 123.94 kgf, with no significant difference before and after the activities ($p = 0.5279$). The exoskeleton proved to be more effective in lifting, by reducing the muscular demand on trunk extension, than in static flexion postures (Dahmen & Hefferle, 2018).

15.4.3.1 Correlation Between Work Variables

The average strength varied between 116.82 and 123.94 kgf, with no significant difference ($p = 0.5279$). The exoskeleton reduced the muscle load in trunk extension, being more effective in lifting than in static postures (Dahmen & Hefferle, 2018)

The correlation between dynamometry, REBA, and discomfort was not significant ($p > 0.05$), indicating that postural risk affects muscle strength independently of the exoskeleton. Perceived pain may involve factors beyond the metrics assessed. The lack of correlation between discomfort/pain and the other measures suggests that the subjective perception of discomfort may be related to factors not captured by these assessments.

This research confirms previous findings on muscle and postural overload at work (Bevan, 2015; Costa & Vieira, 2010; Punnett & Wegman, 2004; Sultan-Taïeb et al., 2017) and advances by analyzing the use of a lumbar passive exoskeleton in grinding activities. Recommendations include the use of the Corlett map to monitor pain and ensure adequate breaks; task rotation and micro-breaks (REBA); and, according to dynamometry, the adoption of technologies that reduce overload, such as adapted tools or grinding robots.

15.5 CONCLUSION

The research revealed a high physical load during rail grinding activity, particularly on the back, shoulders, and legs. The use of the passive exoskeleton reduced some of the discomfort, but did not eliminate postural risks, indicating the need for broader ergonomic interventions. The lack of correlation between muscle strength and pain suggests that organizational and psychosocial factors may also influence the relationship.

The results reinforce the importance of integrating assistive technologies with human-centered solutions, in line with the principles of Industry 5.0, which proposes more intelligent, collaborative, and healthy work environments.

REFERENCES

ABIFER – Associação Brasileira da Indústria Ferroviária. Indústria Ferroviária. 2023. Disponível em: https://abifer.org.br/institucional/. Acesso em: 16 dez. 2022.

Alves, J.; Lima, T. M.; Gaspar, P. D. Is industry 5.0 a human-centred approach? A systematic review. *Processes*, v. *11*, n. 1, p. 193, 2023.

Andrade, C. G. A utilização do algoritmo de árvore de decisão C4.5 para a gestão do absenteísmo organizacional. Postal Brasil – Revista técnico científica dos Correios, v. 2, p. 61–70, 2018.

Azevedo, J. A. D. Formação de líderes em engenharia: via permanente ferroviária. Santos: [s.n.], 2017.

Balasubramanian, V. A. K. R. Comparing dynamic and stationary standing postures in an assembly task. International Journal of Industrial Ergonomics, v. 39, p. 649–665, 2009.

Bevan, S. Economic impact of musculoskeletal disorders (MSDs) on work in Europe. Best Practice & Research Clinical Rheumatology, v. 29, p. 356–373, 2015.

Bogue, R. Exoskeletons – a review of industrial applications. Industrial Robot: An International Journal, 2018.

Bogue, R. Exoskeletons–a review of industrial applications. *Industrial Robot: An International Journal*, v. *45*, n. 5, p. 585–590, 2018.

Bosch, T.; van Eck, J.; Knitel, K.; de Looze, M. The effects of a passive exoskeleton on muscle activity, discomfort and endurance time in forward bending work. *Applied ergonomics*, v. *54*, p. 212–217, 2016.

Bräunig, D. *Work-related musculoskeletal disorders: challenges and prevention*. Berlin: Federal Institute for Occupational Safety, 2013.

Breque, M.; De Nul, L.; Petridis, A. *Industry 5.0: Towards a sustainable, human-centric and resilient European industry*. European Commission, 2021.

Cardoso, M. *Avaliação ergonômica em atividades industriais*. São Paulo, 2019.

Claus, A. et al. Human–robot collaboration in physical tasks. *Ergonomics*, v. 62, n. 3, p. 327–344, 2019.

Costa, B. R.; Vieira, E. R. Risk factors for work-related musculoskeletal disorders: A systematic review of recent longitudinal studies. American Journal of Industrial Medicine, v. 53, p. 285–323, 2010.

Crea, S. et al. Occupational exoskeletons: A roadmap toward large-scale adoption. Methodology and challenges of bringing exoskeletons to workplaces. *Wearable Technologies*, v. *2*, p. e11, 2021.

Dahmen, T.; Hefferle, R. Comparing industrial tasks with and without passive exoskeletons. Procedia Manufacturing, v. 3, p. 456–463, 2018.

De Looze, M. P.; Bosch, T.; Krause, F.; Stadler, K. S.; O'sullivan, L. W. Exoskeletons for industrial application and their potential effects on physical work load. *Ergonomics*, v. *59*, n. 5, p. 671–681, 2016.

Demir, K.; Cicibas, H. Human–machine collaboration in Industry 5.0. *Journal of Industrial Integration*, v. 14, n. 2, p. 133–148, 2021.

Dias, A.; Pinho, M.; Amaral, F. (Ergonomic risks in railway maintenance. *Journal of Safety Engineering*, v. 8, n. 1, p. 22–30, 2021.

Diniz, R. L. Avaliação das demandas física e mental no trabalho do cirurgião em procedimentos eletivos. Tese (Doutorado em Engenharia de Produção) – Universidade Federal do Rio Grande do Sul, 2003.

European Commission. Industry 5.0: Towards a sustainable, human-centric and resilient European industry. Brussels, 2021.

Frost, D.; Abdoli-Eramaki, M.; Stevenson, J. Effect of a passive exoskeleton on EMG activity. *Ergonomics*, v. 52, n. 8, p. 985–998, 2009.

Graham, R. Exoskeletons in automotive manufacturing. *Automotive Ergonomics Review*, v. 4, n. 2, p. 101–118, 2009.

Guimarães, L. B. M. *Abordagem ergonômica: análise macroergonômica do trabalho – AMT*. In: Iida; Guimarães (org.). *Ergonomia de Processo*. Porto Alegre, 1999.

Hasani, M.; Sadeghi, M.; Khosravi, Y. Participatory ergonomics and worker involvement. *Safety Science*, v. 150, p. 105–115, 2022.

Hignett, S.; Mcatamney, L. Rapid entire body assessment (REBA). Applied Ergonomics, v. 31, n. 2, p. 201–205, 2000.

Instituto de Pesquisa Econômica Aplicada (IPEA). *Logística ferroviária e mineração no Brasil,* 2018.

Ippolito, D.; Constantinescu, C.; Riedel, O. Holistic planning and optimization of human-centred workplaces with integrated exoskeleton technology. Procedia CIRP, v. 88, p. 214–217, 2020.

Javaid, M.; Haleem, A. Industry 5.0: Potential applications in medical field. Current Medicine Research and Practice, v. 10, n. 4, p. 201–202, 2020.

Kinebot [Livro Eletrônico]: a inteligência artificial na ergonomia / Alison Alfred Klei ... [et al.]. 2 ed. Curitiba, PR: Kinebot, 2023.PDF

Kong, Y. K. et al. Ergonomic assessment of a lower-limb exoskeleton through electromyography and anybody modeling system. *International Journal of Environmental Research and Public Health*, v. *19*, n. 13, p. 8088, 2022.

Koopman, A. et al. Effects of a passive back exoskeleton. *Journal of Biomechanics*, v. 83, p. 97–105, 2019.

Kunda, R.; Frantz, J.; Moyo, M. Musculoskeletal disorders in mining. *African Journal of Mining Health,* v. 2, n. 1, p. 11–19, 2013.

Lee, J.; Park, J.; Kim, A. Human–robot cooperation. *International Journal of Robotics Research,* v. 31, n. 12, p. 1553–1571, 2012.

Looze, M. P. d. et al. Exoskeletons for industrial application and their potential effects on physical workload. Ergonomics, v. 59, n. 5, p. 671–681, 2016.

Lowe, B.; Powell, D.; Chaffin, D. History of military exoskeleton development. *Military Biomechanics Journal,* v. 9, n. 3, p. 101–119, 2019.

Massad, M. D.; Leite, A. M. P.; Dutra, T. R. Fatores ergonômicos relacionados à saúde e à segurança em trabalhadores de um viveiro florestal. In: Workshop de Análise Ergonômica do Trabalho, 5.; Encontro Mineiro de Estudos em Ergonomia, 2., 2011, Viçosa. Anais [...]. Viçosa: UFV, 2011.

Maurice, P. et al. Objective and subjective effects of a passive exoskeleton on overhead work. *IEEE Transactions on Neural Systems and Rehabilitation Engineering*, v. *28*, n. 1, p. 152–164, 2019.

McFarland, T.; Fischer, S. Ergonomic limitations of exoskeletons. *Ergonomics Review*, v. 45, n. 4, p. 225–240, 2019.

Messing, K.; Stock, S.; Tissot, F. Standing work and musculoskeletal symptoms. *Journal of Occupational Health*, v. 50, n. 4, p. 321–330, 2008.

Nahavandi, S. Industry 5.0: A human-centric solution. *IEEE Engineering Management Review,* v. 47, n. 3, p. 6–12, 2019.

Neto, C. B. Manual didático de ferrovias – Primeira parte. Curitiba: [s.n.], 2012.

Parent-Thirion, A. et al. *6th European Working Conditions Survey*. Eurofound, 2017.

Pauluk, D.; Michaloski, A. O. Análise ergonômica do trabalho nas atividades de preparo do solo com trator agrícola. *Revista ESPACIOS| V. 37, n. 04 Año 2016.*

Paz, J. M. Riscos ergonômicos em trabalhadores no setor de mineração e propostas de intervenção: uma revisão integrativa. 2021.

Possebom, G.; Moreira, A. R.; Carpes, D. P.; Francetto, T. R.; Zart, B. C. C. R.; do Amaral Alonço, P.; dos Santos Alonço, A. Avaliação ergonômica em um viveiro florestal de Santa Maria, RS. *Tecno-lógica*, v. *21*, n. 1, p. 30–36, 2017.

Punnett, L.; Wegman, D. H. Work-related musculoskeletal disorders: the epidemiologic evidence and the debate. *Journal of electromyography and kinesiology*, v. *14*, n. 1, p. 13–23, 2004.

Reardon, C.; Kessler, B.; Hester, L. Ergonomic factors in mining operations. *Mining Health Review,* v. 19, n. 2, p. 77–85, 2014.

Rega Barca, L. Lesiones musculoesqueléticas: exoesqueletos como medida de prevención, 2020.

Russo, F.; D'Agostino, R.; Vicentini, M. Exoskeletons in the era of Industry 4.0. *Procedia Engineering*, v. 204, p. 153–160, 2021.

Santander, T.; Ruiz, E. Correlation methods in ergonomic research. *Ergonomía*, v. 17, n. 2, p. 45–53, 2004.

Schnieders, J.; Stone, R. Military exoskeleton developments. *Defense Technology Journal*, v. 23, n. 1, p. 41–53, 2017.

Schreiber, F. Exoskeleton adoption in smart factories. *Journal of Advanced Manufacturing*, v. 12, n. 3, p. 55–67, 2020.

Selig, E. T.; Waters, J. M. *Track Geotechnology and Substructure Management.* London: Thomas Telford, 1994.

Serranheira, F.; Pereira, M.; Santos, C. S.; Cabrita, M. Auto-referência de sintomas de lesões músculo-esqueléticas ligadas ao trabalho (LMELT) numa grande empresa em Portugal. *Rev port saúde pública*, v. *2*, p. 37–48, 2003.

Sousa, V.; Almeida, S.; Braga, L. Ergonomic risks in underground mining. *Revista de Mineração,* v. 48, n. 1, p. 23–31, 2015.

Sultan-Taïeb, H. et al. Economic burden of musculoskeletal disorders. *Annals of Occupational Hygiene*, v. 61, n. 1, p. 111–122, 2017.

Torricelli, D. et al. Exoskeletons: benefits and challenges. *Frontiers in Robotics and AI*, v. 7, p. 1–19, 2020.

Ulrey, B.; Fathallah, F. Passive back-assist exoskeletons. *Applied Ergonomics*, v. 44, n. 4, p. 728–736, 2013.

Veronesi, L. *Dynamometry and lumbar strength in industrial workers.* São Paulo: Senac, 2020.

Zander, J. E.; King, P. M.; Ezenwa, B. N. Influence of flooring conditions on lower leg volume following prolonged standing. International Journal of Industrial Ergonomics, v. 34, p. 279–288, 2004.

Zhang, T.; Huang, H. A lower-back robotic exoskeleton: industrial handling augmentation used to provide spinal support. IEEE Robotics & Automation Magazine, v. 25, n. 2, p. 95–106, 2018.

16 Ergonomic Challenges and Health Risks in Patient Handling

Addressing Musculoskeletal Disorders Among Healthcare Personnel in India

Sanjeet Kumar and Pradip Kumar Ray

16.1 INTRODUCTION

In the dynamic environment of healthcare, patient handling jobs are fundamental but often overlooked components of care delivery. These jobs, ranging from lifting and transferring patients to repositioning and assisting with ambulation, are not only physically demanding but also among the most ergonomically hazardous activities performed by healthcare personnel (Callison & Nussbaum, 2012; Lunde et al., 2014). Nurses, ward assistants, physiotherapists, and caregivers routinely engage in these manual handling activities, placing them at heightened risk for developing work-related musculoskeletal disorders, particularly affecting the back, shoulders, and neck (Akbari et al., 2017; Hignett et al., 2007; Roffey et al., 2010).

Globally, healthcare personnel face higher rates of occupational injuries than workers in manufacturing, construction, or mining (O. Aluko et al., 2016; Choi & Brings, 2016; Neumann et al., 2018; J. Skotte & Ergonomics, 2008). In India, these challenges are further compounded by systemic issues such as patient-to-staff ratios, aging infrastructure, improper use of assistive devices, and limited access to ergonomic training (Koshy et al., 2020; Rodrigues et al., 2019; Thornton et al., 2004). These factors collectively amplify the risk of injury and reduce the quality of care. Consequently, improving ergonomics in patient handling is both a workforce safety issue and a patient care imperative (Gimeno et al., 2005; Mahajan et al., 2023; Pal Singh, 2018).

The current study investigates the nature and extent of ergonomic challenges encountered in patient handling activities within Indian healthcare systems. Building on primary data collected through ergonomic assessments, subjective workload evaluations, and workplace observations, the study identifies high-risk activities, explores root causes of discomfort and injury, and proposes actionable redesign strategies.

 DOI: 10.1201/9781003559399-18

The goal is to create a comprehensive intervention framework aligned with the principles of "Ergonomics 5.0", an emerging paradigm that promotes human-centric, intelligent, and adaptive workplace systems.

This chapter is structured as follows: Section 2 presents a review of relevant literature and theoretical grounding; Section 3 outlines the methodological framework; Section 4 discusses results from ergonomic and workload assessments, interprets findings, and proposes interventions; Section 5 discusses their alignment with Ergonomics 5.0 principles; and Section 6 concludes with recommendations for policy, training, and practice.

16.2 BACKGROUND AND THEORETICAL FRAMEWORK

16.2.1 Manual Patient Handling: A Global Occupational Health Concern

Manual patient handling jobs are recognized internationally as one of the most physically demanding and hazardous occupational activities in healthcare systems (O. O. Aluko et al., 2016; Callison & Nussbaum, 2012; Lunde et al., 2014; Mehrdad et al., 2012). These jobs commonly involve lifting, transferring, repositioning, and supporting patients who often have limited mobility, obesity, or critical medical conditions (Ghesmaty Sangachin & Cavuoto, 2016; Yavorovsky et al., 2020). Unlike inanimate loads, patients are unpredictable in movement, further complicating the biomechanics of handling jobs (Gagnon et al., 1988; Kumar & Ray, 2018; Lindbeck & Engkvist, 1993; Parida & Ray, 2015). As a result, healthcare personnel frequently experience overexertion injuries, particularly to the lower back, shoulders, neck, and upper limbs (Coury et al., 2009; Pheasant & Stubbs, 1992).

Multiple studies have reported that nurses and other caregivers rank among the highest occupational groups for musculoskeletal injuries, especially in environments lacking mechanical aids or standardized patient handling protocols. According to the U.S. Bureau of Labor Statistics (2020) (U.S. Bureau of Labor Statistics, 2020), nursing assistants experience the highest rate of overexertion injuries of any occupation. Similar trends are observed in European and Asian healthcare systems (Hignett et al., 2007; Yang et al., 2020).

16.2.2 Indian Healthcare Context

The Indian healthcare systems with varieties of patient volumes at different departments of the healthcare systems, patient-staff ratios, infrastructural investment, and distribution of assistive technologies are the key factors for the ergonomic assessment of the patient handling jobs (Singh & Lillrank, 2017). These systemic challenges may create a high-pressure environment where manual handling remains the default mode of patient mobilization (Pal Singh, 2018; Ray et al., 2015).

Furthermore, the lack of ergonomics education in nursing and allied health curricula means that caregivers often employ improper body mechanics and inefficient handling strategies (Saini et al., 2021). Repetitive exposure to biomechanical stress, without adequate recovery or support, leads to chronic musculoskeletal disorders

(MSDs), absenteeism, job dissatisfaction, and increased healthcare costs (Castaneda & Scanlan, 2014; Felipe et al., 2017; Labrague et al., 2020).

16.2.3 Risk Factors in Patient Handling Jobs

The ergonomic risks in patient handling are multifactorial. The major contributors to risks in patient handling jobs are as follows:

- Load Characteristics: patients represent dynamic, often unbalanced loads, with weights beyond safe manual lifting thresholds (Cooper & Ghassemieh, 2007; J. H. Skotte et al., 2002).
- Postural Strain: Frequent bending, twisting, and overreaching during transfers and repositioning increase spinal load and joint stress (Genaidy et al.,1994; Granström et al., 1985; Schoenfisch et al., 2019; Thornton et al., 2004; Zhou & Wiggermann, 2021).
- Frequency and Duration: Repetition of strenuous tasks with inadequate rest intervals accelerates fatigue accumulation (Morales et al., 2016).
- Environmental Constraints: Space limitations, poor lighting, and cluttered surroundings reduce the ability to perform safe maneuvers (Holman et al., 2010; Rogers et al., 2013; Springer, 2007).
- Organizational Factors: High workloads, time pressures, low staffing levels, and lack of standardized procedures exacerbate injury risks (Hazirbulan, 2019; Schwartz et al., 2019; Visser et al., 2014).

16.2.4 Evolution of Ergonomics in Healthcare

Historically, ergonomic interventions in healthcare have focused on equipment design and job standardization. While these remain important, the rise of smart technologies, artificial intelligence, and human-robot collaboration has ushered in a new era, Ergonomics 5.0 (Chikwendu et al., 2023; Rijwani et al., 2024). This paradigm emphasizes a shift from reactive to proactive ergonomics, aligning with Industry 5.0's focus on human-centric, sustainable, and resilient systems (Li & Duan, 2025; Nair et al., 2024; Sheikh et al., 2024).

Ergonomics 5.0 promotes the co-existence of humans and intelligent systems. In healthcare, this implies the use of sensor-based monitoring for real-time risk detection, smart assistive devices for patient transfers, personalized workload management using AI-based analytics, and ergonomically optimized workflows based on dynamic data (Katuk et al., 2024; Vidhyatharan, 2023; Vijay Sampat, 2023).

Although the concept is in its nascent stages in India, integrating Ergonomics 5.0 into patient handling design promises transformative potential for healthcare personnel's health and operational efficiency.

16.2.5 Gaps in Existing Research

Despite growing recognition of the risks, research in Indian healthcare systems has primarily focused on general occupational health or broad nursing issues, with limited in-depth ergonomic analyses. The few existing studies either lack quantitative risk assessments or are restricted to observational surveys. There is a significant research gap that includes empirically evaluating ergonomic risks in specific patient handling jobs, developing cost-effective, context-specific intervention strategies, validating the effectiveness of ergonomic redesigns in healthcare systems, and bridging the theory-practice divide through real-world implementation.

This study seeks to address these gaps by conducting a detailed patient handling job-based ergonomic analysis, applying both objective and subjective evaluation tools, and proposing validated redesign strategies tailored to Indian healthcare environments.

16.3 METHODOLOGY

16.3.1 Study Design

This research adopted a mixed-methods design, combining quantitative ergonomic risk assessment techniques with qualitative workplace observations and interviews. The primary focus was on evaluating physical demands and risk exposures associated with patient handling tasks in Indian healthcare systems.

16.3.2 Study Sites and Participants

The study was conducted in two major government hospitals in eastern India, selected for their high patient inflow, representative infrastructure, and workforce diversity. A total of 107 participants were selected using stratified random sampling from key healthcare personnel categories as nurses, ward attendants, physiotherapists, and housekeeping personnel. Participants had a minimum of two years of job experience, ensuring patient handling job familiarity.

The demographic characteristics of the participants are summarized below:

- Mean age: 32.7 years
- Gender: 62% female, 38% male
- Average work experience: 7.3 years
- Work shifts: Rotational (day, night).

16.3.3 Jobs Selection and Classification

Patient handling jobs were categorized into five core operations based on job descriptions, frequency, and criticality: Lifting and transferring patients; Repositioning in bed; Assisting in ambulation; Supporting during toileting; and Handling patients during diagnostic procedures.

These jobs were selected through preliminary job analysis and expert consultations with hospital administrators and senior nursing personnel.

16.3.4 Data Collection Tools and Techniques

To evaluate ergonomic risks, The Logistic regression (Lee & Lee, 2017; Tissot et al., 2009; Yu et al., 2012) and CART (Classification and Regression Trees) regression are invaluable tools in this endeavor (Garg et al., 2014; Ray et al., 2015). These analytical methods allow for the identification and quantification of the relationships between various factors, such as ergonomic challenges, occupational risks, and the effectiveness of interventions. Logistic regression enables the modeling of the probability of specific outcomes, such as the likelihood of musculoskeletal injuries among healthcare personnel based on various factors. CART regression helps create decision trees to better understand the complex interplay of risk factors and their impact on patient handling activities.

To capture perceived exertion and musculoskeletal symptoms:

- Borg Rating of Perceived Exertion: Rated by participants post-jobs on a 6–20 scale. Body Part Discomfort Map: Used to localize and rate discomfort intensity across different body regions. Modified Nordic Musculoskeletal Questionnaire: Collected self-reported data on MSD symptoms over the past 12 months and 7 days.
- 16.3.5 Environmental and Contextual Observations:Observational checklists were used to evaluate environmental conditions (e.g., space constraints, bed height, floor type), tool and equipment availability (wheelchairs, stretchers, transfer boards), and organizational support systems (staffing, training, supervision).

16.3.6 Ethical Considerations

The study received institutional ethics clearance. No patients were exposed to harm during observation of the patient handling jobs and collection of data related to patient handling activities.

16.4 RESULTS AND FINDINGS

This section presents the statistical modelling results derived from ergonomic evaluations of patient handling jobs in Indian healthcare systems. Ordinal Logistic Regression and CART were employed to identify the criticality of musculoskeletal discomfort severity among healthcare personnel.

16.4.1 Ordinal Logistic Regression Analysis

An ordinal logistic regression model is developed to analyze the influence of ergonomic risk factors on the severity of reported discomfort. While performing ordinal logistic regression, it is observed that other than the frequency of carrying out the patient handling activities such as pulling, pushing, lifting, lowering, shifting from wheelchair to bed, shifting from bed to wheelchair, shifting from wheelchair to toilet and vice versa, and shifting from wheelchair to stretcher and vice versa

TABLE 16.1(A)
Correlation table

Factors	FOC	FOPL	FOPS	FOLI	FOLO	FOSBW	FOSWB	FOSWT	FOSWS
FOC	1.00	0.21	0.21	–0.01	0.16	0.01	–0.11	–0.25	–0.08
FOPL	0.21	1.00	1.00	0.77	0.80	0.77	0.79	0.67	0.83
FOPS	0.21	1.00	1.00	0.77	0.80	0.77	0.79	0.67	0.83
FOLI	–0.01	0.77	0.77	1.00	0.77	0.97	0.88	0.70	0.88
FOLO	0.16	0.80	0.80	0.77	1.00	0.84	0.80	0.64	0.80
FOSBW	0.01	0.77	0.77	0.97	0.84	1.00	0.91	0.69	0.87
FOSWB	–0.11	0.79	0.79	0.88	0.80	0.91	1.00	0.88	0.97
FOSWT	–0.25	0.67	0.67	0.70	0.64	0.69	0.88	1.00	0.89
FOSWS	–0.08	0.83	0.83	0.88	0.80	0.87	0.97	0.89	1.00

are highly correlated (refer to Table 16.1(a)), which in turn brings up the issue of multicollinearity. In this regard, principal component analysis (PCA) has been used to reduce the dimensions and eliminate the correlation. After PCA, the variables frequency of pulling (FOPL), frequency of pushing (FOPS), frequency of lifting (FOLI), frequency of lowering (FOLO), frequency of shifting from wheelchair to bed (FOSWB), frequency of shifting from bed to wheelchair (FOSBW), frequency of shifting from wheelchair to toilet and vice versa (FOSWT), and frequency of shifting from wheelchair to stretcher and vice versa (FOSWS) are combined into a single variable and named as "activity". In addition to the frequency of patient handling activities, variables such as age, height, gender, workload, tiredness, work environment (WENV), past health problem (PSTHLTPR), nature of work (NATWRK), repetitive work (REPWK), and work beyond shift (WBYSFT) are taken as control variables in the analysis. Separate analyses are conducted for all joint pains such as neck pain, left shoulder pain, right shoulder pain, left elbow pain, right elbow pain, left wrist pain, right wrist pain, left knuckle pain, right knuckle pain, spine pain, hip pain, knee pain, ankle pain, and body discomfort.

In the present research, the model has been evaluated based on the test for all slopes equal to zero, goodness of fit statistic, and measures of association (Somers' D, Goodman-Kruskal Gamma, and Kendall's Tau-a). The results of the test for all slopes equal to zero are presented in Table 16.2(b), and those for the goodness of fit test are presented in Table 16.2(c). The results corresponding to the measure of association test are presented in Table 16.2(d).

Table 16.2(b) shows the model fitting (Test of All Slopes equal to Zero) for body discomfort, pain in the neck joint, left shoulder joint, right shoulder joint, left elbow joint, right elbow joint, left wrist joint, right wrist joint, left knuckle joint, right knuckle joint, spine joint, hips joint, knee joint, and ankle joint. From the test results, it is observed that the initial and final log likelihood (-2 Log Likelihood) values for body discomfort are 306.366 and 113.934. The difference between these two values is presented as a G-statistic that follows a chi-square distribution. The G-statistic value is 113.934, and the corresponding p-value is less than 0.05, indicating a statistically significant relationship between body discomfort and patient handling activities. For the neck joint pain, the initial and final log likelihood values are 171.750 and

TABLE 16.2(B)
ANOVA test results

Risk Factor	-2 Log Likelihood (Intercept only)	-2 Log Likelihood (Final)	DF	G-statistic (Chi-square Distribution)	p-Value
Body discomfort	306.366	113.934	16	192.432	0.000
Neck	171.750	0.000	16	171.750	0.000
Left shoulder	169.131	0.000	16	169.131	0.000
Right shoulder	241.131	0.000	16	243.131	0.000
Left elbow	110.161	0.000	16	110.161	0.000
Right elbow	157.385	0.000	16	157.385	0.000
Left wrist	169.248	0.000	16	169.248	0.000
Right wrist	236.500	0.000	16	236.500	0.000
Left knuckle	125.319	0.000	16	125.319	0.000
Right knuckle	125.319	0.000	16	125.319	0.000
Spine	323.013	123.122	16	199.891	0.000
Hip	219.102	17.72	16	201.382	0.000
Knee	293.752	0.000	16	293.752	0.000
Ankle	274.63	0.000	16	274.263	0.000

0.000. As presented in the same table, the difference between these two values is 171.750. The corresponding p-value is less than 0.05, indicating a statistically significant relationship between neck joint pain and patient handling activities. Similarly, the p-values corresponding to the pain in the left shoulder joint, right shoulder joint, left elbow joint, right elbow joint, left wrist joint, right wrist joint, left knuckle joint, right knuckle joint, spine, hips joint, knee joint, and ankle joint are also less than 0.05, indicating a significant relationship between these pains and patient handling activities.

Table 16.2(c) presents the goodness of fit test for all the joint pains. The Pearson and Deviance chi-square values corresponding to body discomfort are 210.237 and 113.934, respectively. The corresponding p-values are 1.0 (much larger than 0.05), which support the null hypothesis. It indicates that the model is close to the saturated model, hence, an adequate model. The Pearson and Deviance chi-square values corresponding to the neck are 0.0000035 and 0.0000070, respectively. The corresponding p-values are again 1.00, indicating the model to be adequate. Similarly, the models corresponding to pain in the left shoulder joint, right shoulder joint, left elbow joint, right elbow joint, left wrist joint, right wrist joint, left knuckle joint, right knuckle joint, spine, knee joint, and ankle joint are adequate since the p-value is either 1.00 or very close to 1.00. Unlike the aforementioned models, the model corresponding to the hip pain is not found to be adequate as the p-value is equal to 0.001 (<0.05).

For the measure of association, Somers' D, Goodman-Kruskal Gamma, and Kendall's Tau-a tests are performed. Somers' D value ranges between −1 and 1, where −1 indicates strong disagreement and 1 indicates strong agreement in association

TABLE 16.2(C)
Goodness-of-fit test results

Risk factor	Method	Chi-square	DF	p-Value
Body Discomfort	Pearson	210.237	352	1.000
	Deviance	113.934	352	1.000
Neck	Pearson	0.0000035	168	1.000
	Deviance	0.0000070	168	1.000
Left Shoulder	Pearson	0.0000002	168	1.000
	Deviance	0.0000002	168	1.000
Right Shoulder	Pearson	0.0000025	260	1.000
	Deviance	0.0000050	260	1.000
Left Elbow	Pearson	0.0000005	268	1.000
	Deviance	0.0000009	268	1.000
Right Elbow	Pearson	0.0000002	268	1.000
	Deviance	0.0000004	268	1.000
Left Wrist	Pearson	0.0000005	260	1.000
	Deviance	0.0000010	260	1.000
Risk Factor	Method	Chi-Square	DF	p-Value
Right Wrist	Pearson	0.0000012	260	1.000
	Deviance	0.0000023	260	1.000
Left Knuckle	Pearson	0.0000018	260	1.000
	Deviance	0.0000036	260	1.000
Right Knuckle	Pearson	0.0000018	260	1.000
	Deviance	0.0000036	260	1.000
Spine	Pearson	315.698	352	0.918
	Deviance	123.123	352	1.000
Hips	Pearson	230.271	168	0.001
	Deviance	17.719	168	1.000
Knee	Pearson	0.0000294	352	1.000
	Deviance	0.0000588	352	1.000
Ankle	Pearson	0.0001267	260	1.000
	Deviance	0.0002533	260	1.000

among the variables. Goodman-Kruskal Gamma value also ranges between −1 and 1, where −1 means complete negative association and 1 means complete positive association. Similarly, Kendall's Tau-a ranges between −1 and 1, where −1 indicates perfect disagreement and 1 indicates perfect agreement. Table 16.2(d) presents the results for Measures of Association: (between the response variable and predicted probabilities) for body discomfort. The computed Somers' D, Goodman-Kruskal Gamma, and Kendall's Tau-a values are 0.9, 0.9, and 0.65, respectively. All the results indicate a significant positive association between body discomfort and patient handling activities. Similarly, the results of Somers' D, Goodman-Kruskal Gamma, and Kendall's Tau-a tests corresponding to other pains pertaining to the left shoulder joint, right shoulder joint, left elbow joint, right elbow joint, left wrist joint, right wrist joint, left knuckle joint, right knuckle joint, hips, spine, knee joint, and ankle

TABLE 16.2(D)
Measures of association test results (between the response variable and predicted probabilities)

Body	Pairs	Numbers	Percent	Summary Measures	Value
Discomfort	**Concorda**	3888	95.1	Somers' D	0.9
		201	4.9	Goodman-Kruskal	0.9
	Discordant			Gamma	
	Ties	1	0.0	Kendall's Tau-a	0.65
	Total	4090	100		
Neck	Pairs	Numbers	Percent	Summary Measures	Value
	Concorda	2596	100.0	Somers' D	1.0
	Discordant	0.0	0.0	Goodman-Kruskal Gamma	1.0
	Ties	0.0	0.0	Kendall's Tau-a	0.46
	Total	2596	100		
Left Shoulder	Pairs	Numbers	Percent	Summary Measures	Value
	Concorda	2576	100.0	Somers' D	1.0
	Discordant	0.0	0.0	Goodman-Kruskal Gamma	1.0
	Ties	0.0	0.0	Kendall's Tau-a	0.45
	Total	2576	100.0		
Right Shoulder	Pairs	Numbers	Percent	Summary Measures	Value
	Concorda	3591	100.0	Somers' D	1.0
	Discordant	0.0	0.0	Goodman-Kruskal Gamma	1.0
	Ties	0.0	0.0	Kendall's Tau-a	0.63
	Total	3591	100.0		
Left Elbow	Pairs	Numbers	Percent	Summary Measures	Value
	Concorda	1511	100.0	Somers' D	1.0
	Discordant	0.0	0.0	Goodman-Kruskal Gamma	1.0
	Ties	0.0	0.0	Kendall's Tau-a	0.27
	Total	1511	100.0		
Right Elbow	Pairs	Numbers	Percent	Summary Measures	Value
	Concorda	2336	100.0	Somers' D	1.0
	Discordant	0.0	0.0	Goodman-Kruskal Gamma	1.0
	Ties	0.0	0.0	Kendall's Tau-a	0.41
	Total	2336	100.0		
Left Wrist	Pairs	Numbers	Percent	Summary Measures	Value
	Concorda	2301	100.0	Somers' D	1.0
	Discordant	0.0	0.0	Goodman-Kruskal Gamma	1.0
	Ties	0.0	0.0	Kendall's Tau-a	0.41
	Total	2301	100.0		

(Continued)

TABLE 16.2(D)
(Continued)

Right Wrist	Pairs	Numbers	Percent	Summary Measures	Value
	Concorda	3456	100.0	Somers' D	1.0
	Discordant	0.0	0.0	Goodman-Kruskal Gamma	1.0
	Ties	0.0	0.0	Kendall's Tau-a	0.61
	Total	3456	100.0		
Left Knuckle	Pairs	Numbers	Percent	Summary Measures	Value
	Concorda	1541	100.0	Somers' D	1.0
	Discordant	0.0	0.0	Goodman-Kruskal Gamma	1.0
	Ties	0.0	0.0	Kendall's Tau-a	0.27
	Total	1541	100.0		
Right Knuckle	Pairs	Numbers	Percent	Summary Measures	Value
	Concorda	1541	100.0	Somers' D	1.0
	Discordant	0.0	0.0	Goodman-Kruskal Gamma	1.0
	Ties	0.0	0.0	Kendall's Tau-a	0.27
	Total	1541	100.0		
Spine	Pairs	Numbers	Percent	Summary Measures	Value
	Concorda	4133	94.6	Somers' D	0.89
	Discordant	228	5.2	Goodman-Kruskal Gamma	0.9
	Ties	6	0.1	Kendall's Tau-a	0.69
Hips	Total	4367	100		
	Concorda	3498	99.4	Somers' D	0.99
	Discordant	20	0.6	Goodman-Kruskal Gamma	0.99
	Ties	2	0.1	Kendall's Tau-a	0.61
	Total	3520	100		
Knee	Pairs	Numbers	Percent	Summary Measures	Value
	Concorda	4017	100.0	Somers' D	1.0
	Discordant	0.0	0.0	Goodman-Kruskal Gamma	1.0
	Ties	0.0	0.0	Kendall's Tau-a	0.71
	Total	4017	100.0		
Ankle	Pairs	Numbers	Percent	Summary Measures	Value
	Concorda	4026	100.0	Somers' D	1.0
	Discordant	0.0	0.0	Goodman-Kruskal Gamma	1.0
	Ties	0.0	0.0	Kendall's Tau-a	0.71
	Total	4026	100.0		

joint indicate a significant positive association. The results regarding the critical risk factors are discussed in the next section.

16.4.2 Classification and Regression Tree Analysis

It is reported in (R Parida, 2013) that CART analysis is preferred when analyzing critical risk factors. The influence of factors as well as the complex interactions among them, which are not captured by logistic regression, are captured by CART (Forde et al., 2005; Tissot et al., 2009). In this regard, CART is used for further analysis in the present research to assess and prioritize the criticality of the patient handling activities and control variables that are responsible for the severity of joint pain and body discomfort.

There are several other reasons by which CART is differentiated from Ordinal Logistic Regression: Firstly, its ability to handle large numbers of independent variables well in comparison to ordinal logistic regression (Mahajan et al., 2023; Parida et al., 2016). Secondly, different classes may be weighted by it, where compensation is based on their proportion in the population to bring consistency in the classification procedure. Thirdly, its relevance in stratified samples. Fourthly, continuous independent variables may be handled by it as easily as categorical ones without any kind of transformation. Fifthly, computational efficiency and finding quasi-efficient factors for classification are its attributes (Breiman, 1998, 2002).

In the CART analysis, the objective function is minimized at each node. In this analysis, input variables, which determine the split at each node, are selected in a greedy manner. In the process, a number of cases with different features are considered, and the split is selected based on the minimum value of the Gini Index (cost function) (Rickard & Fallon, 2004). This process is a recursive process for all the sub-nodes in the tree.

Tables 16.3(a)–Table 16.3(d) summarize the results of the CART analysis. Table 4.3(a) presents those corresponding to body discomfort, neck pain, left, and right shoulder pain. In the analysis presented in this research, it is observed that the variables such as pulling, pushing, lifting, lowering, shifting from bed to wheelchair, shifting from wheelchair to bed, wheelchair to toilet, and vice versa, shifting from wheelchair to stature, and vice versa are highly correlated, which is supposed to bring the issue of multicollinearity. To eliminate the multicollinearity, PCA is done, and these variables are clubbed into one named activity. From Table 16.3(a), it is apparent that the factors such as patient handling activity (1.000), workload (0.822), patient carrying (0.775), work beyond shift (0.731), healthcare personnel's height (0.632), and tiredness (0.616) are critical factors for body discomfort. Similarly, the factors patient handling activities (1.000) along with the carrying (0.575), and tiredness (0.502) are the critical factors for neck pain. The important factors for left shoulder pain are height (1.000), age (0.942), frequency of carrying (0.936), activity (0.869), work beyond shift (0.719), work environment (0.544), and past health problem (0.520), whereas critical factors for right shoulder pain are height (1.000), activity (0.623), work beyond shift (0.545), workload (0.522), and past health problem (0.518).

From Table 16.3(b), it is evident that the critical factors for left elbow pain are height (1.000), activity (0.615), and repetitive non-repetitive work (0.597), whereas

TABLE 16.3(A)
CART scores for body discomfort, neck, left shoulder, and right shoulder

Significant risk factors for body discomfort with CART score	Significant risk factors for neck with CART score	Significant risk factors for the left shoulder with CART score	Significant risk factors for the right shoulder with CART score
Activity (1.000)	Activity (1.000)	Height (1.000)	Height (1.000)
Workload (0.822)	Frequency of Carrying (0.575)	Age (0.942)	Activity (0.623)
Frequency of Carrying (0.775)	Tiredness (0.502)	Frequency of Carrying (0.936)	Work Beyond Shift (0.545)
Work Beyond Shift (0.731)		Activity (0.869)	Workload (0.522)
Height (0.632)		Work Beyond Shift (0.719)	Past Health Problem (0.518)
Tiredness (0.616)		Work Environment (0.544)	
		Past Health Problem (0.520)	

TABLE 16.3(B)
CART scores for left elbow, right elbow, left wrist, and right wrist

Significant risk factors for the left elbow with CART score	Significant risk factors for the right elbow with CART score	Significant risk factors for the left wrist with CART score	Significant risk factors for the right wrist with CART score
Height (1.000)	Work Environment(1.000)	Activity(1.000)	Height(1.000)
Activity(0.615)	Height(0.947)	Height(0.721)	Frequency of Carrying(0.960)
Repetitive Non-repetitive Work(0.597)	Frequency of Carrying(0.907)	Frequency of Carrying(0.687)	Activity(0.892)
	Repetitive Non-repetitive Work(0.830)	Work Beyond Shift (0.530)	Age(0.551)
	Nature of Work(0.686)		Workload(0.542)
	Activity(0.683)		
	Workload(0.610)		
	Tiredness(0.521)		

the critical risk for right elbow pain is work environment (1.000), height (0.947), frequency of carrying (0.907), repetitive non-repetitive work (0.830), nature of work (0.686), activity (0.683), workload (0.610), and tiredness (0.521). Furthermore, the critical risk factors for left wrist pain are activity (1.000), height (0.721), frequency of carrying (0.687), and work beyond shift (0.530), and those for right wrist pain are height (1.000), frequency of carrying (0.960), activity (0.892), age (0.551), and workload (0.542).

Similarly, referring to Table 16.3(c), it may be noted that the critical factors causing left and right knuckle pain are patient handling activities (1.000), work environment (0.924), workload (0.884), height (0.868), age (0.623), and frequency of carrying (0.538), and those corresponding to spine pain are height (1.000), tiredness (0.674), activity (0.667), repetitive non-repetitive work (0.585), and frequency of carrying (0.548). Patient handling activities (1.000), frequency of carrying (0.581), workload (0.562), and height (0.520) are critical for hip pain.

TABLE 16.3(C)
CART scores for left knuckle, right knuckle, spine, and hips

Significant risk factors for the left knuckle with CART score	Significant risk factors for the right knuckle with CART score	Significant risk factors for the spine with CART score	Significant risk factors for hips with CART score
Activity(1.000)	Activity(1.000)	Height(1.000)	Activity(1.000)
Work Environment(0.924)	Work Environment(0.924)	Tiredness(0.674)	Frequency of Carrying(0.581)
Workload(0.884)	Workload(0.884)	Activity(0.667)	Workload(0.562)
Height(0.868)	Height(0.868)	Repetitive Non-repetitive Work(0.585)	Height(0.520)
Age(0.623)	Age(0.623)	Frequency of Carrying(0.548)	
Frequency of Carrying(0.538)	Frequency of Carrying(0.538)		

TABLE 16.3(D)
CART scores for knees and ankle (for PHA)

Significant risk factors for knees with CART score	Significant risk factors for ankle with CART score
Work Environment(1.000)	Workload(1.000)
Height(0.974)	Age(0.951)
Activity(0.843)	Nature of Work(0.944)
Frequency of Carrying(0.838)	Repetitive Non-repetitive Work(0.935)
Workload(0.714)	Activity(0.925)
Age(0.681)	Height(0.691)
Tiredness(0.542)	Frequency of Carrying(0.653)
	Tiredness(0.561)

From Table 16.3(d), it is evident that work environment (1.000), height (0.974), activity (0.843), frequency of carrying (0.838), workload (0.714), age (0.681), and tiredness (0.542) are critical factors causing knee pain. Similarly, workload (1.000), age (0.951), nature of work (0.944), repetitive non-repetitive work (0.935), activity (0.925), height (0.691), frequency of carrying (0.653), and tiredness (0.561) contribute to ankle pain.

16.4.3 Summary of Statistical Findings

Both ordinal regression and CART analysis highlighted the criticality of patient handling activities, job frequency, and lack of ergonomic training as primary contributors to elevated discomfort risk.

CART analysis provided interpretable decision rules, while ordinal regression offered statistically validated weightings.

The combination of these tools strengthens the ergonomic assessment process and lays the foundation for data-driven ergonomic redesign in healthcare systems.

16.5 DISCUSSION AND INTERVENTION STRATEGIES

The findings of this study reveal a complex but interpretable relationship between job-specific ergonomic factors and the severity of musculoskeletal discomfort among healthcare personnel engaged in patient handling jobs. The application of ordinal logistic regression and CART analysis provided not only statistical validation but also actionable insights to guide ergonomic interventions.

16.5.1 Interpretation of Results in the Context of Ergonomics 5.0

The study's results align with the core principles of Ergonomics 5.0, which emphasize human-centered system design in complex environments like healthcare. The strong association between patient handling activities and discomfort severity reaffirms the need for task reengineering, especially in high-load patient transfers and repositioning jobs.

The CART analysis highlighted that high patient handling loads, training, and job frequency are critical behavioral and organizational modifiers. These findings suggest that Ergonomics 5.0 must address both the physical and cognitive-administrative dimensions of healthcare jobs, combining redesign with systemic behavioral improvements.

16.5.2 Proposed Ergonomic Interventions

Based on the data-driven insights and contextual realities of healthcare systems, the following multidimensional intervention strategy is proposed:

a. Engineering Controls
   i. Redesign of Equipment and Workspaces: Introduce adjustable-height beds and ceiling-mounted hoists where feasible. Redesign ward layouts to reduce unnecessary bending and twisting.
   ii. Provision of Handling Aids: Encourage the use of wheel stretchers, slide boards, and transfer belts. Equip high-dependency units with mechanized patient lifts.

b. Administrative Controls
   iii. Job Rotation and Break Scheduling: Introduce rotating roles among nursing staff to limit cumulative load exposure. Design duty rosters with structured micro-breaks to reduce fatigue.
   iv. Workload Management: Use CART-derived decision rules to identify high-risk shifts and allocate manpower accordingly.

c. Training and Awareness
   v. Mandatory Ergonomic Training: Conduct onboarding and refresher training on safe patient handling techniques.
   vi. Simulation-Based Learning: Employ low-cost training simulators and dummies for repeated practice of lifting, transferring, and repositioning tasks.
   vii. Visual Reminders and Posters: Display posture checklists and lifting techniques at key patient handling zones.

Implementing these interventions will not only reduce injury rates but also enhance healthcare personnel morale, reduce absenteeism, and improve patient safety. As healthcare moves toward technology-integrated systems, embedding proactive ergonomic design will be central to ensuring long-term sustainability and workforce resilience.

16.6 CONCLUSIONS AND FUTURE SCOPE

16.6.1 Conclusion

In this research, the study investigates how patient handling activities in hospitals cause MSDs in healthcare personnel. Data are collected from a reputed government hospital to examine the issue. Although the data originate from the hospital, the methodology and model may be generalized to other similar setups. The analysis results recommend that hospitals need to pay attention to designing body postures during patient handling. Since the height of healthcare personnel is also found to be critical, it is recommended to provide height adjustments in patient handling equipment. Optimizing workload and discouraging overtime work are also advisable. The survey reveals that healthcare personnel often work continuously with less than or equal to 30 minutes of rest during their shifts, so hospitals may provide intermittent rest periods. Considering the impact of patient handling activities on joint and body pain, hospitals may prepare a list of critical activities and introduce automation to some extent. This approach will not only improve the overall health and safety of healthcare personnel but also enhance their productivity and reduce occupational hazards.

Aligned with the ethos of Ergonomics 5.0, which advocates for inclusive, adaptive, and data-driven workplace design, the study offers a path forward for ergonomic redesign and workload optimization in healthcare. The integration of quantitative risk assessment tools and decision-support frameworks (like CART) has the potential to transform injury prevention strategies from reactive to proactive.

16.6.2 Contributions of the Study

This study identifies key ergonomic risk factors in patient handling through a data-driven approach and applies ordinal and CART regression models for effective risk assessment. It proposes integrated interventions combining engineering, administrative, and training-based strategies, and offers scalable solutions tailored for low-resource healthcare systems.

16.6.3 Future Scope

While this study lays a strong foundation for ergonomic intervention in healthcare, future research can build upon this work in several ways:

- Longitudinal Impact Assessment: Implement the proposed ergonomic interventions and track their effectiveness over time in reducing injury incidence and improving caregiver wellbeing.

- Integration with Digital Tools: Develop a mobile-based ergonomic risk assessment app using the CART model to provide real-time risk prediction and task recommendations to healthcare personnel.
- Ergonomics in Smart Healthcare Systems: Investigate how emerging technologies (AI, IoT, and robotics) can be used to automate risk-prone tasks and enhance safety in a post-Ergonomics 5.0 framework.

REFERENCES

2020 Home: U.S. Bureau of Labor Statistics. (2020). Retrieved May 7, 2025, from https://www.bls.gov/opub/mlr/2020/

Akbari, H., Akbari, H., Bagheri Hossein Abadi, M., Gholami Fesharaki, M., & Ghasemi, M. (2017). Assessing the risk of manual handling of patients and its relationship with the prevalence of musculoskeletal disorders among nursing staff: Performance evaluation of the MAPO and PTAI methods. *Iranian Red Crescent Medical Journal*, *19*(2), 39860. https://www.ircmj.com/article_194854_0d2eb0726c0af0eee7daa1c4691c036e.pdf

Aluko, O. O., Adebayo, A. E., Adebisi, T. F., Ewegbemi, M. K., Abidoye, A. T., & Popoola, B. F. (2016). Knowledge, attitudes and perceptions of occupational hazards and safety practices in Nigerian healthcare workers. *BMC Research Notes*, *9*(1). https://doi.org/10.1186/S13104-016-1880-2

Breiman, L. (1998). Arcing classifiers. *Annals of Statistics*, *26*(3), 801–849.

Breiman, L. (2002). Arcing classifier (with discussion and a rejoinder by the author). *The Annals of Statistics*, *26*(3), 849. https://doi.org/10.1214/aos/1024691079

Callison, M. C., & Nussbaum, M. A. (2012). Identification of physically demanding patient-handling tasks in an acute care hospital. *International Journal of Industrial Ergonomics*, *42*(3), 261–267. https://doi.org/10.1016/j.ergon.2012.02.001

Castaneda, G. A., & Scanlan, J. M. (2014). Job satisfaction in nursing: A concept analysis. *Nursing Forum*, *49*(2), 130–138. https://doi.org/10.1111/nuf.12056

Chikwendu, O., Ezeanyim, O. C., & Igbokwe, N. C. (2023). Human-robot interaction enhancement through ergonomics and human factors: Future directions. *Hal.Science*, *19*(6), 34–40. Retrieved from https://hal.science/hal-04348897/

Choi, S. D., & Brings, K. (2016). Work-related musculoskeletal risks associated with nurses and nursing assistants handling overweight and obese patients: A literature review. *Work*, *53*, 439–448. https://doi.org/10.3233/WOR-152222

Cooper, G., & Ghassemieh, E. (2007). Risk assessment of patient handling with ambulance stretcher systems (ramp/(winch), easi-loader, tail-lift) using biomechanical failure criteria. *Medical Engineering and Physics*, *29*(7), 775–787. https://doi.org/10.1016/j.medengphy.2006.08.008

Coury, H., Moreira, R., Physical, & Dias, N. B. (2009). Evaluation of the effectiveness of workplace exercise in controlling neck, shoulder and low back pain: A systematic review. *SciELO Brasil*. Retrieved from https://www.scielo.br/j/rbfis/a/GVZBFxm4ysJkYCyN3TXqXZb/?lang=en

Felipe, F., Junior, S., Andrés, E., Merino, D., Reiter, C., David, J., … Ea, M. (2017). Proposal for management of absenteeism among hospital nurses: A systematic review. *Acta Paulista de Enfermagem*, *30*(5), 546–553.

Forde, M. S., Punnett, L., & Wegman, D. H. (2005). Prevalence of musculoskeletal disorders in union ironworkers. *Journal of Occupational and Environmental Hygiene*, *2*(4), 203–212. https://doi.org/10.1080/15459620590929635

Gagnon, M., Roy, D., Lortie, M., & Roy, R. (1988). Examination of biomechanical principles in a patient handling task. *International Journal of Industrial Ergonomics*, *3*(1), 29–40. https://doi.org/10.1016/0169-8141(88)90005-4

Garg, A., Waters, T., Kapellusch, J., & Karwowski, W. (2014). Psychophysical basis for maximum pushing and pulling forces: A review and recommendations. *International Journal of Industrial Ergonomics*, *44*(2), 281–291. https://doi.org/10.1016/j.ergon.2012.09.005

Genaidy, A. M., Al-Shedi, A. A., & Karwowski, W. (1994). Postural stress analysis in industry. *Applied Ergonomics*, *25*(2), 77–87. https://doi.org/10.1016/0003-6870(94)90068-X

Ghesmaty Sangachin, M., & Cavuoto, L. A. (2016). Obesity-related changes in prolonged repetitive lifting performance. *Applied Ergonomics*, *56*, 19–26. https://doi.org/10.1016/j.apergo.2016.03.002

Gimeno, D., Felknor, S., Burau, K. D., & Delclos, G. L. (2005). Organisational and occupational risk factors associated with work related injuries among public hospital employees in Costa Rica. *Occupational and Environmental Medicine*, *62*(5), 337–343. https://doi.org/10.1136/oem.2004.014936

Granström, B., Kvarnström, S., & Tiefenbacher, F. (1985). Electromyography as an aid in the prevention of excessive shoulder strain. *Applied Ergonomics*, *16*(1), 49–54. https://doi.org/10.1016/0003-6870(85)90146-2

Hazirbulan, Y. (2019). *Risk assesment for musculoskeletal disorders in an international manufacturing firm by rapid entire body assessment method*. Retrieved from https://search.proquest.com/openview/09406bb23e6d1918645540fde75fa914/1?pq-origsite=gscholar&cbl=2026366&diss=y

Hignett, S., Fray, M., Rossi, M. A., Tamminen-Peter, L., Hermann, S., Lomi, C., … Johnsson, C. (2007). Implementation of the manual handling directive in the healthcare industry in the European Union for patient handling tasks. *International Journal of Industrial Ergonomics*, *37*(5), 415–423. https://doi.org/10.1016/j.ergon.2007.01.003

Holman, G. T., Ellison, K. J., Maghsoodloo, S., & Thomas, R. E. (2010). Nurses' perceptions of how job environment and culture influence patient handling. *International Journal of Orthopaedic and Trauma Nursing*, *14*, 18–29. https://doi.org/10.1016/j.joon.2009.03.004

Katuk, N., Vergallo, R., & Sugiharto, T. (2024). The future of human-computer integration. In *The future of human-computer integration: Industry 5.0 technology, tools, and algorithms*. https://doi.org/10.1201/9781003479727

Koshy, K., Syed, H., Luckiewicz, A., Alsoof, D., Koshy, G., & Harry, L. (2020). Interventions to improve ergonomics in the operating theatre: A systematic review of ergonomics training and intra-operative microbreaks. *Annals of Medicine and Surgery*, *55*(February), 135–142. https://doi.org/10.1016/j.amsu.2020.02.008

Kumar, S., & Ray, P. K. (2018). *A comprehensive framework for ergonomic evaluation of patient handling jobs in healthcare systems* (pp. 97–113). https://doi.org/10.1007/978-981-10-5631-4_8

Labrague, L. J., Nwafor, C. E., & Tsaras, K. (2020). Influence of toxic and transformational leadership practices on nurses' job satisfaction, job stress, absenteeism and turnover intention: A cross-sectional study. *Journal of Nursing Management*, *28*(5), 1104–1113. https://doi.org/10.1111/jonm.13053

Lee, S.-J., & Lee, J. H. (2017). Safe patient handling behaviors and lift use among hospital nurses: A cross-sectional study. *International Journal of Nursing Studies*, *74*(January), 53–60. https://doi.org/10.1016/j.ijnurstu.2017.06.002

Li, L., & Duan, L. (2025). Industry 5.0: Exploring the role of human-centric design in transformative learning factories. *Politesi.Polimi.It*. https://doi.org/10.1080/00207543.2025.2462657

Lindbeck, L., & Engkvist, I. L. (1993). Biomechanical analysis of two patient handling tasks. *International Journal of Industrial Ergonomics*, *12*(1–2), 117–125. https://doi.org/10.1016/0169-8141(93)90043-D

Lunde, L. K., Koch, M., Knardahl, S., Wærsted, M., Mathiassen, S. E., Forsman, M., … Veiersted, K. B. (2014). Musculoskeletal health and work ability in physically demanding occupations: Study protocol for a prospective field study on construction and health care workers. *BMC Public Health, 14*(1). https://doi.org/10.1186/1471-2458-14-1075

Mahajan, D., Gupta, M. K., Mantri, N., Joshi, N. K., Gnanasekar, S., Goel, A. D., … Bhardwaj, P. (2023). Musculoskeletal disorders among doctors and nursing officers : an occupational hazard of overstrained healthcare delivery system in western Rajasthan, India. *BMC Musculoskeletal Disorders, 24*(1). https://doi.org/10.1186/S12891-023-06457-Z

Mehrdad, R., Dennerlein, J. T., & Morshedizadeh, M. (2012). Musculoskeletal disorders and ergonomic hazards among Iranian physicians. *Archives of Iranian Medicine, 15*(6), 370–374. DOI-012156/AIM.0011

Morales, L., McEachern, B. M., MacPhee, R. S., & Fischer, S. L. (2016). Patient acuity as a determinant of paramedics' frequency of being exposed to physically demanding work activities. *Applied Ergonomics, 56*, 187–193. https://doi.org/10.1016/j.apergo.2016.04.003

Nair, A., Pillai, S. V., & Senthil Kumar, S. A. (2024). Towards emerging Industry 5.0–a review-based framework. *Emerald.Com*. https://doi.org/10.1108/JSMA-04-2023-0067

Neumann, W. P., Steege, L. M., Jun, G. T., & Wiklund, M. (2018). Ergonomics and human factors in healthcare system design – An introduction to this special issue. *IISE Transactions on Occupational Ergonomics and Human Factors, 6*(3–4), 109–115. https://doi.org/10.1080/24725838.2018.1560927

Pal Singh, L. (2018). Prevalence of MSDs among Nurses: A Study in Indian Health Care Industry. *Ergonomics International Journal, 2*(4). https://doi.org/10.23880/eoij-16000149

Parida, R. (2013). *Ergonomic Study of Manual Material Handling Tasks in Construction* (Doctoral dissertation, IIT Kharagpur).

Parida, R., & Ray, P. K. (2015). Biomechanical modelling of manual material handling tasks: A comprehensive review. *Procedia Manufacturing, 3*, 4598–4605. https://doi.org/10.1016/j.promfg.2015.07.539

Parida, R., Sarkar, S., & Ray, P. K. (2016). *Selection of alternate work systems to improve occupational health of Indian construction workers: A design of experiment-based approach*. https://doi.org/10.1007/978-981-287-585-3_11

Pheasant, S., & Stubbs, D. (1992). Back pain in nurses: Epidemiology and risk assessment. *Applied Ergonomics, 23*(4), 226–232. https://doi.org/10.1016/0003-6870(92)90150-T

Ray, P. K., Parida, R., & Saha, E. (2015). Status survey of occupational risk factors of manual material handling tasks at a construction site in India. *Procedia Manufacturing, 3*, 6579–6586. https://doi.org/10.1016/j.promfg.2015.07.279

Rickard, S., & Fallon, M. (2004). The Gini index of speech. *Annual Conference on Information Sciences and Systems*, 5.

Rijwani, T., Kumari, S., Srinivas, R., Abhishek, K., Iyer, G., Vara, H., … Gupta, M. (2024). Industry 5.0: A review of emerging trends and transformative technologies in the next industrial revolution. *Springer*. https://doi.org/10.1007/S12008-024-01943-7

Rodrigues Ferreira Faisting, A. L., & de Oliveira Sato, T. (2019). Effectiveness of ergonomic training to reduce physical demands and musculoskeletal symptoms - An overview of systematic reviews. *International Journal of Industrial Ergonomics, 74*. https://doi.org/10.1016/j.ergon.2019.102845

Roffey, D. M., Wai, E. K., Bishop, P., Kwon, B. K., & Dagenais, S. (2010). Causal assessment of workplace manual handling or assisting patients and low back pain: results of a systematic review. *The Spine Journal: Official Journal of the North American Spine Society, 10*(7), 639–651. https://doi.org/10.1016/j.spinee.2010.04.028

Rogers, B., Buckheit, K., & Ostendorf, J. (2013). Ergonomics and nursing in hospital environments. *Workplace Health & Safety, 61*(10), 429–439. https://doi.org/10.1177/216507991306101003

Saini, S. K., Pandey, V., Kumar, A., & Elhence, A. (2021). Musculoskeletal problems and expressed practices of body mechanics among nursing officers at All India Institute of Medical Sciences Jodhpur, Rajasthan, India. *Journal of Integrative Nursing, 3*(3), 128–134. https://doi.org/10.4103/JIN.JIN_22_21

Schoenfisch, A. L., Kucera, K. L., Lipscomb, H. J., Mcilvaine, J., Becherer, L., James, T., & Avent, S. (2019). Use of assistive devices to lift, transfer, and reposition hospital patients. *Journals.Lww.Com*. https://doi.org/10.1097/NNR.0000000000000325

Schwartz, A., Gerberich, S. G., Kim, H., Ryan, A. D., Church, T. R., Albin, T. J., … Arauz, R. F. (2019). Janitor ergonomics and injuries in the safe workload ergonomic exposure project (SWEEP) study. *Applied Ergonomics*, *81*. https://doi.org/10.1016/j.apergo.2019.102874

Sheikh, R. A., Ahmed, I., Faqihi, A. Y. A., & Shehawy, Y. M. (2024). Global perspectives on navigating Industry 5.0 knowledge: Achieving resilience, sustainability, and human-centric innovation in manufacturing. *Springer*. https://doi.org/10.1007/S13132-024-02498-4

Singh, V., & Lillrank, P. (2017). *Planning and designing healthcare facilities: A lean, innovative, and evidence-based approach*. Retrieved from https://books.google.com/books?hl=en&lr=&id=n449DwAAQBAJ&oi=fnd&pg=PT9&dq=The+Indian+healthcare+s,+patient-staff+ratios,+infrastructural+investment,+and+distribution+of+assistive+technologies+are+the+key+factors+for+the+ergonomic+assessment+of+the+patient+handling+jobs+&ots=Kha4DZEh8G&sig=lRe_mEQ9w3o3px2SGsyXzhvKDZw

Skotte, J., & Ergonomics, N. (2008). Low back injury risk during repositioning of patients in bed: The influence of handling technique, patient weight and disability. *Taylor & Francis*, *51*(7), 1042–1052. https://doi.org/10.1080/00140130801915253

Skotte, J. H., Essendrop, M., Hansen, A. F., & Schibye, B. (2002). A dynamic 3D biomechanical evaluation of the load on the low back during different patient-handling tasks. *Journal of Biomechanics*, *35*(10), 1357–1366. https://doi.org/10.1016/S0021-9290(02)00181-1

Springer, T. (2007). Ergonomics for Healthcare Environment. *Journal of Ergonomics and Health Care* (2), 1–36. Retrieved from https://www.ncbi.nlm.nih.gov/pubmed

Thornton, L. J., Stuart-Buttle, C., Wyszynski, T. C., & Wilson, E. R. (2004). Physical and psychosocial stress exposures in US dental schools: The need for expanded ergonomics training. *Applied Ergonomics*, *35*(2), 153–157. https://doi.org/10.1016/j.apergo.2003.11.007

Tissot, F., Messing, K., & Stock, S. (2009). Studying the relationship between low back pain and working postures among those who stand and those who sit most of the working day. *Ergonomics*, *52*(11), 1402–1418. https://doi.org/10.1080/00140130903141204

Vidhyatharan, V. P. (2023). *Human factors in industry 4.0: A systematic literature review*. Retrieved from https://www.politesi.polimi.it/handle/10589/230027

Vijay Sampat, M. (2023). *Consideration of ergonomic aspects in human-robot task allocation methods* (12202177), 44190. https://doi.org/10.34726/hss.2023.112768

Visser, S., van der Molen, H. F., Kuijer, P. P. F. M., Hoozemans, M. J. M., & Frings-Dresen, M. H. W. (2014). Evaluation of team lifting on work demands, workload and workers' evaluation: An observational field study. *Applied Ergonomics*, *45*(6), 1597–1602. https://doi.org/10.1016/j.apergo.2014.05.009

Yang, S., Li, L., Wang, L., Zeng, J., & Li, Y. (2020). Risk factors for work-related musculoskeletal disorders among intensive care unit nurses in China: A structural equation model approach. *Asian Nursing Research*, *14*(4), 241–248. https://doi.org/10.1016/j.anr.2020.08.004

Yavorovsky, O., Paustovsky, Y., Nikitiuk, O., & Zenkina, V. (2020). *Hygienic description of working conditions and health status of medical workers*. Retrieved from https://doi.org/10.36074/logos-27.10.2023.73

Yu, S., Lu, M. L., Gu, G., Zhou, W., He, L., & Wang, S. (2012). Musculoskeletal symptoms and associated risk factors in a large sample of Chinese workers in henan Province of China. *American Journal of Industrial Medicine*, *55*(3), 281–293. https://doi.org/10.1002/ajim.21037

Zhou, J., & Wiggermann, N. (2021). The effects of hospital bed features on physical stresses on caregivers when repositioning patients in bed. *Applied Ergonomics*, *90*. https://doi.org/10.1016/j.apergo.2020.103259

Index

For Product Safety Concerns and Information please contact our EU representative GPSR@taylorandfrancis.com Taylor & Francis Verlag GmbH, Kaufingerstraße 24, 80331 München, Germany